MAY 0 3 2013

W9-AFK-458

37

14.99

[library stamp]

Principles of
Benthic Marine
Paleoecology

560.45
X
Bou

66.50- 59.85

560.45
B755p
1981

Principles of Benthic Marine Paleoecology

ARTHUR J. BOUCOT

Department of Geology
Oregon State University
Corvallis, Oregon

With Contributions on Bioturbation,
Biodeposition, and Nutrients by
ROBERT S. CARNEY

National Science Foundation
Washington, D.C.

68415

1981

ACADEMIC PRESS

A Subsidiary of Harcourt Brace Jovanovich, Publishers

New York   London   Toronto   Sydney   San Francisco

NC WESLEYAN COLLEGE
ELIZABETH BRASWELL PEARSALL LIBRARY

COPYRIGHT © 1981, BY ACADEMIC PRESS, INC.
ALL RIGHTS RESERVED.
NO PART OF THIS PUBLICATION MAY BE REPRODUCED OR
TRANSMITTED IN ANY FORM OR BY ANY MEANS, ELECTRONIC
OR MECHANICAL, INCLUDING PHOTOCOPY, RECORDING, OR ANY
INFORMATION STORAGE AND RETRIEVAL SYSTEM, WITHOUT
PERMISSION IN WRITING FROM THE PUBLISHER.

ACADEMIC PRESS, INC.
111 Fifth Avenue, New York, New York 10003

United Kingdom Edition published by
ACADEMIC PRESS, INC. (LONDON) LTD.
24/28 Oval Road, London NW1 7DX

Library of Congress Cataloging in Publication Data

Boucot, Arthur James, Date
    Principles of benthic marine paleoecology.

    Bibliography: p.
    Includes index.
    1. Paleoecology. 2. Marine ecology. I. Title.
QE720.B68   560'.45   79–8535
ISBN 0–12–118980–5

PRINTED IN THE UNITED STATES OF AMERICA

81 82 83 84   9 8 7 6 5 4 3 2 1

One principle of research probably comes nearer than any other to being without exceptions: no matter what your problem is, there are not enough data to solve it. That applies with bitter force to paleoecology.

—Simpson, 1969 (p. 163)

Contents

4

Communities and Their Characteristics   177

5

Bioturbation and Biodeposition   357

ROBERT S. CARNEY

6

Sampling   401

Preface

The concept of organic evolution stated so convincingly by Darwin and Wallace was in part generated from their acquaintance with biogeographic and animal community information of the Recent. A thorough understanding of historical biogeography and animal community history could supplement understanding of the evolutionary process, as geologic time provides further information. This hope provided the primary motivation for the preparation of a preliminary account of Phanerozoic historical biogeography and community history. For the geologist, there are practical considerations necessary for the attainment of a better understanding of basin analysis than could be had from the presentation of physical data alone.

This book deals with the paleoecology of the shallow-water marine environment, emphasizing those aspects of both ecology and paleoecology considered to be useful for reconstructing the continental shelf, shallow water environments richly represented in the fossil record. Very little has been written about historical biogeography, and less about animal community history. We can think of no better way to draw attention to these questions than to provide an admittedly deficient account that colleagues, present and future, can enjoy criticizing, we hope in a constructive manner; stimulating them will be profitable to our profession.

Biologists do not study the long-term, irreversible changes in morphology that paleontologists since the time of Louis Dollo have routinely classified as "evolution." Since they are restricted by a geologically instantaneous life span, biological studies have been largely limited to mechanisms of evolution including inheritance, geographic dispersion, and survival in response to environmental stresses. Although they have been successful at documenting morphological variation and manipulation of variation in experimental populations, biologists have not studied change on the time scales available to the paleontologist. Without the increasingly detailed understanding of evolutionary mechanisms gained through biological investigation, however, modern paleontology could resemble a blend of mindless classification and overimaginative polytheism. Without detailed knowledge of how animals exist, reproduce, and vary in the present, a meaningful synthesis of the knowledge gained through study and classification of fossils would be impossible.

If evolution is considered to be the result of processes operating over time periods beyond the reach of the biologist its study should in principle be available to the student of past life, the paleontologist. The paleontologist studies biogenic objects that have been preserved in a time sequence. Although it is most often relative, the time sequence may sometimes be assigned an absolute date through radioisotope dating. Under ideal conditions, the paleontologist studies a time series of morphologically similar objects, time plane by time plane. Patterns of changing similarity are interpreted as having been produced by evolutionary mechanisms in operation over the intervals of time between the time planes from which specimens are available. The paleontologist observes many discrete time planes with greater or smaller morphological gaps in the specimens from each; in some instances there may even be morphological overlaps between time planes. The morphologies seen in each plane are like cinematic frames that merge into a continuum when rapidly viewed. Of course, in the case of a movie, the viewer is not aware that the smoothly flowing film is really a rapid sequence of still pictures. In paleontology, interpretation of apparent evolutionary lineages is carried out in the conscious brain; it draws heavily upon a knowledge of biology, geology, logic, and a reasonable bit of imagination. The paleontologist, as does any scientist, deals with circumstantial evidence, filling in the gaps by means of informed judgment.

The biologist, however, is free to examine in greatest detail biological activity in a single time plane. When appropriate, observed patterns of morphological or even genetic similarity may be interpreted easily and logically as being due to divergence from a common ancestral population. Too often, however, biologists may either fail to recognize or to acknowledge the importance of long-time-span processes because they are unfamiliar with current paleontological knowledge and thinking. Similarly, biologically uninformed (or worse, misinformed) paleontologists are too often unable to see biologically important information in their sequences of fossil material. One purpose of this text is to increase mutual awareness between the two partial studies of evolution, evolutionary biology and paleontology.

In addition to observing discrete changes over periods of time, the paleontologist can calculate rates and rate changes in morphology, size, cladogenesis (both diacladogenesis and metacladogenesis) and anagenesis (phyletic evolution) based on the assumption that the materials involved have been subject to evolutionary changes. The paleontologist can also calculate differing rates of morphologic change affecting the same anagentic lineage to arrive at a better understanding of mosaic evolutionary rates.

If one is willing to admit that the paleontologist can, in principle, estimate rates of varying types of evolution as well as changes in these rates, it is of considerable concern to inquire whether or not the paleontologist can correlate such rate phenonmena with any features of interest to the evolutionary biologist. Biologists have been concerned, in regard to rate control parameters, with such items as radiation flux, intraspecific competition, interspecific competition, interbreeding population size, fluctuating or nonfluctuating (stable) nature of the environment, predictable and nonpredictable factors, and varying nutrient levels, as well as such vague terms as *differing selection pressures*. The paleontologists, and particularly the paleontologist doubling as a geologist, can check some of these possibilities against the geologic record. In any event, the paleontologist has a time-averaged sample that provides a type of information not available to the biologist. The information provided by the biological and paleontological approaches complement and reinforce each other if they are properly integrated. One might think of the common interest in evolution as analogous to the interest shown by the thermodynamicist and the nuclear physicist in the properties of matter: One deals with relatively small-scale, discrete events and the other with time-averaged, statistical entities.

We hope that the interested biologist will realize, after reading this book, that the fossil record is heavily biased toward low trophic-level filter, suspension, and deposit-feeding invertebrates: Low trophic-level organisms predominate (Wigley and Stinton, 1973). Researchers forming generalizations based on the fossil record should keep this fact in mind, as well as the almost total absence of soft-bodied organisms and the lack of attention given to many fossil groups that require extensive preparation techniques for proper study.

We have tried to follow an analytical approach; to suggest how the manifold variables operating in the benthic marine environment can be recognized and how the intensity of their effect could be measured in individual situations. We have tried not to dwell on any one specific environment, but to emphasize characteristics formed from the interactions of many variables. We have not provided a series of vignettes detailing how particular fossil deposits have been interpreted environmentally. The synthetic approach to the study of particular fossil deposits has been carried out with varying degrees of success for almost two centuries. The intensive efforts of Sonderforschungsbereich 53 in this direction (see Seilacher, 1976, as well as Seilacher and Westphal, 1978 for detailed summaries of more recently published work) are examples of varied, and very useful, synthetic efforts. One could, and possibly should, prepare a detailed summary of such syntheses with examples devoted to the myriad environments represented in the fossil record. This type of "case" approach to the problem of benthic marine paleoecology would give the interested student a number of examples to emulate in terms of logic, technique, and conclusions to be drawn from data. We will never have, however, a complete set of consistently high-quality examples to set before the student. Therefore it is critical that the student be provided with the appropriate analytical tools and attitudes with which to work on newly encountered and poorly known examples. It would be poor pedagogy to encourage a "cookbook" approach to the analysis of past environments when no two situations are ever precisely identical. Above all, the student must try to think independently.

It should not be forgotten that the basic tool available to the benthic marine paleoecologist is the careful, combined mapping of community, sedimentary rock, and sedimentary structure data. To be most successful, such mapping should be done on a regional, broad scale and incorporate local, detailed surveys. The scale, or scales, will normally be dictated by the nature of the available exposures and the time available for study. The results of such mapping are presented as community frameworks, combined litho- and biofacies maps, and community diagnoses. They all lead to a more comprehensive basin analysis that provides a better environmental interpretation. A variety of other paleoecologic tools, including an understanding of functional morphology, will supplement and buttress this basic mapping program. But, it should be firmly understood that the basics of benthic marine paleoecology must involve a careful program of mapping. Needless to say, such a mapping program is commonly extended into the time dimension by careful consideration of litho- and biofacies behavior in stratigraphic sections. It must be emphasized that the map data must be interpreted in a manner consistent with that obtained from stratigraphic sections, and vice versa. Major advances in our understanding may be made once specialists in environmental studies of paleontology and sedimentology begin to work jointly. For example, work now being carried out on mid-Paleozoic rocks of Gaspe by Pierre-Andre Bourque, Laval University, in coordination with myself as paleontologist, suggest that, based on the study of sedimentological evidence, Benthic Assemblage 2 may well represent a low shoreface, high subtidal position, rather than the low intertidal.

Many years ago, Cloud (1959a) perceptively reviewed the status of paleoecology and commented on its future directions and possibilities. Many of his single paragraphs, individual sentences, and even clauses have since been transmuted into discrete papers, monographs, and books. This book is one more elaboration of his paper.

Acknowledgments

It is hardly possible to acknowledge all the people who have assisted in the accumulation of the material presented in these pages. My students have certainly played an important role. I began to study fossils seriously in 1948 and have been advised, assisted, and criticized by more people than I could possibly list or really credit for their help. The purpose of this book is to provide a practical guide, and I can only hope that those who have helped over the years will feel that the product has been worth their trouble.

Stephen Shabica, of the National Park Service, Bay St. Louis, Mississippi, deserves my thanks for providing helpful comments on an initial version of the manuscript. Robert Carney, of the Smithsonian Institution, in addition to preparing Chapter 5 and other important sections, has critically reviewed much of the initial manuscript in detail. Jane Gray, of the University of Oregon, Eugene, is involved in the preparation of a parallel treatment of nonmarine paleoecology. Dr. Gray and I have had to cope with similar problems and have used each other for sounding boards for some time. I am deeply indebeted to her for friendly assistance, constant advice, and useful criticism during the preparation of this entire manuscript.

During the final stages of manuscript preparation, the following colleagues have helped me toward an understanding of a number of unfamiliar problems dealing with fishes: Wendy Gabriel, William G. Pearcy, and David L. Stein, of the College of Oceanography, Oregon State University, Corvallis, Oregon; Roland L. Wigley, of the Northeast Fisheries Center, Woods Hole, Massachusetts; Edmund L. Hobson, of the Southwest Fisheries Center, Tiburon, California; John E. Randall, Bernice P. Bishop Museum, Honolulu, Hawaii; C. Lavett Smith, American Museum of Natural History, New York City. Consideration of these problems would have been impractical without their patient assistance.

G. Arthur Cooper, of the National Museum of Natural History, generously provided photographs of pathologic brachiopod specimens.

I am greatly indebted to Nils Spjeldnaes, of the Department of Palaeoecology, University of Aarhus, for giving me an opportunity to spend some weeks in his department during 1978 and for providing me with contacts in the Department of Ecology. The stimulus provided by Spjeldnaes and his colleagues, including Tom Fenchel, was of great value.

Important information is often contained in articles and reports that are, unfortunately, found in difficult to obtain publications. E. Doris Tilles, the interlibrary loan librarian at Kerr Library, Oregon State University, was a marvel at locating libraries in which these publications were available. I want to thank Ms. Tilles and her staff for their unfailing optimism and for their ability to find important papers and to produce copies of them. In addition, I commend the patience and enthusiasm with which the staff of the Science–Technology Division of the Kerr Library met my frequent and sometimes difficult requests for information: Thanks are due Robert Lawrence, Mike Kinch, Molly Goheen, Dave Schacht, Barbara Kienle, Mary Lewis, and all the other members of the staff. My work would have been far more difficult without their very able assistance.

1

Paleoecology

DEFINITION

Simply put, paleoecology is the ecology of the past. In principle, there is no distinction between ecology and paleoecology except for the limitations imposed on the former by the shortness of time available for observing ecological phenomena, and those imposed on the latter by the difficulty of directly measuring or indirectly estimating the many significant parameters. Such parameters as salinity, oxygen tenor, temperature, pressure, nutrient supply, tidal fluctuation, turbulence, biotic character, and abundance can be observed and measured directly by the ecologist, but can only indirectly be studied or inferred with the older materials available to the paleoecologist.

A real understanding of either paleoecologic or ecologic problems can be best realized by a thorough consideration of data available from both the past and present.

Paleoecology has both descriptive and historic aspects, and although they are intimately and inextricably involved with one another, we will consider them separately.

The basic compilation for anyone concerned with paleoecology is the *Treatise on Marine Ecology and Paleoecology* [Vol. 1, Hedgpeth (Ed.); Vol. 2, Ladd (Ed.), 1957]. Despite their publication date these two volumes remain the basic references in the field. During the past few decades the Zoological Record has devoted space to ecologic and paleoecologic materials; this is a rich source of ready reference material. "Textbooks" of paleoecology (see Ager, 1963; Gall, 1976; Hecker, 1965, for examples) commonly divide the field into *autecology* and *synecology*. Autecology may be

defined for the paleontologist as the ecologic information concerning an organism or individual species that can be derived from a study of the rock matrix and of the individual species. Through simple observation and measurement it is possible to gain some knowledge of the relations between species and the physical and biological environment in which they lived. For the ecologist, synecology is defined by Odum (1959) as the, "study of groups of organisms which are associated together as a unit [p. 8]." Attempts to recognize such synecologic groups in nature and to study the groups' responses to sets of physical parameters is perhaps the major activity in present-day ecology. In spite of an increasing application of multivariate statistics to large bodies of faunal and environmental data, synecology remains the most difficult approach to ecologic studies. In what ways can organisms interact among themselves, and with a complex physical environment, so that they behave as a unit? What are the synergistic properties of a group of organisms, and how does the field ecologist measure them? Despite the great difficulties that lie ahead for the synecologist, we think that a synecologic approach has as much value in paleontology as it does in ecology. Indeed, the possibility exists that ecologists may gain a far greater understanding of the types of interactions that could bind several species together into a synecologic unit from observing long time-span animal associations in the fossil record. Paleontological synecology can be seen as a process of varied observation and synthesis. Through observation of fossil assemblages that are sufficiently removed from one another spatially and temporally to be associated with different physical conditions, the paleoecologist is able to suggest rational correlates between the biological assemblages and the physical environment. Synecology and autecology are not mutually exclusive; each depends in part upon the other.

Much of the information in this unit deals with community ecology, or *coenology,* as some term it. Community ecology provides a convenient operational level for the modern-day ecologist as well as for the paleontologist. If the community is thought of as an assemblage of organisms that acts more or less as a unit, community ecology is clearly a subdivision of synecology. Community ecology is obliged to operate simultaneously with all the relevant autecologic and synecologic data available at any one place (see Chapter 4, pages 177–356). Although it is interesting to study communities of the past for their own sakes, it is important to understand that they can be viewed as a building block affording a unique type of data necessary for a more complete understanding of the evolutionary process. Coenology and biogeographic analysis are both methods of analyzing the biota, but they are not unrelated disciplines. The ultimate synthesis consists of a fabric woven from coenologic and biogeographic threads whose individual fibers are the myriad evolving taxa.

It is necessary to introduce the term *taphonomy.* Taphonomy is the study of the various processes (Schäfer, 1972) leading to the fossil record. In simple language, taphonomy[1] consists of those factors that destroy organisms prior to fossilization, alter organisms before fossilization, and partially destroy or alter organisms and/or their component parts previous to fossilization. This term obviously covers a vast array of problems, most of which are little known, little investigated, and at least partially misunderstood. The extent to which organism structure and interorganism relations are destroyed threatens the accuracy with which the paleontologist can reconstruct ecologies of the past. However much of a hindrance information loss upon preservation is, the paleontologist must remember that ecologists are faced with a similar information loss. As will be discussed, when benthic ecologists depend upon gross samplers such as grabs and trawls much of the information about spatial relationships and even morphologies is often destroyed.

TIME ASPECT AND SCALE

The ecologist and paleoecologist must have some comprehension of the different time and spatial scales imposed on one another by the nature of their objects of study. The ecologist can make observations on a human time scale; that is, on a clock and calendar scale ranging from fractions of minutes to at most decades, with the outside possibility that historical data provide information going back hundreds of years. Thus, the ecologist can observe phenomena occurring on a daily, monthly, yearly and decade-long term, with the possibility, in rare cases, for century long or exceptionally millenia-long phenomena. The ecologist can observe directly such events as setting of larvae, feeding behavior, reactions to physical change, death, transportation, locomotion, and reproduction, as well as the interactions of organisms with each other and with the physical environment. Remarkably, many biological processes have not yet been observed over an adequate period of time; many critical data depend on a single day's or single season's observations. The terrestrial and intertidal ecologist has the advantage of relatively easy, direct observation. Unfortunately, the deeper marine ecologist must often rely on samples obtained by a variety of devices unsuited to the spatial scales of the phenomena being studied. In addition, the inherent biases of the samplers are often unknown, and exact design and plan of use varies so much from worker to worker as to make comparison among samples highly doubtful (See Dickinson and Carey, 1975; Holmes and MacIntyre, 1971). Much of the data obtained from subtidal samples has been compiled by counting animals that have been processed through sieves and sorting devices. As with the spatial scale of the gross sampler, the scales chosen for these devices are usually not appropriate to the characteristic scales of the biological phenomena. The use of scuba gear has revealed the shallower subtidal regions and deep-sea photography (Heezen and Hollister, 1971), and deep submersibles (Grassle *et al.,* 1975) have extended visual observation to abyssal depths. Nevertheless, none of these techniques is comparable in ease or expense to the direct observations by terrestrial and intertidal researchers. Despite the technical elegance of some of the deep-water tools, they have not been able to produce a data base comparable in detail to those produced by terrestrial workers or paleontologists.

The difficulties of extracting macrofossils from different types of matrices give rise to a variety of sampling problems. Fossils extracted from beds cut by slatey cleavage present very different aspects (and shelly

[1]In the past decade there has been a sustained effort made toward really investigating taphonomic problems, with emphasis on the marine benthos, by Sonderforschungsbereich 53 Palökologie, and particularly Projektbereich's B (Fossil Lagerstätten) and E (Fossil-Diagenese) under the leadership of Professor Adolph Seilacher, Tübingen. Many results of these efforts have been published in the Monatshefte and Abhandlungen of the *Neues Jahrbuch für Geologie und Paläontologie.* Everyone concerned with taphonomic questions, as a paleontologist must be, should familiarize themselves with the varied materials produced by this group of scientists. See Seilacher (1976) for a summary of SFB-53 activities during the period 1970-1975.

biomass) than do those extracted from weakly lithified matrix. How many samples are extracted for palynomorphs, conodonts, and other microfossils? How many microfossil samples are also studied for macrofossils?

The paleoecologist frequently has the opportunity to observe biotic distributions and associations on a scale of millions, tens of millions, or even hundreds of millions of years. The paleoecologist's time scale is geologic; thus it routinely encompasses evolutionary changes. The paleoecologist may also observe changes on a shorter scale, but has little chance (except in the later Cenozoic, where Carbon-14 [C^{14}] and magnetic reversal stratigraphies are viable) to be certain of absolute time on a short-term basis. Thus the paleoecologist is generally restricted to measuring time in millions of years with the aid of relatively long-lived radiogenic isotopes. The paleoecologist interested in the subtidal environment often can study the spatial relations of organisms living on and in the sea bottom. But the vagaries of terrestrial postmortem transportation and deposition provide the paleoecologist with a far poorer view of the terrestrial environment.

Any discussion of paleoecology stresses the fact that the fossil record is almost totally restricted to organisms having preservable hard parts. The trace-fossil record does record something about certain soft-bodied organisms (chiefly unknown; Chamberlain, 1975), as well as some aspects of the behavior of certain organisms whose hard parts are known. It is sometimes said that the virtual absence of soft-bodied organisms precludes any definitive ecologic deductions. Although some trace of soft-bodied organisms would certainly allow us to form a more complete picture, the view that no deductions can be drawn is akin to claiming that an ecologist's failure to study the entire biota from place to place invalidates all ecologic conclusions. However, it is true that our deductions are necessarily limited. For example, attempts to recreate past trophic structure relations are barely caricatures of what must have been, even though trophic structures worked out differently for shelly and trace fossil communities can point out similarities that have persisted over time and changes that have occurred. Likewise, speculations about changes in biomass character as revealed from studies of the fossil record are based on studies of shelly biomass rather than studies of total biomass; but such speculations are nevertheless important, for the reasons just cited under trophic structure study.

Thus the overlap of data available from the present and the past is only partial. The ecologist and the paleoecologist must work together in order to achieve a synthesis that approximates reality, either past or present. The paleoecologist provides historical perspective for the ecologist; the ecologist provides the bulk of the directly observed relation existing between the biotic and the physical. The paleoecologist is concerned with physical events manifest on a geologic time scale such as long-term climatic gradients, transgression–regression, levels of volcanism, changes in land-mass positions, paleogeography, and the like; whereas the ecologist works with events on a time scale appropriate to human events.

2

Environments

In the succeeding sections, those physical variables that have been shown to affect the distribution of organisms in the marine and coastal environment will be discussed, with emphasis on the fact that these variables act in concert, not singly. Traditionally, ecologists have first recognized and mapped various major environments by the fauna and flora associated with a specific location. A careful study of the physical parameters such as temperature, salinity, depth, and so on, of that geographic area were measured later when the biological explanations for the distributions were sought. It is a basic tenet of paleoecology that the physical–geographic conditions of the past can be inferred from the associated fossil assemblage or morphologies of specific organisms. Although the physical variables of the marine ecosystem may grade gradually or abruptly over a geographic region, there are certain associations of physical conditions, geographic location, and associated biota that exist now and can be recognized as units in the fossil record. For the sake of simplicity of terminology, we refer to each of these units as an *environment,* specifying the precise meaning with an appropriate modifier.

MARINE ENVIRONMENTS

INTERTIDAL REGION

As reviewed by Hedgpeth *et al.* (1957b) there is a complete gradient from the land to the sea in both the aquatic environment, with its brackish, estuarine and lagoonal reaches, and the terrestrial environment with its supratidal, salt marsh, mangrove, and intertidal reaches. The fossil record provides many examples of varying ages taken from the brackish, estuarine, and lagoonal, but a smaller number from the supratidal. The salt marsh and mangrove environments have provided very few examples for the geologic record. There are many examples of intertidal level-bottom environments, but very few in the rocky-bottom category.

Hedgpeth's review can be supplemented by Newell's (1970) careful discussion of the physiological aspects of intertidal animal life history to provide insight into the intertidal world. The unique character of this environment includes complex interactions among the parameters of changing temperature, oxygen, salinity, food supply, desiccation, nitrogenous waste disposal, and so on, which have selected for a specialized group of organisms. Two major categories of organisms can be recognized. The first consists chiefly of eurytopic shallow-water marine creatures which do best in the low intertidal. The second group are derived from the larger marine environment, but have evolved into forms that are restricted to the intertidal region. Linguloid brachiopods are an example of the first category; their tolerance of fresh water allowing existence in the lower intertidal (Paine, 1963a). Jackson (1974) has studied the thermal tolerance of some bivalves which

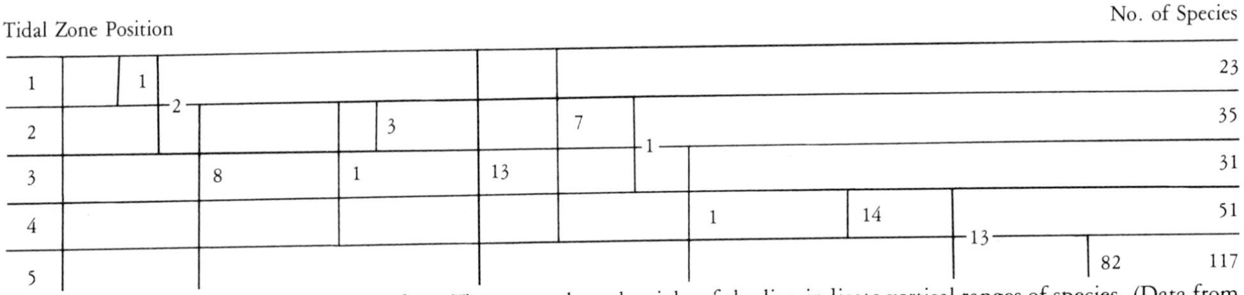

Tidal Zone Position   No. of Species

1		1							23
			2						
2				3		7			35
							1		
3		8	1	13					31
4						1	14		51
								13	
5								82	117

FIGURE 1. The intertidal–subtidal boundary. The numerals to the right of the line indicate vertical ranges of species. (Data from A. D. McIntyre and A. Eleftheriou, The bottom fauna of a flatfish nursery ground, 1968, *J. Mar. Biol. Assoc. U.K.*, 48, pp. 113–142, Table 2. Copyright © 1968 by Cambridge University Press.)

can survive intertidal conditions. The common bivalve *Donax* is a conspicuous example of the first category. Shore crabs such as the ghost crab *Ocypode* and the fiddler crab *Uca* are excellent examples of the second category. *Uca* contains species that live at different heights in the intertidal (Crane, 1975), and *Ocypode* has become quite terrestrial in its habits.

McIntyre and Eleftheriou (1968) have provided some taxic diversity data for macroscopic benthos across the intertidal and well down into (10 m) the subtidal (Figures 1–4). Their figures show a pronounced increase in number of taxa at about the mean low-water position. They list 117 species in the subtidal position (down to 10 m) and 60 species in the intertidal (their Table 5 presents this information more finely divided). Their Figures 7 and 8 also provide an idea of the amount of taxic discontinuity adjacent to the intertidal–subtidal boundary. Jackson (1972) provides somewhat similar data for a tropical region (Jamaica). Manton and Stephenson (1931) provide similar data for an area of the Great Barrier Reef. Kohn (1971) points out that in some genera (*Conus,* for example) the intertidal is much less speciose than the subtidal. Thus, we have every reason to accept a generalization that states that the intertidal region has a significantly lower number of benthic organisms than the subtidal, even the shallow subtidal. The transition in taxic diversity is not gradual over a fair distance; from the intertidal into the subtidal, there is an abrupt shift.

The intertidal biota is far more than a restricted number of species limited by the severity of the environment. The paleoecologist can make use of the special aspects of intertidal fauna to recognize the intertidal zone in the fossil record. First, there are fewer species found in the intertidal as compared to the adjacent subtidal (Batham, 1956). Second, there is a strong tendency for intertidal species of families, or genera that are both subtidal and intertidal to be larger (Boucot, 1975). Low taxic diversity may be combined with larger size and geologic evidence based on sedimentary structures (Ginsburg, 1975; Klein, 1972) to permit recognition of ancient intertidal deposits.

The burrows of infaunal bivalves tend to be far deeper in the intertidal as compared to the shallow subtidal regions (Figure 5 from McAlester and Rhoads,

1967). Thus, trace-fossil morphologies may serve as paleoindicators of the intertidal–subtidal boundary. However, as pointed out by McAlester and Rhoads, the depth of burrowing may be a response to levels of turbulence and not to position in the tidal range per se, and there may be some overlap in the shallowest subtidal region. Nevertheless, the bivalve taxa actually present should provide information as to the turbulence regime.

Caution should be exercised when the zonation of specific fossil species is used to infer a specific height in the tidal range. The conditions determining vertical zonation of a given species are complex. Foster (1969, 1971) has discussed extensive experimental studies describing the variables that affect the intertidal zonation of barnacle species. He concluded that the intertidal and geographic position of a species was highly correlated with its temperature tolerances, desiccation resistance, and other features of the climate.

When predation becomes an important factor in determining vertical zonation, the vertical position of specific organisms can no longer be taken uncritically as an indicator of a specific physical regime. In early experiments (Warren, 1936) it was found that the typical intertidal mussel *Mytilus edulis* could flourish subtidally when protected by caging from the subtidal starfish predators. Thus, fluctuations of key predators in time and/or space might have a great effect upon vertical zonation, and the relatively rare predator could seldom be preserved in association with its prey species.

The width of the intertidal zone may vary from a few meters to scores of kilometers, depending on slope and tidal amplitude. There is every reason to suspect that, in many places, the broad epicontinental platforms were characterized by correspondingly broad intertidal reaches. Also keep in mind that macrotidal intertidal regimes afford many more niches than do the microtidal. See Chapter 3 (pp. 25–175) for additional criteria distinguishing the shallow subtidal region.

Jackson (1977b) has pointed out the rarity of colonial organisms in the modern rocky intertidal environment; the fossil record is certainly in accord with Jackson's conclusions. In addition, note that hermatypic corals, dasycladacean algae, and crinoids are normally not present in the modern intertidal environment. The forego-

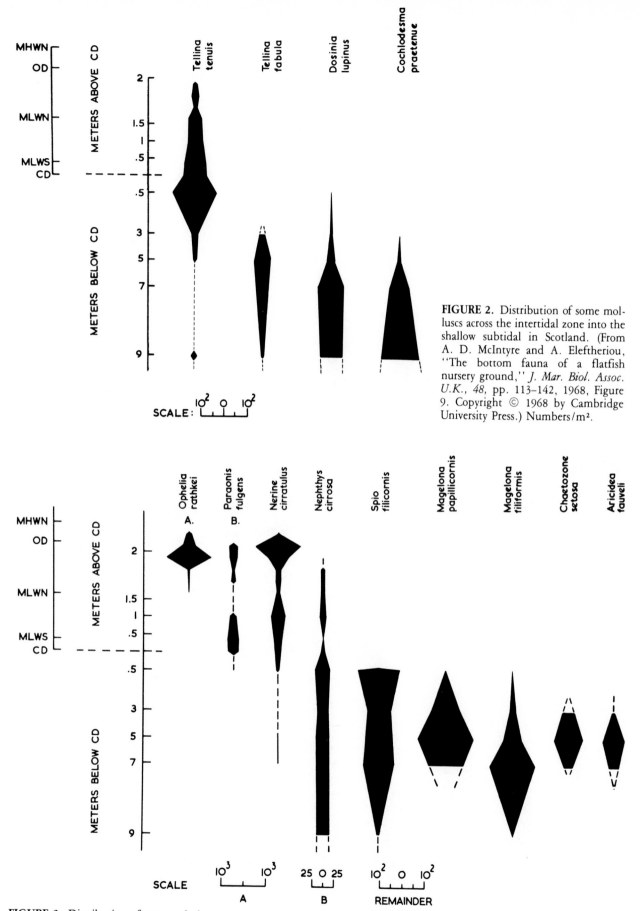

FIGURE 2. Distribution of some molluscs across the intertidal zone into the shallow subtidal in Scotland. (From A. D. McIntyre and A. Eleftheriou, "The bottom fauna of a flatfish nursery ground," *J. Mar. Biol. Assoc. U.K., 48,* pp. 113–142, 1968, Figure 9. Copyright © 1968 by Cambridge University Press.) Numbers/m².

FIGURE 3. Distribution of some polychaetes across the intertidal and shallow subtidal in Scotland. Numbers/m². (From A. D. McIntyre and A. Eleftheriou, 1968, The bottom fauna of a flatfish nursery ground, *J. Mar. Biol. Assoc. U.K., 48,* pp. 113–142, Figure 7. Copyright © 1968 by Cambridge University Press.)

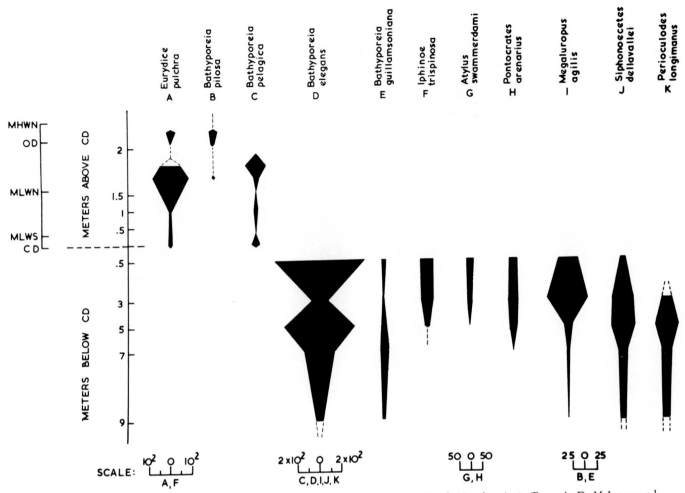

FIGURE 4. Distribution of some intertidal and shallow subtidal crustacea in Scotland. Numbers/m². (From A. D. McIntyre and A. Eleftheriou, The bottom fauna of a flatfish nursery ground, 1968, *J. Mar. Biol. Assoc. U.K., 48,* pp. 113–142, Figure 8. Copyright © 1968 by Cambridge University Press.)

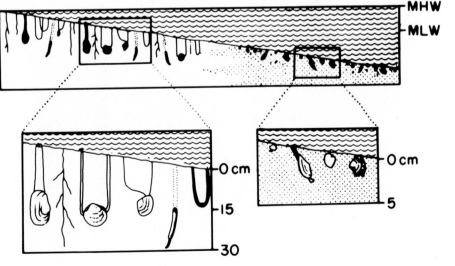

FIGURE 5. Diagram showing relation of bivalve burrow depth to water depth, with intertidal and shallow subtidal region shown enlarged at left and deeper subtidal zone enlarged at right. (From A. L. McAlester and D. C. Rhoads, 1967, *Marine Geology, 5,* pp. 383–388, Figure 2. Copyright © 1967 by Elsevier Scientific Publishing Company.)

ing are a few of the criteria in use by researchers trying to establish a lower limit for the intertidal in many examples from the fossil record.

Different Intertidal Environments Recognized in the Fossil Record

Beaches. Morton and Miller (1968) provide an extensive descripton of the uniqueness of various intertidal environments and transects (Figures 6–8); Stephen (1930, see Figure 9) presents semiquantitative information relative to intertidal transect shell-growth size differences; and Womersley and Edmonds (1958, see Figure 10) give examples of how many different environments with different biotas may be encountered within the intertidal environment.

Hedgpeth (1957a) points out that the intertidal sandy beaches, which grade imperceptably into the muddy beaches, lack large attached plants and support far fewer species than do the intertidal rocky bottom areas, with a greater level of environmental heterogeneity or the muddy intertidal substrates. These intertidal sandy beaches also commonly have a black, reduced layer at depth. Since much detrital food is supplied from the sea, active predation by snails is important, as is mortality due to protracted high temperature coupled with desiccation. Longshore currents help to effect larval distribution, and there are many vertical biotic zones on a beach and into the shallow subtidal sandy environment (see Figures 6–10 and Figure 165, p. 181).

The recognition of the beach-intertidal environment depends on various physical and biotic criteria. DeWindt's (1974) discussion of the ubiquitous calianassid burrows that leave the trace fossil *Ophiomorpha*, a distinctive, easily recognized structure (see Pryor, 1975; Sellwood, 1971; Weimer and Hoyt, 1964) known from the Jurassic to the present, is only one among many such criteria. Each of the varied intertidal environments has its own complex of biotic and physical features by which its presence may be recognized under favorable circumstances.

I have discussed (1975) the recognition of the intertidal environment of the past in terms of the following criteria (*a*) relatively low taxic diversity; (*b*) larger species shells in the intertidal than of the same genus in the shallow subtidal; (*c*) absence of coral reef from the intertidal although present up to mean low tide position; (*d*) the absence of evidence for pelmatozoans which are unable to tolerate desiccation (Brower, 1975); (*e*) the absence of the dasycladacean algae so characteristic of the shallow subtidal regions in the tropical and subtropical regions today (Johnson, 1961) and presumably in the past as well.

Salt Marsh. The salt-marsh environment is characteristic of only the macrotidal regimes (Davies, 1964; Gill, 1973, 1975) and is present in temperate regions; being essentially the homologue in the nontropical regions of the mangrove environments, which also occur in microtidal regions. Salt marshes appear to be absent in the polar regions where ice push would not permit their sediments or taxa to accumulate undisturbed. The taxa unique to the salt marsh have been studied in some detail by ecologists; thus it is surprising that the paleontologists have had such difficulty in recognizing the salt-marsh fauna (the flora does not appear to be preserved in most instances), although Murray (1973) explains that, with the aid of benthic foraminifera, this environment can be recognized in the geologically younger strata where comparisons with the Recent are valid. In some ways the salt marsh and the mangrove environment can be viewed as an intertidal swamp. Defined as such, many of the paralic coals of the Paleozoic, and their associated floral materials, might be seen as representations of intertidal swamps as suggested by the intimate interbedding of marine layers.

Mangrove. Paleobotanical evidence for the existence of mangrove taxa (such genera as *Rhizophora, Sonneratia, Avicennia*, etc., including *Nypa*), if not of the mangrove environment (Chapman, 1976, 1977; MacNae, 1968, for a discussion of some of the actual divisions of the mangrove environments), are available from the Paleocene onward (Muller, 1966). The restriction of the mangrove environment largely to the tropical and subtropical regions, where the conditions of high-level bacterial activity and oxygenation are prevalent, tends to discriminate strongly against preservation of the organic residues.

It is probable that most of the Paleozoic and post-Paleozoic coals represent the results of cool Temperate to Tropical zone activity rather than tropical or subtropical accumulations. Plaziat (1970) provides evidence for the presence of mangroves in the form of the xenomorphic impressions generated by mangrove bark on mangrove oysters both present and past (Figure 11), and Stenzel (1971) provides additional background on the subject of mangrove oysters.

Moorman (1963) discusses the fact that mangrove and other coastal, quiet-water environments in the tropics tend to generate sediment high in iron sulfide and organic residues. This material will ultimately lithify to pyrite, coaly shales, and siltstones. If the material weathers a bit before lithification, "cat-clays" formed from the reaction of sulfuric acid produce bleached spots; resulting in alums and jarosites. Lithified cat-clays might be recognizable in the rock record. The faunal and floral facies present in these environments considered with their lithology may help in definitively recognizing their presence.

Rocky Bottom. We have few examples of intertidal rocky-bottom communities and environments in the fossil record despite the present importance of these environments (See Lewis, 1964; Ricketts and Calvin,

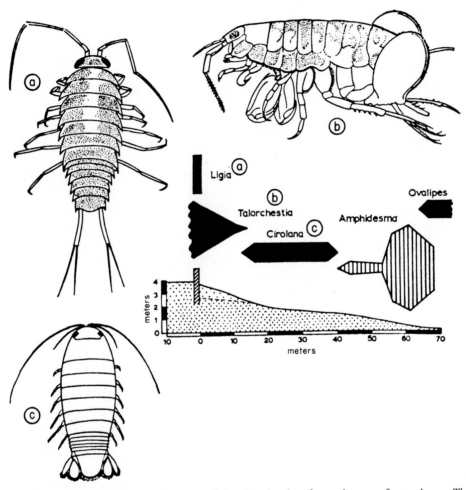

FIGURE 6. Section of the open sand beach in New Zealand showing the abundance changes of several taxa. The profile is based on an east coast beach, showing the successive zones and their typical occupants in relation to the dominant zoning bivalve *Amphidesma subtriangulata*. Illustrated in detail are: (*a*) *Ligia novaezelandiae;* (*b*) *Talorchestia tumida*, with characteristic expansions of posterior pereiopods; (*c*) *Cirolana arcuata*. (From J. E. Morton and M. Miller, 1968, *The New Zealand Sea Shore.* Copyright © 1968, Figure 167, by W. Collins & Sons, Ltd.)

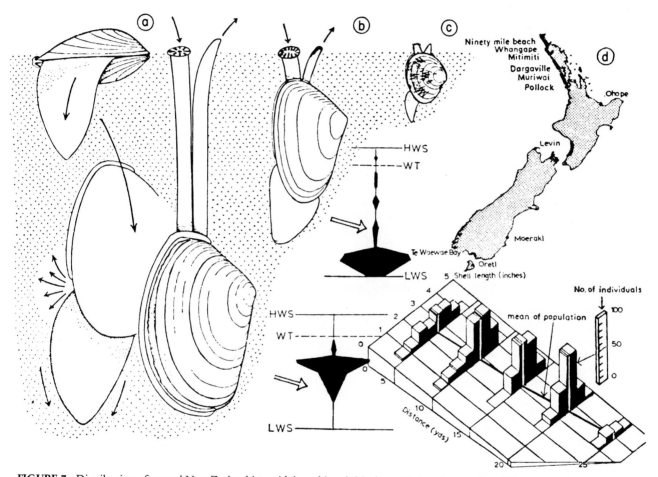

FIGURE 7. Distribution of several New Zealand intertidal sand beach bivalves. (*a*) the toheroa, *Amphidesma ventricosum*, shown in situ with stages in extrusion of the foot and digging (× ⅓). The inset shows the distribution across the beach at Muriwai; (*b*) *Amphidesma subtrianglulata,* shown in situ (× ½). The inset shows the distribution across the beach at Milford (in both insets the letters WT indicate the level at which the water table reaches surface of sand); (*c*) *Tawera spissa* shown in situ (× ½); (*d*) the location of the principal toheroa beds. (From J. E. Morton and M. Miller, 1968, *The New Zealand Sea Shore,* Figure 170. Copyright © 1968 by Wm. Collins & Sons Ltd.) Lower right: size distribution of the toheroa down a beach transect at Muriwai. (After R. M. Cassie.)

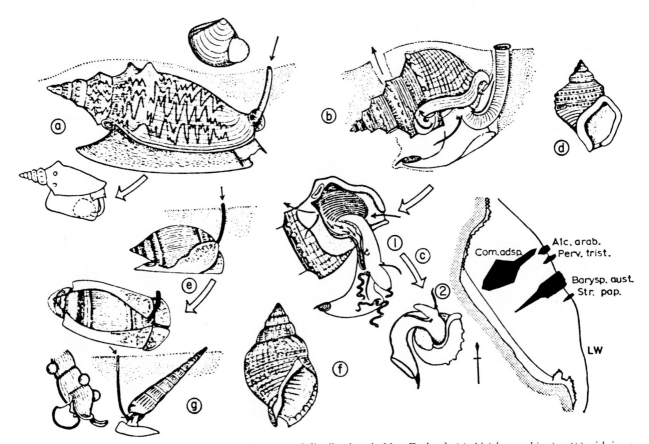

FIGURE 8. Some intertidal protected sandy beach gastropod distributions in New Zealand. (*a*) *Alcithoe arabica* (× ½) with inset sketches of egg capsule and of an animal feeding on a bivalve; (*b*) *Struthiolaria papulosa* (× ½), animal in position in the sand. (*c*) *Struthiolaria papulosa*: (1) animal extended from shell to display ctenidium and food groove, (2) use of the operculum in righting movement. (*d*) *Struthiolaria (Pelicaria) vermis* (× ⅔); (*e*) *Baryspira australis* (× ¾), the animal in situ in the sand and crawling upon the surface. (*f*) *Cominella adspersa* (× ⅔) (*g*) *Pervicacia tristis* (× 1) burrowing and (left) detail of shell with egg capsules of *Baryspira australis*. Lower right: distribution of five species across the lower beach at Cheltenham. (From J. E. Morton and M. Miller, 1968, *The New Zealand Sea Shore*, Figure 182. Copyright © 1968 by Wm. Collins & Sons Ltd.)

FIGURE 9. Diagram outlining the relation between species abundance with size distribution within the intertidal zone in Scotland (*see facing page*).
TOP: West Sands, St. Andrews (2/9 natural size). (*a*) Low Water Mark (¼ m²); (*b*) one-third way up the beach from Low Water Mark (¼ m²); (*c*) two-thirds way up the beach from Low Water Mark (½ m²).
BOTTOM LEFT: Sands at the Bar, Nairn (2/9 natural size). (*a*) Low Water Mark (¼ m²); (*b*) one-third way up the beach from L. W. M. (¼ m²); (*c*) two-thirds way up the beach from L. W. M. (¼ m²); (*d*) near High Water Mark (¼ m²).
BOTTOM RIGHT: Sands at Aberlady (2/9 natural size). (*a*) Low Water Mark (¼ m²); (*b*) one-third up the beach from Low Water Mark (¼ m²); (*c*) two-thirds up the beach from Low Water Mark (¼ m²); (*d*) near High Water Mark (¼ m²).
(From A. C. Stephen, 1930, *Trans. Roy. Soc.* Edinburgh, *56*, pp. 521–535, Figures 1–3.)

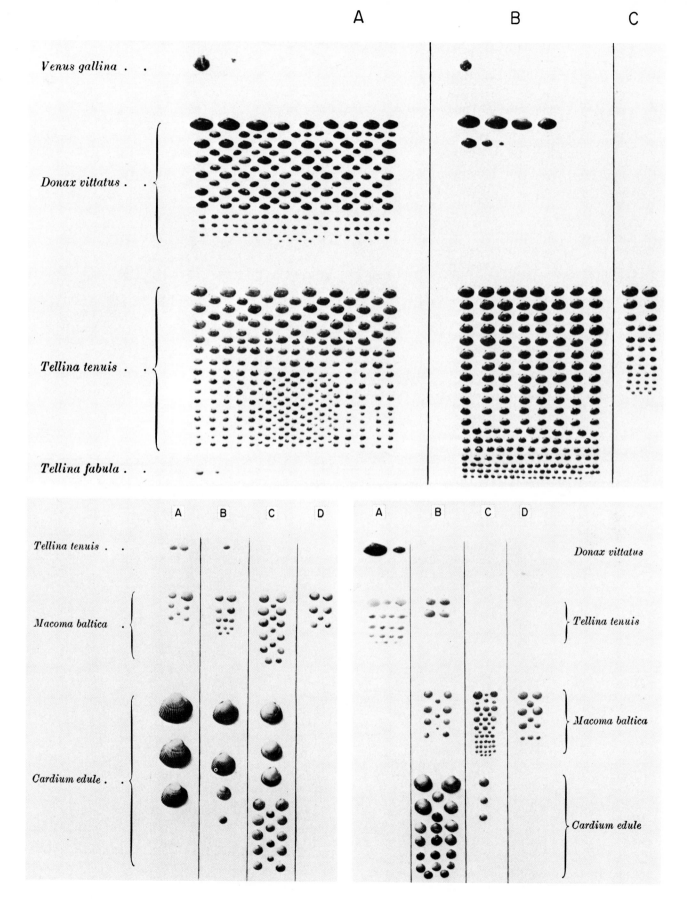

Zone	Western and central coasts					South-eastern coast
	Extreme-to-strong wave action			Moderate wave action	Slight wave action	Extreme-to-strong wave action
	Steeply sloping Palaeozoic rock	Horizontal rock platforms	Sand beaches	Sheltered coasts	Sandy or muddy flats	Horizontal rock platforms
Supralittoral	*Calothrix* · · · · · · · *Melaraphe*	*Melaraphe Lichina Verrucaria*	*Talorchestia*	*Melaraphe Lichina*	Relatively bare when rocky Samphires	*Melaraphe Lichina Verrucaria*
Upper littoral	*(Chthamalus) Chamaesipho*	*Chthamalus Chamaesipho*	Isopods	*Chthamalus Chamaesipho*	Samphires, mangroves, or *Bembicium*	
Mid littoral	*Catophragmus* (Molluscs, *Galeolaria*, blue-green algae)	*(Catophragmus)* Molluscs, *Galeolaria*, blue-green algae *Brachyodontes rostratus*		Molluscs, *Galeolaria*, blue-green algae	*Bembicium, Modiolus inconstans*, mangroves, *Enteromorpha*	Molluscs, *Galeolaria*, blue-green algae
Lower littoral	*Balanus*-coralline mat	*Hormosira* Other algae	Bivalves	*Corallina–Gelidium, Hormosira*	*Hormosira Brachyodontes erosus, Katelysia, Pinna*	*Hormosira*
Sublittoral fringe	*Cystophora intermedia*	*Cystophora intermedia*	Bivalves	*Cystophora* sp. ↓	*Hypnea, Spyridia,* and other algae	*Durvillea* ↓
Upper sublittoral	Fucoid algae etc.	Fucoid algae, *Ecklonia*, red algae		Other algae	*Zostera* ↓ *Posidonia* ↓	*Macrocystis*

FIGURE 10. Correlation of taxa and environments on some South Australian Coasts. (From H. B. S. Womersley and S. J. Edmonds, 1958, *Austr. J. Mar. Freshwater Res.*, 9, pp. 217–260, Table 2.)

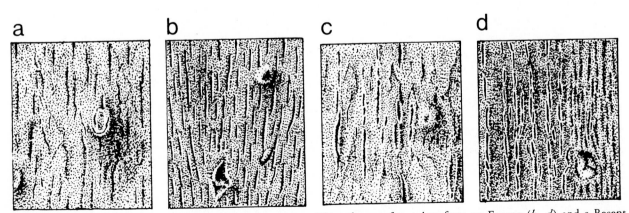

FIGURE 11. Impressions (positive and negative) of mangrove *Rhizophora* surface taken from an Eocene (*b, d*) and a Recent mangrove oyster (*a, c*). (From J. C. Plaziat, 1970, *Geobios, 3*, pp. 7–27, Figure 4.)

1939; Stephenson and Stephenson, 1972, for good summaries). Reasons for this apparent lack include the predominance of epicontinental sea deposits in much of the fossil record. These deposits are known for an absence or relatively low level of shoreline deposits compared to the present. Their physiographies are characterized by very low relief; virtual peneplanation, unlikely to be rich in the rocky-intertidal environment. In addition to the epicontinental sea problem, much of which occurs in regions of tropical weathering that tends to quickly reduce areas of bold relief, is the inherent self-destructive effect of the rough-water, rocky-bottom intertidal environment on its fauna (although not on the quiet-water, rocky-bottom intertidal environment). The enigma is well illustrated by the fossil record of the common rocky-bottom intertidal barnacles, which have an almost complete gap from the Cretaceous (their first appearance) to the present. This situation is puzzling, as the paleontologist has no difficulty in recognizing level-bottom intertidal regimes from the Cambrian to the present in some abundance; the facies change from nonmarine to marine is common knowledge to just about every biostratigrapher using all types of physical and biotic evidence. Where then are the intertidal rocky bottoms? A similar absence is encountered in the subtidal with the exception of the carbonate-cemented hard bottom. Perhaps the prominence of the rocky bottom today is largely a function of the overall continentality and high stand of the continents, a situation that has apparently been most uncommon during the rest of the Phanerozoic, although the possibility that paleontologists have been negligent in searching for and recognizing the rocky intertidal environment cannot be completely dismissed.

Woodring (1928, 1931) has emphasized the rarity of the shell types that are known to have lived attached to and under rocks. Why, for example, should abalones be so uncommon in the fossil record compared to the present? Again we may appeal to the relative scarcity of rocky substrates as compared to level bottoms both today and in the past. The shells referred to by Woodring do not always inhabit unusually turbulent epifaunal environments so that elimination from the fossil record by postmortem breakage and abrasion is not a reasonable chief explanation.

BRACKISH ENVIRONMENTS
AND ESTUARINE ENVIRONMENTS.

The brackish environment, whether intertidal or subtidal, is commonly recognized by a combination of biological and physical criteria. First, the physical criteria suggesting intimate facies relations with nonmarine beds are important when combined with a paleogeography consistent with separation from truly marine deposits. Excellent examples for both the intertidal and subtidal are provided for the Black Sea–Cas-

pian–Pontic region and for the Baltic region by Caspers (1957); Zenkevitch (1963), and Runnegar and Newell (1971) for the South American Permian. The biotic evidence depends on lowered diversity at the class and ordinal level, accompanied by an overall lowered diversity at the specific and generic levels (see Kauffman, 1973, for a Cretaceous example). In other words, the brackish environment is generally unfavorable to many groups of organisms. Following Gunter's (1956) lead (see also Emery and Steveson 1957) it would appear that the lower salinity biotas of the brackish and estuarine environments represent areas where euryhaline taxa and taxa derived from the marine (not from the freshwaters) environment predominate. In many ways the brackish environment is a world unto itself rather than a busy halfway station. It is obvious that the aquatic taxa of the nonmarine environment have largely found their way through the brackish and estuarine environments, but their numbers have been very few. Their subsequent wildly rich diversification on land should not blind one to the fact that very few taxa survived to the land before being subject to this ecologically rich, biogeographically controlled, adaptive radiation diversification that was not possible in the aquatic environment (either fresh or salt) because of the relative scarcity of niches. Murray (1973) has summarized much of the literature devoted to using foraminifera in delineating the brackish and estuarine environments. Lohman (1957) provides examples of the use of diatoms for the same purpose. Tarlo (1967) has summarized evidence from the early Mesozoic devoted to brackish-water reptiles. A summary of the distribution of manatees and dugongs of the Cenozoic, as well as Mesozoic to Recent crocodiles, would presumably produce a similar result. Runnegar and Newell (1971) comment on the famous little Permian reptile *Mesosaurus,* the presumably fish-eating, brackish water, endemic of the Gondwana region. Hecker *et al.* (1963) describe a Paleogene impoverished fauna from Central Asia.

Land-locked bodies of brackish water tend to develop highly endemic faunas as is pointed out for the Gondwana Permian (Runnegar and Newell, 1971; including Caspers, 1957; Zenkevitch, 1963) and also for the Paratethys Pontic regions during the later Cenozoic. However, little attention has been devoted to the patterns of evolution present in the brackish-water regions more intimately connected with the seas as in coastal lagoons and estuaries. It would be most illuminating to try to trace the modern biotas of the coastal lagoons and estuaries back into the Mesozoic and to ascertain their overall patterns of stability and exchange with the fully marine environments. We need to evaluate the extent to which the coastal-brackish environments as opposed to the adjoining fully marine faunas tend to be globally isolated from each other in a reproductive sense.

The brackish and brackish–estuarine environments are characterized by a sharply reduced number of taxa

TABLE 2.1
Salinity Distribution and Faunal Distribution along the Knysna Estuary, South Africa[a]

Percentage	Mouth (%)	Lagoon (%)	Middle reaches (%)	Head (%)	Total species (%)
Salinity range	34.5–35.7	29.1–34.8	18.9–26.5	1.1–14.0	
Fresh water	0	0	0	7 = 18.7	7 = 2.3
Estuarine only	3 = 1.7	20 = 10.0	12 = 18.7	12 = 36.4	27 = 8.6
Euryhaline and marine	80 = 44.6	108 = 54.2	46 = 71.9	14 = 42.4	137 = 44.1
Stenohaline marine	96 = 53.7	71 = 35.8	6 = 9.4	0	139 = 45.0
Total species	179	199	64	33	310

[a] From J. H. Day, *In* G. H. Lauff, 1967, Estuaries, *AAAS Pub. No. 83*, Table 1. Copyright © 1967 by American Association for the Advancement of Science.

including a few euryhalines derived from the marine environment, a few tolerant freshwater forms, and a few truly endemic brackish-water forms. Even the brackish-water plant population is reduced in number of species (see Table 3.11, p. 127) as compared with either the fresh or the marine environment (Gessner and Schramm, 1971). Table 2.1 (Day, 1967); Figures 111 and 112 (p. 125) (Emery and Steveson, 1957, modified from Remane, 1934); and Table 2.2 (Remane, 1934) provide some idea of the reduced taxic composition of the brackish-water and brackish-estuarine environment.

The truly reduced number of taxa which are restricted to the brackish and brackish–estuarine environments probably reflects the geologically transitory and ephemeral nature of that environment. Only when such an environment is permitted to remain fixed for a significant length of time can cladogenetic development of a rich and varied brackish fauna take place. Those taxa having reproductive characteristics that permit them to spread easily from one estuarine region to another will, because of the large interbreeding population size, be the ones that will tend to evolve most slowly.

TABLE 2.2
Distribution of Species Belonging to Several Major Animal Groups for the Marine through the Brackish Water Environments of the North Sea and Eastern Baltic Sea[a]

Group	I	II	III	IV
Porifera	15	13	0	0
Hydroidpolypen	49	26	7	3
Scyphozoa	8	5	2	1
Anthozoa	12	5	2	0
Ctenophora	3	2	1	1
Nemertini	?	ca.25	3	2
Archianneliden	12	7	1	0
Polychaeta	>100	66	13	5
Cumacea	19	5	2	0
Mysidacea	9	7	5	4
Amphipoda	?	36	13	8
Decapoda	49	13	5	0
Amphineura	3	1	0	0
Lamellibranchia	69	34	7	5
Opisthobranchia	—	23	6	4
Echinodermata	27	8	1	0
Ascidiae	16	6	1	0

[a] From A. Remane, 1934, *Zool. Anz., verh. Deutsch. Zool. Gesellsch. Supplementband*, 7, pp. 34–74, Tabelle 1.

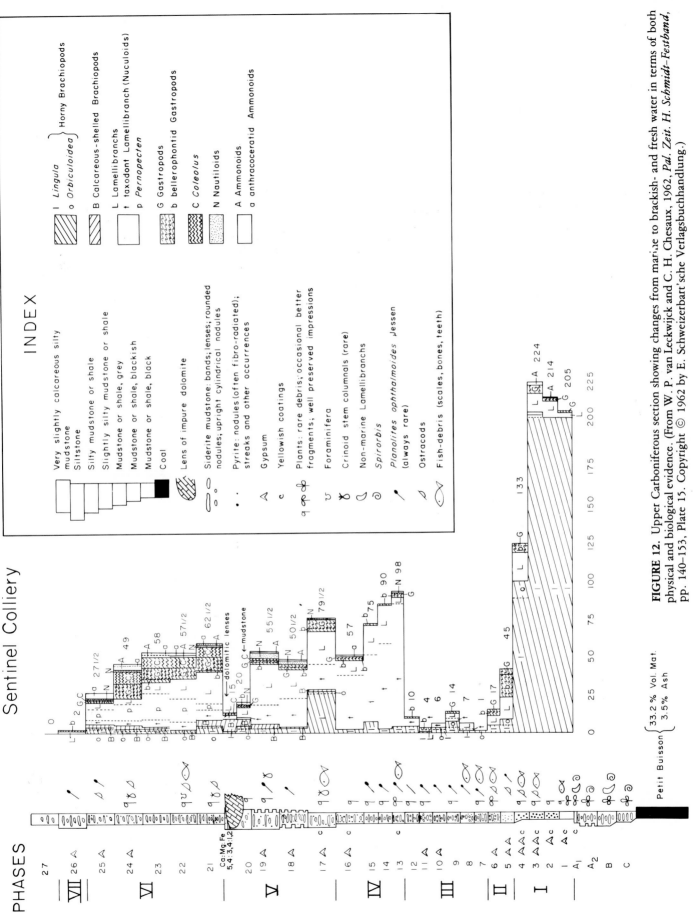

FIGURE 12. Upper Carboniferous section showing changes from marine to brackish- and fresh water in terms of both physical and biological evidence. (From W. P. van Leckwijck and C. H. Chesaux, 1962, *Pal. Zeit. H. Schmidt-Festband,* pp. 140–153, Plate 15. Copyright © 1962 by E. Schweizerbart'sche Verlagsbuchhandlung.)

Additional evidence suggesting that there may be more vacant niches within the estuarine and brackish-water region is Carlton's (1975) conclusion that exotics are far more easily established within existing estuarine and brackish systems than they are in the fully marine.

Rhoads (1975) shows that larger numbers of infaunal burrowers (Figure 115, p. 128), in contrast to nonburrowers, go into the brackish environment because of the more constant salinity. Calder (1976) maintains that there are relatively fewer epifaunal estuarine species than infaunal estuarine forms in those estuaries where the water salinity is subject to erratic fluctuations, since changes in the salinity of the infaunal environment are smaller in magnitude, and occur less frequently. Calder's point could be used to distinguish estuarine environments in the fossil record subject to fluctuating conditions of salinity from those possessing an unchanging infaunal taxic diversity contrasted with high and low epifaunal diversities.

The recognition of the brackish, estuarine transition situated between the fully marine and nonmarine environments is commonly very difficult to establish. In most instances, the use of purely physical criteria is not sufficient. However, a number of studies have combined animal and plant fossil information, including community-type data, with physical information to provide a very sound approximation to actual conditions. Examples worth noting may be found in Belt, 1975; Boger, 1964; Calver, 1968a,b; Hudson, 1963; van Leckwijck and Chesaux, 1962; see Figures 12 and 113, page 126.

Many of the examples are taken from the Carboniferous and the Permian. This is due partly to the extent of the coal workings and also to the intensive study to which Carboniferous and Permian cyclothems have been subjected, as contrasted with many other parts of the stratigraphic column.

SUBTIDAL REGION

The subtidal region is more extensive than the intertidal and has presumably been so through geologic time. It encompasses the sublittoral, the continental shelf, bathyal, abyssal, and hadal zones in the open ocean. For paleontological purposes, this varied environment can be divided into two zones, the photic–phytal and subphotic, which may be easily recognized in the fossil record. Nothing can be said here concerning paleoecological recognition on a finer scale of the bathyal, abyssal, or hadal environments because the fossil record available to us today provides little information.

Subtidal Environments Recognizable in the Fossil Record

Photic–Phytal Zone. Both in the present environment and in the past, it is easy to see the importance of a division into photic and subphotic zones. Production of plant food whether it be phytoplankton or bottom-attached algae is restricted to the photic zone. The bottom-attached, subtidal marine plants provide a variety of environments impossible in the subphotic zone and, consequently, greatly enrich the environmental and taxic diversity of the shallower subtidal region. The importance of lime secreting algae at present and in the past is clear, and their skeletal debris provides us with the possibility of paleontologically recognizing the photic zone. Various boring algae do the same (Golubic *et al.*, 1975). The probability that hermatypic coral taxa of the post-Paleozoic were dependent on zooxanthellae (Yonge, 1963a) gives significance to the presence of hermatypics as indicators of the photic zone, as does their absence in the intertidal region. Tridacnids may be employed in the same manner. The inference that some reef organisms of the Paleozoic, including many of the tetracoral and stromatoporoid taxa, signify photic zone conditions is in agreement with cooccurrence of these taxa with various types of calcareous algae (see Wray, 1971, for a summary of calcareous algae possibilities).

The overall tendency for plant-derived food to diminish from the shoreline region is partly a function of the primary productivity of the photic zone, partly a function of the more abundant nutrient supply in the shoreline region as a result of influx of dissolved nutrients from the land and of the high productivity of brackish areas including mangrove, salt marsh (Odum and de la Cruz, 1967), and estuary.

The overall decrease in biomass found as one moves away from the shoreline (Rowe, 1971) is well known in the modern environment. Virtually no work has been done in ancient environments attempting to estimate shelly biomass as a function of shoreline proximity, but one does observe over the years that shells become rarer as the facies becomes more distant from the shoreline region and approaches the shelf margin.

The interbedding of photic–phytal zone materials with subphotic zone materials should not cause confusion. Clearly, near or adjacent to any boundary, there is apt to be an interlayering of debris derived from either side of the boundary. Minor changes in bottom topography, transparency of the water, and even seasonality will permit such a boundary to fluctuate a bit—creating a transitional zone. The existence of a messy transitional zone should be favorably regarded as evidence for the presence of photic–phytal and subphotic zones on either side.

Subphotic Zone. The subphotic zone today is, of course, divided into outer shelf (or deeper shelf), shelf margin, bathyal, abyssal, and hadal regions—each with its characteristic fauna. Paleontologically, the subphotic zone is recognized for the most part as the result of paleogeographic studies indicating a departure from the region where calcareous algae and reef deposits occur in growth position, segregated from intertidal region and nonmarine deposits. Recognition of photic zone materials rolled into or transported into the subphotic zone will always be a problem, but at worst, it leads to the existence of a transitional area unless there are regions of excessively steep bottom slope. (See Boucot [1975] and Shabica and Boucot [1976a,b] for a review of some of the factors that permit recognition of the shelf margin region.)

The absence of an adequate abyssal fossil record (see Boucot, 1975, for a brief discussion of the problem) is puzzling as the presence of bivalves, varied echinoderms, and fishes of all sorts would lead one to expect a reasonable diversity, low population density, set of collections from the varied dredge hauls, JOIDES cores, and the like. Menzies *et al.* (1973) discuss some of the characteristics of the modern abyssal fauna. It may be, however, that the abyssal infauna as deposit feeders, as well as scavengers, are so unusually efficient that virtually nothing is left to the fossil record except a few ostracodes, plus Foraminifera, and that the solution of calcium carbonate beneath the carbonate compensation depth (CCD) is also a factor. The deep bathyal and abyssal macrurid fish scales described by David (1946a,b) are, unfortunately, a Paleogene exception that proves the rule. Benson's work on the abyssal ostracodes (1975) is another exception, but it provides far too little to permit many generalizations.

TERRESTRIAL ENVIRONMENTS

It is one of the paradoxes of ecology and paleoecology that the ecologist has focused attention almost totally on the terrestrial environment (see Figures 13–15 for examples of some of the possibilities for subdivision) whereas the paleoecologist has riveted attention on the marine environment. This paradox makes sense in terms of the more ready access and economic importance of the terrestrial for the ecologist and of the marine for the paleoecologist. Therefore, any treatment of the terrestrial environment of the past in terms of ecology is likely to be more a recital of possibilities for future study than a compilation of what we now know.

The best known environment on the land during the past is the lacustrine; because it provides, at least in the case of large bodies of water, a far better site for deposition of organic materials in a relatively undisturbed condition than does the fluviatile environment; also an environment favoring the reducing conditions usually necessary for preservation, so commonly absent in the subaerial environment. Despite this introduction there are few paleoecologic studies dealing with lacustrine materials.

Studies of the fluviatile environment are chiefly accounts of bits and pieces of log and bone buried here and there; hardly what could be called paleoecology. Important exceptions are Matthews (1970a,b, 1974a,b). The subaerial environment has provided information largely preserved in either fluviatile or lacustrine deposits complicated by the problem of post–mortem transport.

The paleobotanists have been concerned chiefly with ecology on the large scale; little attention has been given to defining recurring associations. However, from the Carboniferous to the present, a certain amount of attention has focused on floral differences that probably correlate with proximity to shore line and to differences in elevation.

Cryptic environments of the past, with the exception of Quaternary deposits which have afforded much information, are essentially unknown on land although there is reason to believe that they existed and that the fauna they contained was as significant as present.

It is important to understand that the fossil record for the terrestrial environment provides a fair amount of data concerning regions of low relief, particularly those that were generated near sea level, but exponentially fewer in high-relief regions and in low-relief regions that are removed from sea level. The rarity or absence of high-relief and high-elevation biotas in the fossil record should not be interpreted as indicating their recent evolution any more than should the virtual absence of desert biotas from the fossil record (no mention need be made of the small chances for preservation in the pitiless high oxidizing desert environment subject to violent erosive activities during the rare intervals of catastrophic rainfall).

Nonmarine hypersaline conditions are well known in the Cenozoic (see Palmer, 1957, for a Miocene example) as evidenced by the presence of actual evaporites. Further back in the column, few evaporite deposits have survived. However, the presence of crystal voids representing the solution of evaporite minerals (see Figure 16 for a Triassic example) do provide a means of obtaining positive evidence favoring hypersaline conditions; evidence which can aid in the interpretation of biotic data found in association.

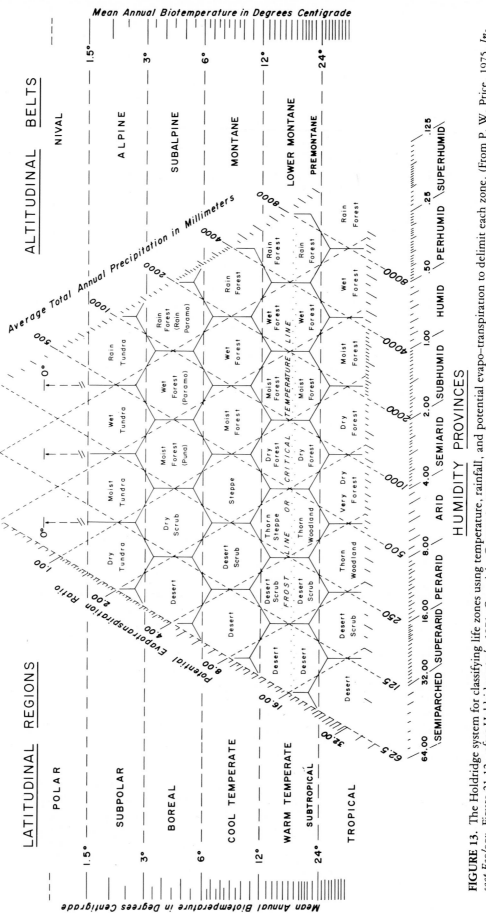
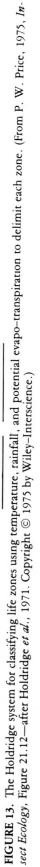

FIGURE 13. The Holdridge system for classifying life zones using temperature, rainfall, and potential evapo-transpiration to delimit each zone. (From P. W. Price, 1975, *Insect Ecology*, Figure 21.12—after Holdridge *et al.*, 1971. Copyright © 1975 by Wiley–Interscience.)

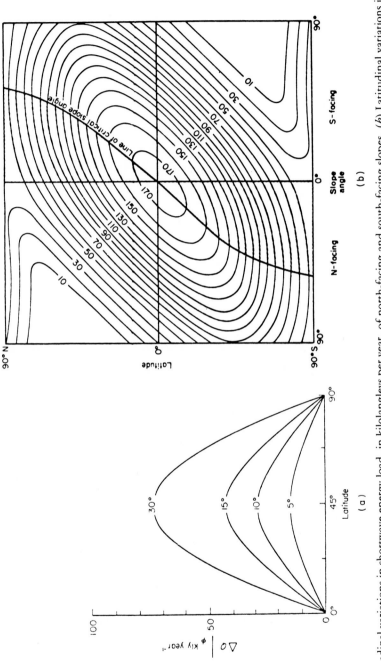
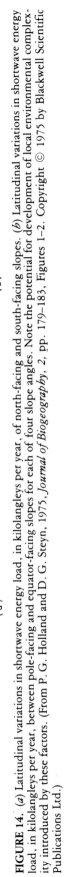

FIGURE 14. (*a*) Latitudinal variations in shortwave energy load, in kilolangleys per year, of north-facing and south-facing slopes. (*b*) Latitudinal variations in shortwave energy load, in kilolangleys per year, between pole-facing and equator-facing slopes for each of four slope angles. Note the potential for development of local environmental complexity introduced by these factors. (From P. G. Holland and D. G. Steyn, 1975, *Journal of Biogeography*, 2, pp. 179–183, Figures 1–2. Copyright © 1975 by Blackwell Scientific Publications Ltd.)

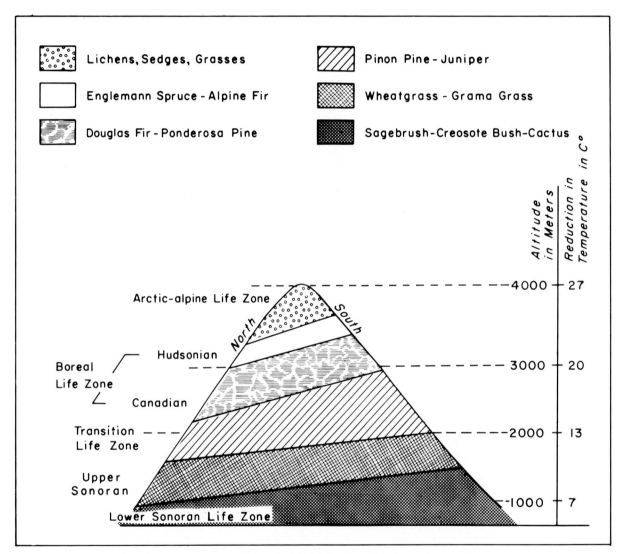

FIGURE 15. Life zones recognized by Merriam (1931) in the Rocky Mountains. South slopes in the Northern Hemisphere receive more insolation and thus ecotones are at slightly higher altitudes than on north slopes. (From P. W. Price, 1975, *Insect Ecology*, Figure 21.11 after Boughey, 1968. Copyright © 1975 by Wiley-Interscience.)

FIGURE 16. Glauberite crystal cavities in Upper Triassic Shale from Pennsylvania. (From E. T. Wherry, 1916, *American Mineralogists* (× 2).)

3

Variables of the Paleoenvironment

———————— INTRODUCTION TO ECOLOGIC ATTRIBUTES ————————

ECOLOGIC ATTRIBUTES

Kinne (1970, 1971a, b, 1972, 1975) has provided a comprehensive view of the characteristics known to be of ecological importance in the marine environment. This excellent summary of data and interpretation can serve as a guide to paleoecologists who must make use of every bit of evidence that can be gleaned from fossil material and its rock matrix. In subsequent sections, we will discuss those parameters which may be measured or reasonably inferred by the paleoecologist, weigh their importance, suggest additional parameters that might be worth studying, and give examples from the literature of successful attempts at extracting useful information about physical parameters from the fossil material.

Kinne (1970, 1971a,b, 1972, 1975) has emphasized that the distribution of marine organisms is not so much the result of any single factor, either physical or biological, as it is the complex interaction of **all** factors. This understanding is illustrated by the example shown in Figure 17, in which each parameter of the physical environment is represented as an orthogonal dimension in a hyperspace. It is Hutchinson's (1944) ecologic hyperspace concept in practice. In the example presented in Figure 17, the values representing temperature, oxygen, and salinity describe a solid in three-dimensional space. Within this solid of actual environmental conditions, there is a surface that corresponds to conditions that the American lobster can tolerate. The complexity of the survival surface serves as a warning that the simplistic, yet common, assumption

that a single factor or even a few factors can explain an organism's presence or absence is unrealistic and probably wrong. Although Figure 17 clearly illustrates the complexity of the interaction of physical factors, it takes a misleading, simplistic attitude toward the species. If one were concerned only with the animal's growth and reproduction rather than with total survival, separate surfaces in the environmental space could correspond to each of these biological processes.

It will be emphasized again and again that the distribution of organisms, both on land and sea, should be viewed as **correlated with** rather than **caused by** any of the variety of factors to be discussed. It is critical not to interpret a high correlation between or among variables as causal, despite the probability, in many instances, that high correlation may be synonymous with cause. Paleoecology (and ecology as well) is a field in which conservatism and reluctance to draw final conclusions are virtues. Nature is almost always more complex than we are given to expect. Once a physical parameter in the paleoenvironment has been inferred or accurately measured, the interpretation of its significance must rely heavily upon biological knowledge. If the assumed relation between any physical factor and the biota is misunderstood in terms of paleoecology, the interpretation will suffer. For those physical factors that have extensively-studied physiological impact, such as temperature, light, salinity, and pressure, interpretation may be straightforward. However, when parameters such as organic carbon or environmental nutrients are involved, the actual manner in which the factor influences the biota must

25

NC WESLEYAN COLLEGE LIBRARY
ROCKY MOUNT, NC 27804

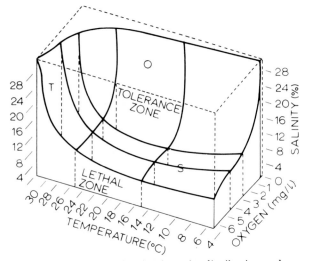

FIGURE 17. Diagram showing how the distribution and occurrence of an organism (e.g., the American lobster *Homarus americanus*) is controlled by the combined interaction of *all* variables to which that organism is sensitive. (From O. Kinne, Ed., 1972, *Marine Ecology*, vol. 1, part 3, Figure 12-16, p. 1684. Copyright © 1972 by Wiley–Interscience.)

be known. Too often, biology only provides assumptions as to interactions: These are not immediately useful to the paleoecologist. A prime example is the lack of available information about how sediment-feeding animals extract nutrient from the sediment.

There is no doubt that the interacting complexities of the marine environment as controls for organic distributions lack terrestrial counterparts. The terrestrial realm is far more complex, and affords much greater environmental heterogeneity than the marine realm. The restrictiveness of the terrestrial realm excludes most major groups of organisms (Nicol, 1972, Table 3.1), but those few that have adapted to it maintain distributions correlated with a bewilderingly complex set of interacting variables. Because of the relatively poor fossil record provided by terrestrial environments, the attention of the paleoecologist is largely directed toward the marine environment.

ATTRIBUTES AVAILABLE TO THE ECOLOGIST

Kinne (1970, 1971a, 1972, 1975) encyclopedically reviews the variables available to the marine ecologist. In addition to substrates, such items as pressure, light, seasonality, temperature, salinity, nutrients, oxygen, suspended sediment, turbulence, tidal phenomena, and the like must be considered. The number of variables is limited only by lack of ingenuity in the investigator. The extensive discussions of the ecologic importance of physical parameters pertain to autecology because they are based upon known effects of certain parameters upon specific organisms. There are a host of biological parameters that must be considered as well,

TABLE 3.1
Occurrence of Phyla in Marine, Freshwater, and Land Environments[a,b]

Phylum	Marine	Freshwater	Land
Protozoa	x	x	x
Porifera	x	x	
Mesozoa	x		
Monoblastozoa			x
Coelenterata	x	x	
Ctenophora	x		
Platyhelminthes	x	x	x
Rhynchocoela	x	x	x
Acanthocephala	x	x	x
Rotifera	x	x	x
Gastrotricha	x	x	
Kinorhyncha	x		
Priapuloidea	x		
Nematoda	x	x	x
Gordiacea	x	x	x
Calyssozoa	x	x	
Bryozoa	x	x	
Phoronida	x		
Brachiopoda	x		
Mollusca	x	x	x
Sipunculoidea	x		
Echiuroidea	x		
Myzostomida	x		
Annelida	x	x	x
Tardigrada	x	x	x
Pentastomida			x
Onychophora			x
Arthropoda	x	x	x
Chaetognatha	x		
Pogonophora	x		
Echinodermata	x		
Pterobranchia	x		
Enteropneusta	x		
Planctosphaeroidea	x		
Tunicata	x		
Cephalochordata	x		
Vertebrata	x	x	x
Totals 37	34	17	15

[a]From D. Nicol, 1972, *Quart. Jour. Florida Acad. Sci. 3*, pp. 191–194, Table 1. Copyright © 1972 by the Florida Academy of Science.

[b]Note how few living phyla occur on land as contrasted with the aqueous, particularly the marine, environment. Consider how ecologically significant is the far greater environmental heterogeneity of the terrestrial environment (possibly 800,000 land; 166,677 marine; 33,333 freshwater living species according to Nicol (1972) p. 191).

because these may determine the organization of individual animals into a synecologic unit. Of particular importance is the behavior of individual organisms with respect to other organisms, and the response of one organism to a modification of the environment caused by another organism.

Despite the abundance and diversity of literature dealing with ecology, both marine and terrestrial, the paleoecologist and his ecological colleague are aware of the sparcity of available knowledge compared to potential understanding of the relations of organisms to their environment. The paleoecologist continually reconsiders data provided by the fossil record in light of data continually provided by the biologist. Long favored conclusions, deductions and correlations must be discarded as additional information, and with it understanding, appears. We proceed, by a series of successively more meaningful approximations to a valid interpretation.

Community Parameters

There are many phenomena in nature that may be modeled as relatively simple processes, but that would yield a complicated set of data if carefully recorded by an observer. This is commonly the case in many fields of science. Analysis of data proceeds by the adoption of a model that allows reasonably accurate representation of the entire data set by a small number of sufficient parameters. When data are normally distributed, the mean and the standard deviation describe the data set for many purposes. Ecologists have attempted to find a sufficient number of statistical parameters to allow them to reduce the complexity of the species rank of abundance data produced in their field surveys. Often the parameters have been adopted without reference to an underlying model, whereas at other times a parameter suggested by a model has proven to be insufficient for its intended purpose. The interested reader is encouraged to read the review by May (1975) and to consult the many papers cited by May.

Diversity indices are the most commonly used "community parameter" in ecology today. Although new ones are constantly being published, suggestions of meaningful interpretations are rare (see Pielou, 1975, for a review). In spite of the variety of indices, most of the calculations are intended to summarize information on the relative abundance of different species in a sample, and by extension, in the sampled environment. Throughout this book, we use the term *taxic diversity*, rather than *species diversity*, out of paleontological necessity. The ecologists are in an enviable position in that they are able to infer from morphologic variation that the animals in their collections are distinct species with unique ecologies and a unique genetic character. The paleontologists are restricted. They must work with relatively few specimens and, thus, are unable to conclude that the range of morphologic variation shown by their material is found within a single species. Conversely, they may erroneously consider ecophenotypes to be distinct species. Without genetic information, we cannot assume that the species of the paleontologist represent the same level in a taxonomic hierarchy as the species of the neontologist. However, this is not a severe limitation! Diversity may be assessed even when the hierarchical level of the faunal identifications is not

well known; the practice is quite common in ecology. If a benthic ecologist computed a diversity index from the species rank of abundance data of a collection of echinoderms, bivalves, fishes, polychaete worms, and nematodes, what assurance is had that the species recognized by specialists are in any way comparable in terms of an ecologic or genetic definition of species?

ATTRIBUTES AVAILABLE TO THE PALEOECOLOGIST

In the physical realm, the paleoecologist is concerned with the nature of the sedimentary rock in which organisms are entombed. The depositional environment of most marine organisms usually appears to approximate closely the locale preferred in life; this is particularly true in the case of infaunal organisms, somewhat less so for bottom-epifaunal organisms (see Chapter 4, pp. 323–331). The high correlation observed between marine organisms and sedimentary rock (substrate of the ecologist) is soon obvious to any student of fossils. Study of sedimentary rock provides insight about many of the physical variables easily measured by the ecologist, but elusive for the paleoecologist.

Such study must be combined with a careful exploration of regional relations, that is, lithofacies and biofacies studies joined with paleogeographic synthesis. We cannot directly measure light penetration in the Lower Cambrian sea, but we can infer its extent from consideration of the occurrence of calcareous algae in combination with marine fossils on a regional scale. Regional biofacies studies, added to lithofacies studies, can reveal much about past ecologic complexity. Such variables as depth, temperature, light penetration, turbulence, tidal magnitude, salinity, oxygen tenor may be reasonably inferred.

It is necessary to study the taxonomic contents of communities; to differentiate between benthic infaunal and epifaunal, plus pelagic (including both nektic and planktic communities). Study of the taxic content of communities generates information about taxic diversity. The relative abundances of various taxa belonging to a community may be recorded to be employed for various purposes at a later time. The stratigraphic and geographic position of communities and taxa relative to each other and to various physical features may be noted both in the same time interval and during different time intervals. Trends in the size of closely related taxa may be ascertained in transects normal to shoreline (i.e., depth-related) and parallel to shoreline (i.e., nondepth-related). Hard-part population density and shelly biomass may be studied or approximated in a variety of ways; community successions may be recorded; phenotypic gradients are also available for study.

In addition to these obvious, noninterpretive factors,

items such as biogeographic relations, community evolution, community replacement in the long term, and the basics of taxonomy and organic evolution may all be deduced.

The above items can be synthesized and attempts made to explain the observed correlations. Each of the factors mentioned involves a degree of technical expertise and training in methodology. The ultimate synthesis will be a better understanding of organic evolution; one which will explain morphology, taxonomy, modern biology, biogeography, community ecology and ethology, past and present. The inclusion of ethology among the aforementioned disciplines is valid since we can routinely study the closed-system type behavior involved in substrate selection by benthic creatures through time in terms of repetitive associations between the organisms and the lithologies with which they are associated. It is also possible to study the spacing characteristics of various species; spacing of individuals belonging to the same species is certainly a form of behavior.

SUBSTRATE

INTRODUCTION

The entombing sedimentary rock is probably the prime evidence available to the paleoecologist for paleoenvironmental interpretation. Without it the investigator could not arrive at significant conclusions concerning animal–environment interactions. When the sedimentary matrix and the enclosed fossil material are studied together, the assembled information is many times more informative than when either is studied alone. Whereas the fossil material provides information about the biota, the sedimentary rock provides information as to the physical environment of deposition. Sedimentary rock must act as depth gauge, salinometer, photometer, thermometer, pH meter, and so on.

As with any basically descriptive science, the study of sedimentary rock begins with classification of the various rock types according to some chosen set of attributes. For example, grain-size classifications, sorting and textural classifications, mineralogical classifications, inorganic–organic content classifications, chemical classifications, environmental classifications are all possible categories. A number of classificatory schemes are available to the paleoecologist; a "perfect" substrate classification does not exist. This last statement is true for the ecologist as well as for the paleoecologist. Ideally, the ecologist bases classifications of the environment upon attributes of direct importance to the associated fauna. Yet the parameters measured are often defined by nonbiologists who, for various reasons, study portions of the environment. For example, rather than determining the exact aspects of a soft sediment responded to by the whole fauna, the benthic ecologist measures those aspects mentioned in basic sedimentology texts and those which are easily available through a soils-testing laboratory. The purpose of classifying the entombing sedimentary rock is to assess environmental similarity, as perceived by the biota, over space and through time. Although some given classificatory scheme may indicate that two environments are similar, there is no guarantee. It is necessary to consider more than one attribute and more than one scheme. We must exhaustively consider the physical, biotic, and distributional aspects of the substrate–faunal complex.

A useful approach to ecologically meaningful classification for both ecologist and paleoecologist is to divide substrates into those in which physical properties predominate in determining faunal composition and distribution, and those in which biological factors predominate. Since there are obvious gradations between these two extremes in nature, there will also be controversy about correct classification of certain environments such as the deep-sea. There are also fairly clearcut examples. Today the bulk of shelf depth or shallower, level-bottom substrates are probably physically dominated. There are environments where physical topography conditions the various sloping, cryptic, and crevice environments. These are the sloping and level rocky-bottom environments. The obviously biologically dominated environments are those in which topography is biologically conditioned; the so-called reef (including "mound," bioherm, and "buildup") environments of the past and present. On a finer scale there are the massive, stony carbonate reefs, the serpulid worm reefs, and shelly-bottom environments (oyster banks, hippuritid reefs, crinoid and bryozoan thickets, etc.). There are, on an even smaller scale, a multitude of epibiotic environments from the surfaces of attached marine algae to the shells of living scallops. Pelagic and planktic biological substrates have undoubtedly existed, but have been seldom recognized in the fossil record. Algal substrates, except those provided by the calcareous algae, decompose upon death, and the hard-shelled epizoans are indeterminantly mixed with the benthic fauna. An extreme case of a biologically dominated substrate and its associated fauna is the host–parasite relationship. The ecologic importance of parasite infestation is undoubted, yet there are few examples in the fossil record due to the soft bodies of most parasites, and absence of host skeletal deformation.

As indicated, the distinction between physically and

biologically dominated substrates is not absolute in many cases. This is especially true when the physical environment is considered over a hierarchy of spatial scales. Both biologically generated reefs and rock reefs offer a variety of sloping, crevice, and cryptic environments of different sizes. The fauna within a particular crevice of a coral reef may be dominated by the effects of tidal exposure, whereas the fauna of a crevice in a rock face may be dominated by a layer of encrusting bryozoa. Thus, it is important to determine the spatial extent of a given substrate–faunal assemblage when classifying it as to dominating factors. Another problem area is that in which the fauna is modifying the physical substrate. An obvious example is the case of rock-boring fauna that radically alter the topography of an otherwise physically dominated environment. Far less obvious is the situation in soft level-bottoms where physical factors such as temperature and salinity are thought to have a dominating effect. In some instances, the level bottom consists largely of biologically generated debris that does not serve as a substrate for epifauna, but alters the characteristics of the sediment substrate. Then there is the possible effect of bioturbation, present in most soft bottoms, upon the fauna found in the sediment. The importance of such biological modification of a presumably physically dominated environment is now an active research area, of interest to the paleoecologist as well as to the marine ecologist.

Cutting across this classification, which is based upon a gross assessment of the dominance of biological or physical factors, are classifications based upon specific physical and biological aspects. These are predominantly physical, such as: tidal–subtidal, photic–subphotic, shelf–subshelf, brackish–fresh, hyposaline–hypersaline, tropical–subtropical, temperate–polar, macrotidal–microtidal, turbulent–quiet, hypobaric–hyperbaric, warm–cool, high oxygen–low oxygen. There are a few biological items such as suspension feeding–sediment feeding, oligotrophic–eutrophic, high diversity–low diversity, cyclic–steady state. Again, there is no "perfect" substrate classification that can be taught in a classroom and outlined in a handbook as an infallible technique. We must remember that the real world is complex and has been complex through time. A simplistic one-or two-factor explanation has little place in the realm of ecologic hyperspace. Indeed, the increasingly popular notion of the importance of multicomponent, linear factors borrowed from the statistics of psychology may prove to be pointlessly simplistic. In an environment that is changing through time, with an associated fauna evolving through time, there must be a large number of variables interacting in a continuously changing (and seldom linear) pattern. In subsequent discussion of substrate analysis by the paleoecologist, keep in mind that many factors are considered simultaneously, and that the same factor is considered in combination with others from different viewpoints. In the following sections, some of the interactions between biota and substrate that are observed in present environments and have been inferred to have existed in the past are discussed.

SUBSTRATE SELECTION

The high correlation between substrate type (Powell, 1937, Figure 18) and benthic organisms, both sessile and vagrant as well as epifaunal and infaunal, has been long recognized by ecologists (see Purdy, 1964, for review [Figures 19–21]; Toulement, 1972; Farrow, 1971 [Figure 22 and Table 3.2]). This common correlation is not restricted solely to the larger fauna; it is found for the meiofauna (Wieser, 1960, see Figure 23), as well as traces of animal activity such as fecal pellets (Bandel, 1974) and trace fossils (Fürsich, 1975a). There is even evidence that the larvae of some fish show distinct substrate preferences, and this might correlate in some instances with the adult community position (Marliave, 1977).

Of primary importance in terms of substrate association is the division into hard substrate benthos, including rocks, cemented hard substrates, and so on, and soft bottoms in all size grades from clay through granules and pebbles, plus the various boring groups restricted to substrates of varying types. Meadows and Campbell (1972b) provide a useful summary of much of the experimental and observational data in this area. In dealing with fossil assemblages, it must be remembered that preserved fauna may be an accumulation of both benthic and pelagic organisms. If benthic, the paleoecologist must consider the animal–sediment correlations; if pelagic, one must be concerned about correlations existing between the organisms and the containing water, which may be thought of as a "watery substrate." Little is known of these relationships because of the considerable difficulties inherent in pelagic studies (McGowan, 1971; Peterson and Miller, 1975, 1976).

Concern has been expressed as to whether these commonly observed organism–substrate correlations are caused by effects of the bottom upon the fauna, or are due to dependence on unmeasured factors affecting both. Figures 18–20 and 22 show that various measures of the fauna correlate with sediment parameters such as percentage of gravel, percentage of clay, and so on. The great difficulty in assessing causality is illustrated in Figure 20. The various sedimentary parameters are generally interrelated among themselves and with other environmental variables (see Parker, 1975, for an exhaustive treatment of this subject). In deciding whether or not the correlation between a sediment type and a given organism reflects a causal agent, it is necessary to determine if the relationship indicates a purposeful search programmed by an innate, or closed,

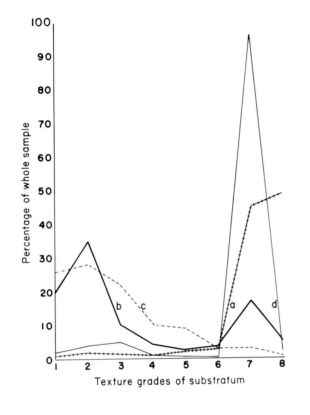

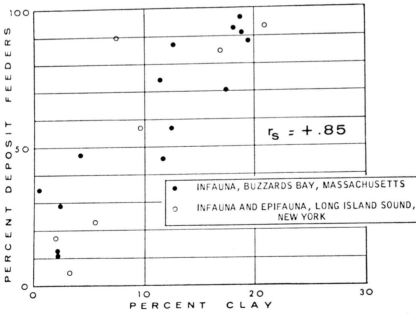

FIGURE 18. Diagram showing the correlation between sediment grain size and benthic animal communities. *Texture grades of Substratum:* 1. Material left on sieve with round holes of 15 mm diameter; 2. Material left on sieve with round holes of 5 mm diameter; 3. Material left on sieve with round holes of 2.5 mm diameter; 4. Material left on sieve with round holes of 1.5 mm diameter; 5. Material left on sieve with round holes of 1 mm diameter; 6. Material left on sieve with round holes of .5 mm diameter; 7. Material that passes sieve 6 and settles in 1 min (fine sand); 8. Material that passes sieve 6 and does not settle in 1 min (silt). (*Percentages* are percentage of each grade to the whole sample). *The Animal Community of the particular Substratum-curve it appears against:* (*a*) *Echinocardium* formation. (*b*) *Maoricolpus* formation. (*c*) *Tawera* + *Glycymeris* formation. (*d*) *Arachnoides* formation. (From A. W. B. Powell, 1937, *Trans. & Proc. Roy. Soc. New Zealand*, *65*, pp. 354–401, Diagram 1.)

FIGURE 19. Diagram indicating the correlation between percent of deposit feeding organisms and percent of clay in the associated sediment. (From E. G. Purdy, 1964, *In* J. Imbrie and N. Newell, Eds., *Approaches to Paleoecology*, Figure 7. Copyright © 1964 by John Wiley & Sons.)

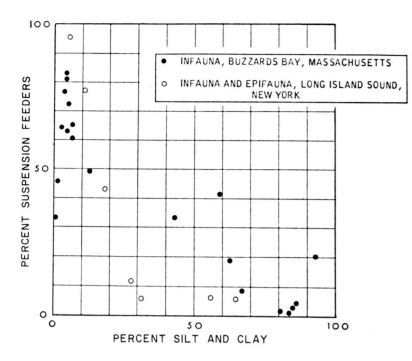

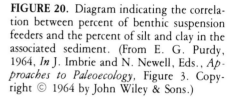

FIGURE 20. Diagram indicating the correlation between percent of benthic suspension feeders and the percent of silt and clay in the associated sediment. (From E. G. Purdy, 1964, *In* J. Imbrie and N. Newell, Eds., *Approaches to Paleoecology*, Figure 3. Copyright © 1964 by John Wiley & Sons.)

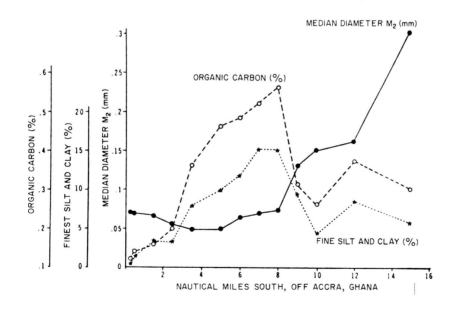

FIGURE 21. Diagram illustrating the correlation between organic carbon content and percent of silt plus clay, as well as the poor correlation with median diameter of the sediment particles. Keep in mind that figure 20 indicates the positive correlation of deposit feeders with percent of silt and clay also correlated with total organic carbon. (From E. G. Purdy, 1964, *In* J. Imbrie and N. Newell, Eds., *Approaches to Paleoecology*, Figure 4. Copyright © 1964 by John Wiley & Sons.)

31

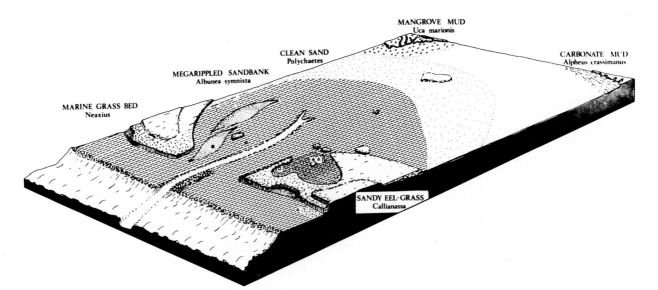

FIGURE 22. Burrowing, intertidal, largely crustacean biofacies on Aldabra. (From G. E. Farrow, 1971, *Symp. Zool. Soc. London, 28,* pp. 455–400, Figure 4. Copyright © 1971 by Zoological Society of London.)

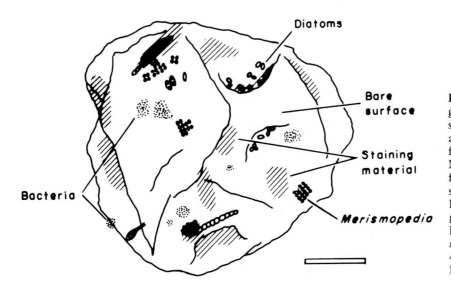

FIGURE 23. Diagram indicating the great environmental complexity of a single 100μ diameter sand grain, from an intertidal beach, insofar as meiofauna and microfauna are concerned. Note also the potential availability of food from the surface of such a grain insofar as a deposit feeder is concerned. Bar = 100μ; *Merismopedia* is a blue-green alga. (From P. S. Meadows and J. I. Campbell, 1972b, *Advances in Marine Biology, 10,* pp. 271–382, Figure 4. Copyright © 1972 by Academic Press Inc. [London] Ltd.)

Environment	Dominant burrowing crustacean	Burrow type	Burrow density	Sediment grain size	Sorting	Sedimentary structures
Leeward Oceanic Coasts						
Upper beach	*Ocypode ceratophthalma*	Spiral "U" "J"	Moderate	Coarse-fine sand	Bimodal	Parallel lamination
Marine grass bed	*Neaxius* sp.	Simple chambered	Low	Conglomeratic fine sand	Poor	Massive with rhizomes
Back-reef sandbar	*Albunea* sp.	Vertical deep	Very low	medium-fine sand	Good	Massive with air heave and thixotropic structures
Channel Platforms						
Megarippled sandbank	*Albunea* sp.	Vertical deep	Very low	coarse sand	Good	Partially erosive cross ripples
Migrating sandsheets	*Callianassa* sp.	Branching horizontal	Moderate	coarse sand	Good	Cross ripples
Sandy eel-grass or stable sandsheets	*Callianassa* sp	branching galleries	High	coarse-fine sand	Bimodal	Massive with rhizomes
Lagoon Platforms						
Polychaete sandflats	(*Arenicola* sp.)	"U" with surface trials	High	medium-very fine sand	Fair	Small oscillation ripples
Carbonate mudflats	*Alpheus crassimanus*	extensive horizontal networks	Very high	silt	Fair-poor	Destroyed
Mangrove muds	*Uca marionis*	simple, "U" or "Y"	High	highly organic mud		Algal mats, mangrove roots

Environment	Tidal range (springs) (m)	Tidal height	Wave action	Current activity	Associated fauna[b]
Leeward Oceanic Coasts					
Upper beach	3.0	HWOT	Moderate	Swash	*Ocypode kuhli*
Marine grass bed	2.8	LWOT	Moderate to high	Slight	*Erycina*[b] *Capulus*[b] *Codakia tigerina*
Back-reef sandbar	3.0	LWNT–LWOT	High	Slight	*Oliva espiscopalis* Maldanid polychaetes *Ochetostoma erythrogramma* *Polinices* sp. *Calappa* sp.
Channel Platforms					
Megarippled sandbank	2.1	LWNT–LWST	High	High	*Mitra mitra* *Oliva* sp. *Metalia* sp.
Migrating sandsheets	1.6	LWNT–LWST	Low	Moderate	*Terebra* spp. *Conus tessulatus* *C. arenatus* *Laevistrombus* sp.
Sandy eel-grass or stable sandsheets	1.1–2.5	LWOT	Low	Moderate	*Laevistrombus gibberulus* *Conus* spp. *Fragum fragum* *Balanoglossus studiosorum*
Lagoon Platforms					
Polychaete sandflats	1.6	LWNT–LWST	Slight	Slight	*Cypraea* spp.
Carbonate mudflats	1.2	ELW	Nil	Nil	*Macrophthalmus* Gobioid fishes[b]
Mangrove muds	1.0	HWOT	Nil	Nil	*Terebralia palustris* *Cerithium morum*

(Cont.)

bivalves. Pumping rate decreased and production of pseudofeces increased in turbid water as compared to clear. This finding is consistent with the observation of slower growth for filter-feeders in silty environments as opposed to sandy environments. Another good example, in which there is a connection between animal–substrate relations and predation, is the study of Tenore *et al.* (1968). Their findings suggested that specimens of *Rangia cuneata,* a bivalve, that lives in sand lives longer, grew faster, and suffered less predation than specimens living in shallower, finer-grained substrates.

Not only animals are affected by sedimentary factors. Neushul (1967) documents preferences for rocks as contrasted with muddy substrates for certain subtidal algae and their motile zygotes. In an experimental study, Hartog (1972b) points out the preference of certain attached algae for differently textured rock surfaces (haptophytes). Concrete blocks were accepted by these attached algae, whereas smooth stones were not, and crumbly rocks were little used. Some taxa (rhizophytes) were found to prefer muddy or sandy bottoms to stone.

There are many ways to portray graphically the correlation between substrate and taxic associations (see Figure 25, McNulty *et al.,* 1962; and Figure 26, Franz, 1967), and there are many statistical methods of assessing the linear correlation between two things, such as Pearson's product moment correlation. However, unless there is a logical reason for seeking linear correlations between the sedimentary parameter and the fauna, graphic techniques are preferable. If, after plots of variables have been examined, linear correlations seem appropriate, statistical analyses may be informative and meaningful. But if observations are obvious from the data, overall conclusions should remain the same no matter how the data are analyzed.

Substrate selectivity is not a topic for the biologist only. The work of Thomsen (1976) clearly indicates that careful analysis of substrate relations among a group of Paleocene encrusting bryozoans showed a high level of substrate selection—this selection was at the genus and species level and involved both the settling species and the favored substrate species. Figure 27 indicates graphically the high correlation observed by Ross (1961) between fusuline species and carbonate grain size and percentage of silt–clay "impurities" in the sediment.

Antibiosis

Although the surface of an organism may serve as an excellent colonization site for another organism, an encrusting epizoan may compete with its host for food or be otherwise detrimental to it. In order to prevent fouling of their surfaces, many marine organisms demonstrate antibiosis. In simple terms, antibiosis is the generation of substances unfavorable to others coming in contact with the generator (Burkholder, 1973). Needless to say, such an antibiotic generating capacity will influence the composition of the recurring association of taxa.

The fact that certain fouling organisms attach to one organic substrate but not to another may be related to antibiotic substances. Burkholder (1973) cites the example of a dead gorgonian fan quickly becoming encrusted by a fouling organism, whereas a living fan remained pristine for some time. This is an example of a potential absence. Conover and Sieburth (1964) provide additional examples from the oceanic *Sargassum* community. Presences may also be generated in terms of such examples as the venomously hemolytic holothurine (Bakus, 1974), characteristic of many tropical and subtropical holothuroids. Jackson and Buss (1975) provide further examples taken from sponges and other groups in the reef environment. Antibiosis is well known among the land plants, where the generation of deleterious substances acts to keep competitors away from the same ground.

To what extent benthic organisms may introduce antibiotic substances into their environment is not known, nor is what effect any such substances may have upon the associated biota. However, this may be a very important factor in determining animal–sediment correlations. Proctor (1957) demonstrated that cultures of freshwater phytoplankton produced fatty acids that hindered or promoted growth of other algal species. Hardy (1956) suggested that antibiotic excreted by plankton into the water might be a partial cause of patchy plankton distributions. There have been some experimental results suggesting that there are animal–plant antibiotic effects in the sea. However, the ecologic importance of antibiosis is not universally accepted (Fogg, 1965, and Provasoli, 1963). If low concentrations of plankton-produced antibiotics do affect plankton communities despite the tremendous dilution possible in the open sea, greater effects might be possible in sediments or on rocks. The reduced capacity for diffusion due to entrapment of interstitial water, and the absorptive capacity of many sediments would tend to keep antibiotic materials restricted to the environment into which they were generated.

In terms of special examples of substrate selectivity, Ross (1967) indicates how certain anemones employ a chemical factor present in molluscan shell material in establishing themselves on hermit crab-occupied shells. This is an example of selection by an adult on the basis of chemical clues, and is akin to antibiosis, but in a reverse order. So the possibility of substrate selection on the basis of biologically produced substances is just as likely as substrate avoidance (Crisp, 1974). The clumping characteristic of many sessile benthos is possibly based on such a selection mechanism by the larvae.

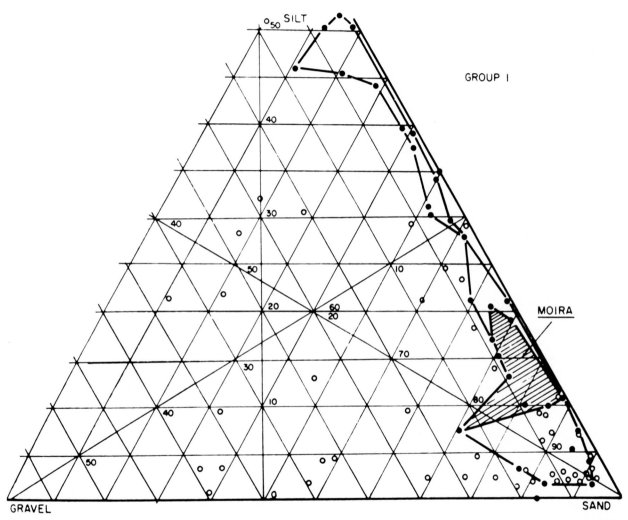

FIGURE 25. Correlation between benthic animals (Group 1, solid dots) and sediment grain size. (From J. K. McNulty, R. C. Work, and H. B. Moore, 1962, *Bull. Mar. Sci. Gulf & Caribbean, 12,* pp. 322–332, Figure 5. Copyright © 1962 by Rosenstiel School of Marine & Atmospheric Sciences, University of Miami.)

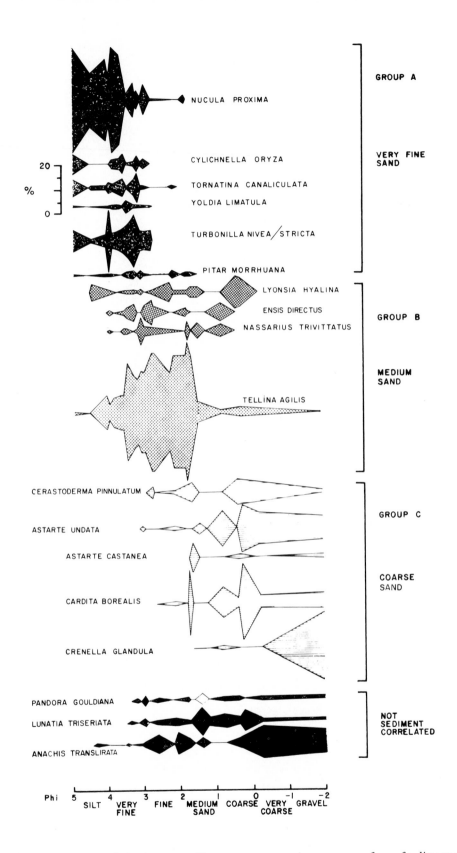

FIGURE 26. Diagram showing the correlation between sediment type—even using a very gross form of sediment analysis—and both relative abundance and presence of various benthic species. (From D. Franz, 1976, *Malacologia, 15,* pp. 377–399, Figure 6. Copyright © 1976 by Philadelphia Academy of Natural Sciences.)

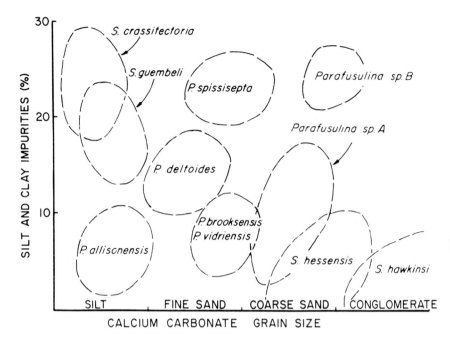

FIGURE 27. Schematic showing the high correlation between grain size and sorting, and species—fusuline foraminifera (species of *Schwagerina* and *Parafusulina*) from the Permian of Texas. (From C. A. Ross, 1961, *J. Paleo.*, 35, p. 399, Figure 1.)

Changes in Substrate during Ontogeny

The life histories of most marine invertebrates contain developmental sequences in which different environments are occupied by a single species at different stages of growth. The distinctions between juvenile and adult environments may be as subtle as slightly different salinities, or as radical as pelagic versus benthonic. In ecologic studies where both the physical environment and the biota can be jointly surveyed, recognizing growth stages as reflections of differing environments is relatively simple. Work with living benthos has shown that many shelled animals have the ability to change their substrate orientation during life, and that some actually change their substrate. The most obvious example is that of bivalve mollusk spatfalls, larval settlings. The spat of many infaunal organisms begin benthic life at a very shallow level of the sediment–water interface; perhaps even attached as part of the meioepibenthos. With growth, the animal may outgrow its holdfast and assume the deeply burrowed posture associated with the adult form. Medcof (1950) relates how a young clam begins its settled existence attached to plants and only later drops to the bottom to take up an infaunal existence.

Possibly the classic brachiopod example of organisms changing their substrate has to do with certain productids that possess an early growth stage in which a pair of hinge spines wrap about a crinoid stem, and a later growth stage in which the organism outgrows this lifestyle and drops to the bottom, where it thrives (Figures 28 and 29).

Although a few examples of organisms changing their substrate during ontogeny have been recognized

in the fossil record, little serious attention had been given to this possibility prior to the publication of Surlyk's (1972) paper on Danish Chalk paleoecology. Surlyk made a very strong case for a substantial number of level bottom brachiopods associated with muddy bottoms having begun life as small shells attached to small biogenic fragmentary substrates scattered about on the mud, followed by a later growth stage in which the delthyrium becomes filled by some material resulting in an atrophied pedicle opening coincident with the swelling-out of the pedicle valve into a hemispherical structure: The process can be conceived of as rocking back and forth on the bottom under the influence of weak currents, or of having been only half-buried so as to permit the smaller, flattish brachial valve and commissure to remain free of the substrate mud. Such creatures have a mortality curve indicating very high mortality in the early growth stages. The presence in many of the extinct strophic brachiopod shells of a pedicle tube may be interpreted as evidence of an early attached stage followed by a free-living later stage.

LEVEL-BOTTOM ENVIRONMENTS AND GENERAL CONSIDERATIONS OF ANIMAL–SUBSTRATE RELATIONSHIPS

The seas of the present and past have been dominated in terms of area by the sediment-covered level bottom. This is generally true from the epicontinental-intertidal zone and hypersaline lagoons to the hadal and central abyssal regions. It is this type of environment with which the paleoecologist is most often concerned. Obvious physical parameters available to the

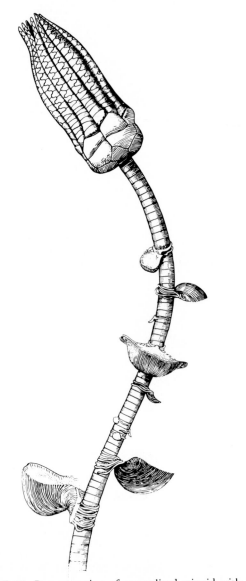

FIGURE 28. Reconstruction of generalized crinoid with attached shells of *Linoproductus angustus* in various stages of growth. Note that larger specimens break off under their own weight to take up positions on the level bottom beneath. (From R. E. Grant, 1963, *Jour. Paleontol.*, 37, pp. 134–140, Figure 1. Copyright © 1963 by Society of Economic Paleontologists & Mineralogists.)

paleoecologist for study are those of the sedimentary petrographer including grain size, sorting, texture, mineralogy, chemistry, organic content (both organic carbon and organic structures), and various others. The overwhelming bulk of ecologic and paleoecologic work has considered these parameters using techniques borrowed from the geologist–sedimentologist. McNulty *et al.* (1962) present a wealth of data showing high correlations between sediment grain size and taxa (Figure 25), which latter aggregate naturally into communities.

Franz (1976, Figure 26) provides more data showing the high correlation between taxa and sediment type, but he emphasizes that this is by no means a correlation of 1.0. However, it is common knowledge that obviously differing benthic communities occur on what the sedimentary analysis suggests is the same type of sediment. This apparently contradictory finding has been explained by an appeal to other, unmeasured, variables.

The word *substrate* can assume a variety of meanings, two of which reflect basic differences in the ecologists' or paleoecologists' conceptualization of the organism–substrate interaction. *Substrate* is most often defined as a foundation; something wholly apart from the biota. The preponderance of this reductionistic approach can be seen in many marine ecology texts, and even appears in portions of this work, where it is most common to find substrate discussed in a series of chapters on physical environment; salinity, temperature, substrate, and so on. An alternate definition is common in biochemistry, and should be adopted by ecologists and paleoecologists. A substrate is that upon which an enzyme acts, and indeed may control the reaction depending upon the kinetics of the system. For our purposes the important thing is that the organisms interact with the substrate; they do not simply sit upon it or within it. Animals and sediment compose a dynamic system that must be studied conjointly and in detail. Certainly, grain size as such correlates logically or statistically with many taxic phenomena, but the truly difficult problem is to understand why the correlation exists and how might the sedimentary environment, as experienced by the fauna and preserved in the rock, be better measured. There are obviously many factors correlated with any given sedimentary parameter that may be of great importance in determining organism distributions; not all of them vary uniformly with substrate type. When an ecologist or paleoecologist has studied some aspect of the physical environment, the traditional approach has been to borrow the techniques of a physical scientist working in that area. But too seldom have the techniques borrowed for convenience been questioned as to their applicability to the study of ecologically meaningful parameters. The paleoecologist must be cautioned against such continued uncritical borrowing, and encouraged to devote considerable thought to the measurement of the environment as sensed by an organism.

Surprisingly, the intimacy of the animal–sediment relationship and the failure of sedimentologic techniques to shed light upon the relationship has been better noted by paleontologists working with living systems than by the ecologists. Johnson (1974) has conclusively shown that the routine analytic procedures of the sedimentologist effectively destroy many attributes of the sediment, particularly those involved with organic material, which either coats or helps to aggregate inorganic particles, as well as other features that are un-

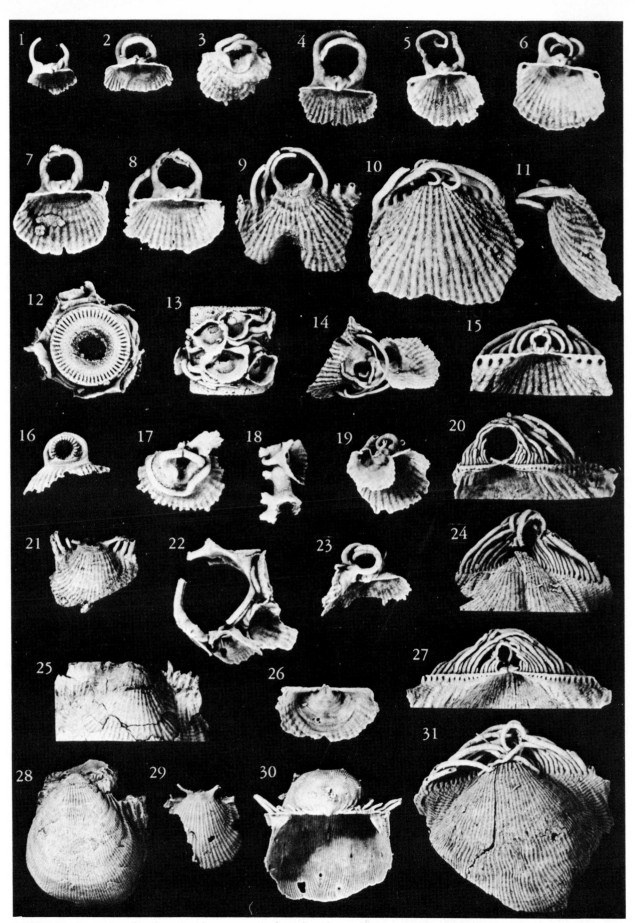

FIGURE 29. Examples of early, crinoid stem attached productid growth stages and later bottom-dwelling forms of the same species. (From R. E. Grant, 1963, *Jour. Paleontol.*, *37*, pp. 134–140, Plate 19. Copyright © 1963 by Society of Economic Paleontologists & Mineralogists.)

$-8 \leq$ Phi $\leq +12,$
forget the tremen
vironments of, for
when the sedime
obscure number.
vironmental diffe
Phi terms) sedim
ing that numeri
quently correlate
sediment-size str
able to try to n
affected by grain
tant, such as geo
gists have little
granulometric
that fine struct
a paleoindicato
the technique

The simpl
substrates is
coarse-grained
perhaps even
environments
ing from com
position that
everything th
leads logicall
The terms *f*
preted to m
they are han
vironments
sediment si
and paleo
sedimentar
eye to be
tual geom
sediments
range of s
and Camp
sedimenta
predomin
would ex
fine- and
any giver
geostatis
sample h
modes,
especial
mended
for the
detailed
more
wise, v
sedime
ment
cation

Con
tion

doubtedly ol
biotic occurr
Bokuniewic;
(1949) em|
measuring '
as another |
metric anc
sedimento!
(quicksanc
properties
measured
obviously
the space
fluid mo
mechani
1941, er

Station no

H
(

12

II

10

9

8

Brady, 1943; Brownell, 1970; Jackson, 1973; Tenore *et al.*, 1968; Theede *et al.*, 1969). Probably, the deposit-feeding infauna is closely related to the thixotropic properties of the sediment. Sedimentary rock types of apparently similar properties, according to the petrographer, commonly present indubitable paleontological differences that frustrate the petrographer's powers of discrimination. The physical properties of the fine-grained sediment that affect support and movement—the "soupiness" of the sediment and the resistance of the sediment to movement by the infaunal organism—although poorly investigated, must be of primary importance. These various properties are complex in terms of grain forms, mineralogy, interstitial water chemistry (see Berner, 1976, for a review of the problems involved), organic constituents, and many other variables, but need to be considered in terms of what may appear to be random occurrences of different infaunal associations.

Epifaunal benthos present the problems of attachment and sediment fouling for the sessile types and of footing for the vagrant. Manton and Stephenson (1931) discuss the great variability of the hermatypic corals' ability to cope with both fine-grained and coarse-grained sediments. At first glance, the fine-grained substrates may be considered unsuitable for reliable attachment, but occurrences in the past and present indicate that this need not be the case. There are a variety of structures (Bacescu, 1972, Figure 31; Bromley and Surlyk, 1973, Figure 32) that find the fine-grained substrate to be a suitable attachment site; in addition, fine-grained substrates have a widely varying span of thixotropic behavior that may not be apparent to the petrographer of ancient rocks (or of modern sediments). Franzen (1977), for example, reviews many of the adaptations possessed by crinoids for dealing with varied hard and soft substrates.

The greater possibilities for deposit feeding in the fine-grained substrates lead to greater possibilities for bioturbation and of biodeposition of sediment. The importance of varieties and volume of fecal materials (Arakawa, 1970), must be kept in mind. Data obtained from the cultivation of oysters and observations of other pseudofeces and fecal-producing organisms make clear that environments within the fine-grained sediment realm may be, in certain cases, highly modified or even controlled by excretory behavior of deposit and filter or suspension feeders operating largely in the fine-grained region (Aller and Dodge, 1974; Pandian, 1975).

A mass of evidence, gleaned chiefly from the recent, indicates that fine-grained substrates are far higher in usable nutrients (Bader, 1954; Longbottom, 1970; Rhoads and Young, 1970) for deposit feeders than are coarse-grained sediments. Corollary to this is the more oxygen-rich nature of coarser-grained sediment as compared to finer-grained sediment.

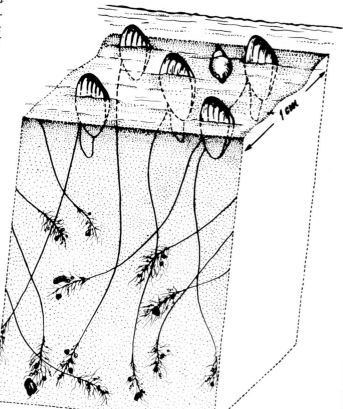

FIGURE 31. Sediment anchoring device composed of byssal threads attached to larger detrital particles employed by *Corbula*, a semiinfaunal bivalve, in an environment of moving sandy material. (From M. C. Bacescu, 1972, *In* O. Kinne, Ed., *Marine Ecology*, vol. 1, part 3, Figure 7-20. Copyright © 1972 by Wiley–Interscience.)

The foregoing observations appear to hold as well for hypersaline, hyposaline and fresh waters as well as for those of the normal marine environment. The record overwhelmingly indicates that fine-grained substrates do support an epifauna and an infauna distinct from that of coarse-grained substrates. But, it should be clearly understood that in many instances not grain size alone, but variables correlating strongly with fine-grained sediment as opposed to coarse-grained sediment may be causal factors. For example, coarse-grained sediments tend to reflect more turbulent, better-oxygenated waters that provide a greater supply of transported food particles than do the quiet waters associated with fine-grained sediments. In addition, fine-grained food particles usable by certain classes of deposit feeders are commonly sedimented with similar-size mineral grains rather than with larger-size mineral grains. The point about correlation versus causation must again be emphasized because when substrate does not correlate with sediment grain size, although an

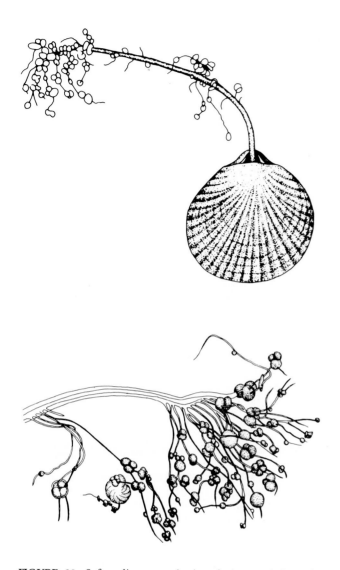

FIGURE 32. Soft sediment anchoring device consisting of modified brachiopod pedicle threads attached to foraminifera occurring buried in the sediment. (From R. G. Bromley and F. Surlyk, 1973, *Lethaia, 6,* pp. 329–365, Figure 1. Copyright © 1973 by Universitetsforlaget.)

organism is highly correlated with grain size, due to the cooccurrence of other variables, there will undoubtedly be occurrences when the critical variables fail to occur with the expected grain size.

Despite the absence of adequate surveys of the matter, one does gain the overall impression that the fine-grained, soft-bottomed substrate benthos gradually changes its predominant lifestyle from filter and suspension feeding in the shallower waters characteristic of the continental shelf to deposit feeding, once the abyssal shades have been reached; the bathyal slopes are intermediate (Shabica and Boucot, 1976a). Atten-

tion must be directed to the fact that hard-substrate, attached benthos, whether deep, intermediate, or shallow, perforce seek their livings through filtering and suspension feeding.

In the carbonate facies rich in limestone and dolomite, the fine-grained facies is characterized by abundant micrite (Wilson, 1975). However, micrite (clay- and silt-size carbonate particles) is an index of relatively quiet water sedimentation, and the percentage found varies according to the amount of larger bioclastic fragments left behind by various organic and physical processes. A large amount of micrite can be assumed to represent a fine-grained substrate formed under conditions of relatively quiet water and to have many of the properties common to a silicate mineral-complex fine-grained substrate.

The fine-grained sediments of many shallow-water tropical and subtropical regions characteristically have a large amount of calcium carbonate in their composition, particularly in regions where low rates of sedimentation predominate, whereas the reverse is the case in more boreal, austral, and polar regions, where calcium carbonate solubility (Alexandersson, 1972) in cooler waters militates against its abundance. However, we have little understanding of the ecologic consequences of lime-rich, fine-grained sediment as contrasted with lime-poor (presumably silicate-rich or organic-rich) sediments in terms of such parameters as thixotropic properties. Therefore, the best that ecologists, and particularly paleoecologists, can do is to carefully observe what we see, and marvel at the awesome complexities and varieties of the natural world while continually searching for correlations between the physical and the biological, thereby extending the hope that experimental and modern observational work can make sense from our observations.

Turbidity. When currents or wave motion have bottom-water velocities sufficient to suspend bottom sediments, the nature of the bottom substrate will have a direct impact upon the turbidity of the overlying pelagic environment. Turbidity, usually associated with fine-grained substrates, is an important environmental variable. This is true for both the pelagic organisms and the benthos that is dependent upon the near bottom-water for nutrition. Pandian (1975, pp. 84–85) explains how certain bivalves such as species of *Musculus* and *Modiolus* have developed modifications of ciliary behavior suited to filter feeding in a highly turbid environment unfavorable for other organisms. Loosanoff (1962) details how turbidity discourages some shelly benthos at the adult stage, whereas other species are discouraged from settling at the larval stage; turbidity is clearly an important factor in determining community composition among the shelly benthos. Moore *et al.* (1961) point out how at least one echinoid will not tolerate turbid water. Kristensen (1957) cites circumstantial evidence suggesting that cockles off the Dutch Coast thrive in a subtidal turbid environment, but not in a clear-water environment; he interprets the distribution as the result of predatory fish being unable to see the cockles in turbid water. Whether or not Kristensen is correct, it is clear that turbid and nonturbid waters form another class of environmental variables that affect the distribution of level-bottom and pelagic organisms in a variety of ways. Therefore, one must be aware that similar lithology shales, mudstones, and claystone may still bear different communities as a result of physical factors that may not be evident from examination of the rock or sediment type. Rhoads (1973) and Rhoads and Young (1970, 1971) point out the possibility that the turbidity generated by organisms living on a level bottom may serve to keep additional taxa out of an area. They showed that oysters ''hung'' **over** a particularly muddy, stirred up substrate, thrived despite the unfavorable bottom turbidity. Thus, the process of bioturbation may be important in controlling the fauna associated with an actively burrowed bottom. Manton and Stephenson (1931), in their review of hermatypic coral reaction to sediment, made it clear that certain taxa are incapable of dealing with fine-grained, sediment-caused turbidity in a quiet-water environment. Pratt and Campbell (1956) indicate how turbid water slows the filtering rate and may retard growth of suspension feeders with consequent increase in production of pseudofeces. Aller and Dodge (1974) report that turbid water conditions decrease coral growth and taxic diversity after artificial perturbation. They also make the important general point that an abundance of suspension feeders should correlate with clear waters, whereas an abundance of deposit feeders may correlate with more turbid conditions.

Coarse-Grained Sedimentary Environments

In general, but not in all instances, coarse-grained sediment correlates highly with more turbulent, better-oxygenated conditions than does fine-grained sediment. However, well-oxygenated waters above the bottom do not necessarily correlate with a higher level of oxygen in the interstitial waters below. Thus, foul sands may underlie clean waters (see Fager, 1968, for a typical example). Coarse-grained sediments tend to provide a far lower level of particulate organic matter to deposit feeders than do fine-grained sediments, a fact that correlates with the observed lower abundance and diversity of deposit feeders in coarse-grained sediments.

Regionally, there is a tendency for coarse-grained sediment to be most abundant in the intertidal and shallow subtidal regions of higher turbulence and proximity to sources of coarser-grained materials. The distinctions between the biotas of coarse- and fine-grained sediments within the intertidal and shallow subtidal region are legion. Such distinctions range from correla-

tions with turbulence to problems of rooting for both plants and animals. Again, the correlation of many other environmental variables with grain-size makes us cautious about misinterpreting grain size as the sole causative variable when studying fossil deposits. A good example of the caution needed is Brown's (1973) account of the high correlation between the mutually exclusive occurrence of two intertidal isopods and grain-size of the sand with which they are associated. The differing sand grain-sizes are controlled themselves by the complex of factors involved in relationships of water of differing velocities and food supply, oxygen, temperature, and motion.

Oolitic sediments, a special case of the coarse-grained variety, generally reflect turbulent, better-oxygenated environments, and tend to be highly correlated with low rates of sedimentation.

Fine-grained substrates tend to be highly correlated with deposit-feeding organisms that thrive on organic detritus that is found in the fine-grained size range and thus in sediments there, whereas the coarser-grained substrates tend to correlate with the suspension- and filter-feeding benthos, which do better in rapidly moving waters where they obtain a more plentiful food supply than is available in the normally slow-moving waters overlying fine-grained substrates. Bloom, Simon, and Hunter (1972) provide data for a situation where a gradient from fine to coarse, and deposit feeders to filter feeders, exists. Boaden's (1962) report of a high correlation between interstitial fauna and grain size is not surprising, although it is of small interest to the paleontologist because of the absence of interstitial fauna from the fossil record.

Buchanan (1963a) has observed two species of *Amphiura* in the North Sea; one modified for deposit feeding in slowly moving water, and the other for suspension feeding in more rapidly moving water. There is a general correlation in their respective areas of occurrence with coarser- and finer-grained sediment, but detailed examination of the present sediment pattern showed conclusively that there was a much better correlation with water velocity than with substrate type (Figure 33), suggesting that the sediment distribution pattern, although obviously a result of a current velocity regime, is not the result of the one obtaining during Buchanan's observational period. This example serves as another warning to the geologist that the organisms may be more reliable indicators of environment for a particular time interval than is the associated sediment. Obviously, quiet-water organisms living attached to boulders were not there when the boulders were implaced, but Buchanan's case is far more subtle. Figure 34 indicates the common inverse correlation between grain-size and organic carbon content, as well as between grain size and organic nitrogen.

Fager's (1964) account of a tube-building polychaete (chiefly hornblende and shell fragments in his exam-

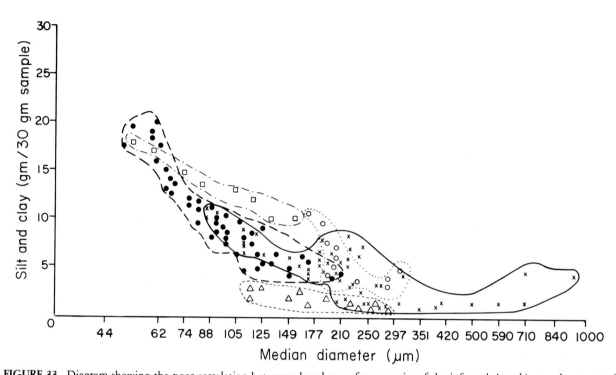

FIGURE 33. Diagram showing the poor correlation between abundance of two species of the infaunal *Amphiura* and amount of silt and clay in the substrate. The number of organisms correlate more highly with bottom current velocity, which suggests that the currents responsible for the deposition of the sediment are not operative at this time. Cross-marks surrounded by solid line = *Amphiura filiformis;* filled-in circles surrounded by broken line = *Amphiura chiajei;* triangles surrounded by dashed line = *Venus;* squares surrounded by dash — and — dot line = *Melinna-Amphipod;* open circles surrounded by dotted line = *Haploops.* (From J. B. Buchanan, 1963a, *Oikos, 14,* pp. 154–175, Figure 4. Copyright © 1963 by *Oikos.*)

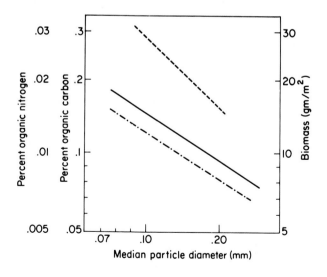

FIGURE 34. Diagram indicating the positive correlation between organic carbon and biomass of *Arenicola,* an infaunal worm, and also of organic nitrogen with median particle diameter. Dashed line shows *Arenicola* biomass; solid line, organic nitrogen; dot–dash line, organic carbon. (From M. R. Longbottom, 1970, *Jour. Exper. Mar. Biol. Ecol., 5,* pp. 138–157, Figure 9. Copyright © 1970 by Elsevier/North–Holland Biomedical Press.)

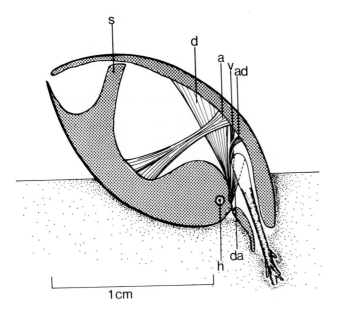

1 cm

FIGURE 35. Pedicle modification in a modern brachiopod permitting a pogo-stick type of lifestyle on moderately turbulent gravel bottoms. The figure shows the modified pedicle withdrawn. The shell is elevated when the pedicle is extended. (From J. R. Richardson and J. E. Watson, 1975, *Science, 189,* 381, Figure 1. Copyright © 1975 by Am. Assoc. Adv. Sci.)

ple) that selects tabular, sand-size grains for its tube (as much as 80–85% hornblende grains from a sand averaging about 4% in one example), and is found in sandy sediments, is an excellent example of a causal relation between sediment characteristic and presence of benthic organisms, but such cause-and-effect relations have been demonstrated in very few cases.

Richardson and Watson (1975a,b) show that their specially modified pogo-stick brachiopod inhabits a coarse-grained, turbulent substrate (see Figure 35), and Neall (1970) draws similar attention to *Neothyris* occurring as a rule on a gravelly, turbulent bottom.

Smith (1933) in his study of the highly turbulent waters surrounding the Eddystone emphasizes that some deposit feeders exist in a highly turbulent environment, characterized by coarse gravels and bare rock, and that there is virtually no upslope movement of shell debris from quieter, deeper-water fine-grained environments. The high correlation he reports between living and dead shells is impressive.

In the carbonate facies rich in limestone and dolomite, the presence of a large amount of sparite (coarsely crystalline calcite) is equivalent to the coarse-grained facies of the silicate-mineral type. The presence of sparite cement between the commonly bioclastic fragments is indicative of a rough-water, better-oxygenated environment than coincides with a large percentage of micrite (see Wilson, 1975, for a summary statement).

Mineralogy

Plant ecologists have always been concerned with edaphic, soil-related, factors that might influence plant distributions. As plants derive their mineral nutrients from the soil, the mineralogy and related soil chemistry has been studied as an important ecologic parameter.

The marine ecologist seldom does more than designate a particular substrate as carbonate- or silicate-rich. Significance is seldom ascribed to biotic differences noted between carbonate- and silicate-rich sediments. One of the disappointing findings of the *HMS Challenger* expedition during the pioneering days of zoogeography was that the deep-sea fauna was not partitioned into faunal zones corresponding to the dominant types of bottom ooze. Nevertheless, the geologically trained paleoecologist is continually tempted to turn the locally strong correlations between sedimentary rock, mineralogy and fauna into causal relations. Thus, the common observation that particular taxa are associated with limestone (or dolomite of replacement origin), whereas others may be found locally with silicate-rich materials, is turned into a causal relation. Paleontologic experience indicates that most of the level-bottom taxa, if studied over a vast region (preferably worldwide), show no significant correlation with substrate mineralogy.

There are certain exceptions to such a generalization; cases in which the mineralogy of the sedimentary rock provides important environmental evidence. If, during deposition, the physical environment affected authigenic and diagenetic processes, this is especially true. For example, glauconite appears to be the result of chemical processes restricted to relatively shallow water and somewhat turbulent, coarse-grained, environments, commonly near shore. Evaporitic minerals certainly indicate that the benthos in question underwent a hypersaline phase. Fenchel (1978) makes it clear that the common geologic misconception that abundant authigenic pyrite in a sediment indicates generally foul conditions within the sediment and the overall environment is likely to be in error. He summarizes data indicating that in the modern near-surface sediment environment, a moderate percentage of iron sulfide is precipitated in the course of the normal recycling ("mineralization") of nutrients occurring in the near-surface sediment. A high percentage of pyrite in a finely laminated, nonbioturbated sediment does, however, suggest the presence of an anoxic-bottom environment, although Kauffman's (in press) studies of the Holzmaden Jurassic beds indicate that conclusions cannot be reached prematurely. The lack of carbonate material may indicate deposition below the calcite compensation depth.

As was mentioned earlier, the commonly measured granulometric parameters of the sedimentologist may not be the most important environmental parameters as far as animals are concerned. In the fine-grained level bottom substrate region, a thorough understanding of sediment mineralogy might tell us something about the thixotropic properties of the sediment. It is too early to report any important relation between the chemistry of interstitial water, the surface properties of the encasing sediment grains, and the biota associated with the whole system. Although such fine-scale structural study is technically difficult, it must eventually be undertaken if the interaction of biological and geochemical processes is to be fully understood.

Limestone Is Not a Marine Substrate. Geologically trained paleoecologists are inclined to assume that the association of a limestone matrix with fossils has ecologic significance. This inclination commonly results in the compilation of faunal data and comparisons of limestone-derived fossils with those derived from noncarbonate shale and quartz-rich sandstones. Such gross categorization of calcium carbonate rocks as limestone, entirely sensible from many geologic viewpoints, leads to ecologic disaster. Limestones may be far more meaningfully categorized into calcarenites having all the textural properties of sandstones; micrites having all the properties of muds possessed of noncarbonate chemistry; subtidally-cemented limestone hard-bottoms having the properties of rocks; limestone-granule conglomerates; and conglomerates having properties appropriate to their grain size. The textural differences alone suggest that these are ecologically significant classifications. Therefore, it is almost meaningless for paleoecologists to consider "limestone" as a useful category with which to correlate faunal and floral associations. The only exception to this statement is the significance of abundant limestone as an indication of tropical or subtropical overall conditions, as limestone is a most uncommon rock under conditions of water temperature low enough (Alexandersson, 1972) to prevent its accumulation. Probably the chief reason why limestone is not a substrate in the sea, in either the chemical or the mineral sense, is the uniform composition of sea water itself. Primary producers in the sea obtain their nutrients from sea water, not from the substrate, as is the case on land.

In the nonmarine environment, however, limestone is definitely a significant substrate category. Boycott (1934, 1936) emphasizes the strong correlation between the presence or absence of certain nonmarine snails and the tenor of calcium. He reports the distributions of both the calcicole and calcifuge snails, as well as of the intermediate types. The same substrate-chemical situation holds for plants; nonmarine botanists are aware that soil chemistry is important, but attached marine algae pay no attention to substrate chemical or mineral composition (Hartog, 1972b).

Shelly Debris

Sedimentary rock classifications such as crinoidal limestone, coquina, shell bed, biostrome, and the like are in common usage among paleoecologists, reflecting the frequency with which a sedimentary matrix composed predominantly of shelly debris is encountered. The ecologist, however, seldom pays much attention to the shelly debris associated with living marine organisms. When shell material is taken in a sediment sample, it is usually classed with the coarse sediment and is seldom considered further. In part, this apparent difference in approach may reflect the present absence of vast epicontinental seas. In the geologic past there may have been far greater accumulation of shell debris in these environments; thus ancient substrates may have contained more shell material. It must be remembered that paleontologists have been drawn to shell beds because of their great abundance of fossil material. Therefore, the extent of nonshelly substrates in the geologic record may be underemphasized.

In addition to forming a unique substrate, shelly debris sometimes provides evidence of the dominant benthos, or information regarding the mixture of taxa living in a high-diversity area. Careful petrographic study of pebble-size and gravel-size shelly debris commonly provides much knowledge about the taxonomic composition of the local flora and fauna; careful study of even sand-size shelly debris can help in the recognition of many taxa. Milliman (1977) has summarized a mass of information regarding the nature of the shelly carbonate debris off the eastern coast of North America. He recognizes a number of major "biolithic" subdivisions that give some idea of the dominant shelly benthos occupying different areas of the shelf. Such subdivisions in themselves do not amount to community analysis in the strict sense, but they may be a useful preliminary to actual community analysis, and they are a useful means of generalization. As pointed out by Milliman (1977; see also Farrow *et al.,* 1978), there are carbonate-rich substrates, although relatively few, except in the warm waters; rich in mollusks, calcareous sponges, echinoids, bryozoans, coralline algae, benthic foraminifera, and in combinations of the aforementioned groups, as well as in other groups. Algal-rich substrates may be divided into those rich in algal fragments, (i.e., sands, gravels and conglomerates), those rich in oncolites, and the hardgrounds in which calcareous algae sometimes play an important constructional role. Pelmatozoan-rich limestones are discussed separately (Chapter 3, pp. 103–104).

There is commonly a tendency on the part of specialists to see only the presence of their own group of organisms. We see specialists interested in brachiopods discussing brachiopod communities occurring in shelly limestone while ignoring the abundance of ostracodes or bryozoans that may dominate both in actual numbers of specimens and shelly biomass. However, we do encounter situations in which individual groups dominate. An effort should be made, when discussing communities derived from shelly carbonate rocks, to distinguish situations in which the community is based on the bulk of the specimens as contrasted with those of interest to the specialist.

There has been little systematic effort to determine whether or not the community characteristics of shelly substrates based on different major groups have different community characteristics for the associated organisms unrelated to the formers of the shelly substrate. It is not clear if shelly gravel looks the same to most associated organisms, regardless of whether they are derived from bivalves, sponges, or calcareous algae. Certainly there are statistical differences in shape for gravels and conglomerates derived from different groups, but just what role these distinctions play in attracting or repelling other groups is uncertain. Studies of successional developments involving shelly substrates might provide at least partial answers to this question.

Shelly substrates should be of greater interest to the ecologist than to other specialists because these substrates often represent a situation in which the benthic biota radically alters its physical environment in a manner that alters, in turn, the subsequent biota. Shell material that accumulates in fine sediments may actually modify underlying sediment sufficiently to make conditions favorable for the establishment of a new biota. Whereas it is common to consider succession as a smooth trajectory through time, much study in shallow water indicates that the route may be almost chaotic. Factors such as seasons and the vagaries of larval dispersion and availability militate against there being any one route from fine sediment to any particular end-product. Shelly substrates may be intermittently covered with blankets of mud that extinguish the biota dependent on the shelly substrate and require that the entire process begin again. Monotonous sequences of shelly debris alternating with fine-grained sediment are present in many parts of the Phanerozoic record in all parts of the world. Bayer (1967) presents an excellent example of this last situation for 15 units in the Upper Ordovician of Minnesota.

Some shelly-debris deposits are the result of significant transport and community mixing. Haas (1940) provides a vivid description of a storm-generated intertidal shell-debris deposit that would challenge to the utmost the paleoecologist assigned the difficult task of making ecologic sense from the taxonomic chaos.

The normal aspect of the Gulf beach on Sanibel Island in both the intertidal and supratidal zones is greatly changed after heavy storms from the Gulf, and, especially after one of the dreadful hurricanes which attain a velocity of more than eighty miles an hour. These storms disturb the water to a depth of about thirty feet, and the waves wash the animals and plants torn loose from the ocean floor onto the shore. Even species of mollusks normally not to be found on the beach are thus thrown up. Such forms include the rare volutid *Maculopeplum junonia* Chemnitz, *Cypraea exanthema* Linnaeus, *Conus proteus* Linnaeus, and the bivalves *Tellina radiata* Linnaeus, *T. alternata* Say, and *Lyonsia floridana* Conrad; they may be collected alive or recently dead, when such a storm is over.

During my stay on Sanibel Island, I witnessed the relatively light storm of June 12–15, 1939, which attained not more than thirty-five miles an hour at its maximum. This gave me an opportunity to observe the changes thus caused in the beach deposits. When the waves had calmed somewhat and the newly built-up shell deposits, which corresponded to the maximum height of the storm flood, were laid bare, the unusually large number of animals quite new to the beach, and especially of gastropod shells, became obvious. Many *Strombus pugilis* had been thrown ashore from the calm depths where they generally live. Those that had not yet died from desiccation were trying to protect themselves from the action of the waves, to which they are unaccustomed, by burrowing in the sand; but most of them perished within a few days by being buried and subsequently suffocated in the sands; from day to day fewer living ones and more dead ones were found on the beach. Living young *Fasciolaria distans* Lamarck were found in the intertidal zone, and their behavior and fate paralleled the case of *Strombus pugilis*. Several species of muricid gastropods, unfit either to withstand the action of the surf or to escape this zone, face the same destiny. *Crepidula fornicata*, whose chains, settled on mussel shells, are thrown up on the beach, perish even more quickly since the animal does not move and cannot escape in any way.

The bivalves washed up by the storm suffer the same fate as the gasteropods. The vast majority of mussels washed ashore consists of the most common species, *Arca ponderosa;* but it is chiefly small specimens, up to three quarters grown, that are thrown up. This species dies in thousands by desiccation in the deposits above the high-water level. In the intertidal zone it attempts to escape the effect of the surf by burying itself in the sand; but since every oncoming wave throws it up again, it generally ends by being buried and choked. That adult *Arca ponderosa* are rare among the victims of the storm is probably due to their sticking deeper in the sea floor than the younger ones, so that even the rough water cannot throw them out. The three other species of *Arca, secticostata, transversa,* and *occidentalis,* behave in quite the same manner as *Arca ponderosa.*

Pinctada radiata Lamarck, *Pteria colymbus, Modiolus tulipa* Linnaeus and *Brachidontes exustus,* all light shells, moored to some foreign body by byssus threads, can frequently be found living or dying on the beach after the storm. They are of special interest as they generally are washed ashore with the base on which they have settled. In most cases this base is a valve of a mussel, usually an *Arca,* resting on the ocean floor on its concave surface and settled on the convex surface by a gorgonid, a hydrozoan, or some other object, on which, in turn, these bivalves have anchored themselves. The only other bivalve thrown ashore alive is *Pinna rigida,* mostly as half-grown specimens.

There are surprisingly few echinoderms and tunicates among the victims of the storms on Sanibel Island, but a few sponges are common on the beach after a storm. Hermit crabs with their shells, and dromiid crabs of the genus *Hypoconcha* (Guerin), which inhabit odd mussel valves, are rather frequent after a heavy gale, but begin to disappear very soon, either because they return to calmer depths or because they have been killed by the surf or by the many herons and terns which watch the seashore for food.

The Gulf beach of Sanibel Island, normally almost destitute of living animals, is overcrowded with animal life the first days after a storm, mostly by gasteropods and bivalves washed up from the greater depths of the sea. All the specimens thrown ashore by the waves at the highest point of the storm lie above the normal high-water mark and die from lack of water, or become the victims of predators like herons, pelicans, cormorants, terns and raccoons. The mollusks deposited within the intertidal belt have a chance of surviving for a short period. Those that are able to retire into the tidal pools behind the low-water mass themselves in these sheltered basins to the degree of literally paving the floor. Such gasteropods and bivalves as are able to bury themselves in the sand hide there during low tide, and thus escape the suction of the receding water. But bivalves unable to make use of the foot, like *Pinctada, Pteria, Pinna,* and the mytilids, are caught by the waves, and continue floating in the water from high tide to low tide and from low tide back to high tide they knock against each other so that their shells grit audibly, they smash each other and expose the animals thus bared to inevitable death, usually by suffocation in the sand. These dead or dying *pinnas,* which constitute such a high percentage of the beach deposits, finally come to rest below high-water mark, and the odor of putrefaction is soon perceivable at a great distance. The tiny crabs of the genus *Pinnotheres,* commensals of the pinnas, desert their dying shell host and roam and swim uneasily in the surf zone, where they are also destined soon to perish.

These victims of the storm floor that find a temporary asylum in the tidal pools, and are able to creep and bury themselves, are:

Bivalves	Gasteropods
Arca ponderosa Say	*Strombus pugilis* Linnaeus
Arca transversa Say	*Murex* spp.
Chione cancellata Linnaeus	*Fasciolaris distans* Lamarck
Venus mercenaria notata Say	*Fasciolaris tulipa* Linnaeus
Venus campechiensis Gmelin	*Fasciolaris gigantea* Kiener
	Busycon spp.

All these mollusks try to find shelter in the sand before the receding flood at low tide can tear them out, and it is a strange sight to see the gasteropods burying themselves with the points of their shells directed downward. When the residual pools that communicate with the open water have emptied at low-tide, occasional movements of the floor or jets of water rising from the bottom betray the presence of a rich fauna hidden under the sand while the tide is out. When the tide returns, it either covers the still-buried animals with sand and consequently chokes them, or the waves tear them out and whirl them around again until they are killed. The fresh, empty shells of the bivalves and those of the gasteropods containing the dead remains of the animal become more common every day, and the number of the living animals in pools diminishes correspondingly. The beach, overcrowded for a short time, is soon again a mortuary. The grinding action of the surf at once begins on the shells of these mollusks. The waves tear off their ligaments, polish their outer surface, grind the valves by mutual friction, and finally break them up. After a short time the beach has regained its usual aspect, revealing only worn and polished mussel valves, with practically no gasteropod shells [pp. 372–374].

Such cataclysmic events as the Sanibel storm undoubtedly have been instrumental in the formation of many shelly substrates. But the paleoecologist must remember that shelly debris does not always reflect turbulent conditions. Many shells living in quiet water may aggregate after death to form shell gravel, coquina, or even a conglomerate. Micrite-rich matrix associated with such aggregates suggests that turbulence was a minor agent in their formation.

The paleoecologist must also be wary of interpreting the mixture of shelly material with gravels as having been invariably due to transportation and winnowing by currents. Figure 35 illustrates an epifaunal organism modified for anchoring to the gravel with a pedicle. The shell morphology alone does not suggest an animal living in gravel, and the unwary paleoecologist might infer that such shells were transported to the site rather than having lived there. However, Richardson (personal communication, 1976) has developed criteria by which study of internal shell morphology leads to correct conclusions.

An appropriate closing for this section is a quote from Cloud (1959b) concerning a biogenic deposit on Saipan where "The small patch of phosphatic gravel on the reef-islet of Agingan point (plate 1235D) was made up largely of human bone from the 1944 military campaign [page 377]."

Shifting-Bottom Sediment Environments

Newell *et al.* (1959) emphasize the various biota encountered on shifting sands as compared with stable sands. Purdy (1964) has reviewed some of the consequences (Figure 36) of stable-sediment environments as contrasted with unstable substrates and formed the logical conclusions that indicate lower taxic diversity in

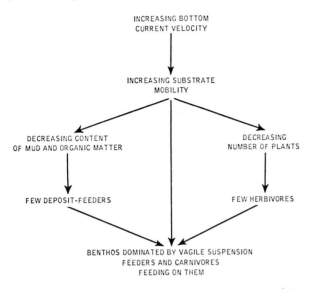

INCREASING BOTTOM
CURRENT VELOCITY

INCREASING SUBSTRATE
MOBILITY

DECREASING CONTENT
OF MUD AND ORGANIC MATTER

DECREASING
NUMBER OF PLANTS

FEW DEPOSIT-FEEDERS

FEW HERBIVORES

BENTHOS DOMINATED BY VAGILE SUSPENSION
FEEDERS AND CARNIVORES
FEEDING ON THEM

FIGURE 36. Diagram illustrating the consequences of increasing substrate mobility. (From E. G. Purdy, 1964, *In* J. Imbrie and N. Newell, Eds., *Approaches to Paleoecology,* Figure 15. Copyright © 1964 by John Wiley & Sons.)

shifting environments; he has also provided observational data that conform to the logical inferences.

Estcourt (1968); Fager (1968); Owens and Sohl (1969); and Purdy (1964) provide information indicating that taxic diversity tends to be significantly lower in shifting-bottom sediment environments than in the normal, less disturbed regimes. Shifting-bottom sediment environments provide still another example of the fact that any departure from the "normal" condition results in lower taxic diversity.

Jumars and Fauchald (1977) provide a regional survey of polychaete diversity and feeding type that indicates that nearshore types are mainly vagrant deposit-feeders, whereas outer-shelf types tend to include a high number of sessile-filter feeders. The authors suspect that this situation relates to the shifting nature of the nearshore sediment as opposed to the more stable nature of the offshore sediment.

ROCKY-BOTTOM ENVIRONMENTS

Much of the marine ecology literature is devoted to the rocky bottom, and specifically directed at description and discussion of the rocky intertidal zone. Why is it that virtually none of the paleontological literature deals with this extensively studied modern environment? Where is the environment of the Lewises and Stephensons in the geologic past? We are faced with three possible answers to these questions: (*a*) an environment similar to the present rocky intertidal may not have existed in the past; (*b*) the remains of such an environment may remain unrecognized by an imperceptive geologist (Glaessner, 1969, p. R424; Ladd,

1966; and Woodring, 1928a,b); and (*c*) it is possible that a hard-shelled, rocky-intertidal fauna is a relatively recent development. Which is it?

The vast epicontinental seas of the past have no modern counterpart, although much of the Arctic shelf and Sunda Sea may be somewhat similar. The bulk of the marine fossil record studied to date appears to represent level bottoms of one sort or another, reflecting the past extensiveness of the epicontinental seas. The overwhelming upheaval of the continents so characteristic of the present provides a rock surface extensively colonized by barnacles, mussels, chitons, and sea weeds, and impeccably studied by the Lewises and Stephensons. The intertidal is a narrow, transient line that must change radically as the sea level fluctuates and the coastal topography is altered by erosional and tectonic forces. It may be inferred that this commonly high-energy intertidal environment is not likely to be preserved in a slowly accumulated fossil record. However, quieter level-bottom intertidal records do appear to have been preserved in some abundance, as have the subtidal reef environments. The absence of the rocky-intertidal fauna in the fossil record may be made striking by thinking, for a moment, of the overall intertidal abundance and ubiquity of rocky-bottom barnacles outside of the Tropics today, and their virtual absence in the Tertiary nontropical fossil record. A corollary to this general rarity of intertidal rocky-bottom is an overall rarity of bouldery environments.

As with intertidal rocky-bottoms, subtidal rocky-bottoms with their abundant sessile faunas find few counterparts in the fossil record. Again, this rarity of Phanerozoic rocky-bottom appears to reflect a norm of low relief over most continental areas during the past, with the additional complication that high-relief areas tend to be worn down and destroyed. The discovery of a genuine fossil rocky-bottom situation is noteworthy. (Surlyk and Christensen [1974] provide a well-documented example from the Cretaceous; Pyanovskaya and Gekker [1966] and Gekker and Uspenskaya [1966] present additional well-documented examples as well as a review of the literature on European Mesozoic and Cenozoic examples; Woodring, 1931, for a California Miocene example.)

Subtidal rocky substrates, in contrast to hard bottoms, are found today on rock exposures and boulders. Boulders today are particularly abundant in areas where glacial ice has melted and dropped its load. Knox (1970) and Foster (1974) discuss some of the Antarctic taxa found attached to such boulders. In the fossil record few such occurrences have been noted, but adjacent to reef-type bodies, the blocky talus has the potential for supplying bouldery substrates similar to those of modern reef-talus aprons.

This overall rarity of preserved rocky-bottom environments is unfortunate for the paleontologist. Consider for a moment how much of our specific knowledge and how many of our general conclusions about marine ecology are based upon the study of the modern rocky bottom. It is interesting to wonder how different our present concept of the evolution and ecology of benthic animals would be if biologists conducted their studies with the full awareness that the extensive rocky environments of today are not typical of the preserved marine ecosystem of the past. The significance to the ecologist of microenvironmental variation giving rise to crevice and cryptic habitats within the rocky environment is obvious. The paleontologist will rarely have access to these habitats. Despite this unfortunate situation, the paleoecologist must always be alert for a preserved remnant. Fossils might have been transported from a rocky substrate into an adjacent level-bottom depositional site. There may even be some examples discovered of rocky-bottom faunas preserved in situ.

In the section on substrate selection (Chapter 3, pp. 29–36), it was mentioned that sedimentary faunas are selective with regard to the sediment on which they settle as larvae. This relation is so well-recognized that virtually all workers are aware of the faunal distinctions between different sediments. In the case of rocks and rocky bottoms, the size of the substrate makes it easy to forget that substrate selection is still operating. One must not take the position that all rocky substrates have the same attraction to settlement, and that all substrates provide equal fitness potential. Hartog (1972b) provides good examples of attached algae that prefer a rough concrete surface to a smooth stone surface. Bennett and Pope (1953) emphasize that weathered rock behaves very differently as a substrate than does fresh rock. Pequegnat (1964) distinguishes rock-types having different substrate properties for organisms. Roughness, smoothness, color (if it affects the amount of heat absorbed in shallow or intertidal environments), weathering properties, water retentiveness of intertidally exposed stone; all these and many other variables affect rocks as substrates. Stephenson and Stephenson (1972) discuss the widely varying attractiveness of different rock types (moisture-retaining rock, hard substrate, easily bored substrates and so on) in the intertidal zone to different flora and fauna.

HARD-BOTTOM ENVIRONMENTS

The foregoing sections have discussed rocky bottoms with the implicit understanding that the rocky substrate will be geologically older, such as the Precambrian boulders in an Upper Cretaceous Sea discussed by Surlyk and Christensen (1974), than the organisms thriving on it. However, there is another category of rocky bottom; the penecontemporaneous hard bottom. These hard bottoms are characteristic of modern and past tropical to subtropical regions where carbonate cementation of clastic debris is an important process. Such hard bottoms are reported from the supratidal,

intertidal, and shelf depth subtidal regions (depths of as much as 600–700 m are known, Neumann et al., 1977). James et al. (1976) provide an introduction to the conditions under which hard bottoms are being formed subtidally at the present time, including data suggesting that they form more rapidly in areas of fine-grained sediment and in areas subject to oxidation. The most comprehensive treatment of hard-bottom character, genesis, and classification is provided by Fürsich (1979). The fossil assemblages found in these hard bottoms may reflect very complex successions as a result of biotic changes made possible by environmental alterations involved in the passage from unindurated to indurated sediment as well as complex conditions of erosion at different stages of the numerous processes. Hard bottoms have many characteristics of ordinary rocky bottoms, but retain their own uniqueness. Think of the hard-bottom biotas as progressing through a series of changes conditioned by a series of physical alterations in the substrate, or the presence of a substrate that is undergoing a series of such changes. Baird and Fürsich (1975); Bromley (1975c); Goldring and Kazmierczak (1974); Kennedy and Klinger (1972); Lindström (1979); and Palmer and Fürsich (1974) provide adequate introduction and good illustrations to the extensive literature on the subject.

In addition to epifaunal organisms occurring on hard substrates, there are boring organisms of varied types and cryptic organisms (such as the lingulids described by Richards and Dyson-Cobb, 1976) living within the burrows, as well as a generally higher level of environmental heterogeneity than that found in the more uniform level-bottom environment. However, hard bottoms occur intimately interlayered with ordinary, contemporaneous unindurated level bottoms. Hard bottoms have commonly been overlooked as distinct environments in the past, and lumped together with the level bottoms that became lithified after burial. Hard bottoms do not form cliffs, steep slopes, or boulders; they are identified only after careful examination and thought (see Boyd and Newell, 1972, for an appropriate example). The individual case histories of the hardgrounds described to date (see references) read like a series of paleontologic Sherlock Holmes adventures. No two tales are alike, although each involves a hardground (Figures 37–50).

A key variable affecting the nature of a hardground is the time after hardening (penecontemporaneous lithification), during which the surface has been exposed to the actions of physical and biological agencies. Hardgrounds exposed for a brief time prior to burial by the deposition of more material present a simpler story than is the case when they are exposed for a lengthy period. Accumulations of glauconite, phosphorite nodules, manganiferous and ferriferous crusts, accumulations of pyrite, intensive differential solution to some depth (giving rise to embedded materials—see Figures 43, 47, and 48) and only partially preserved at

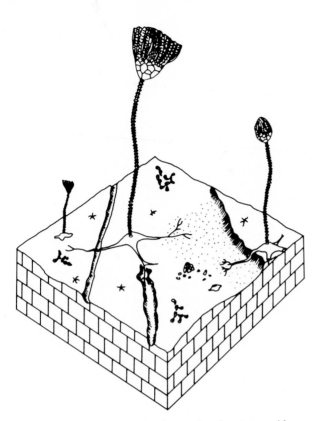

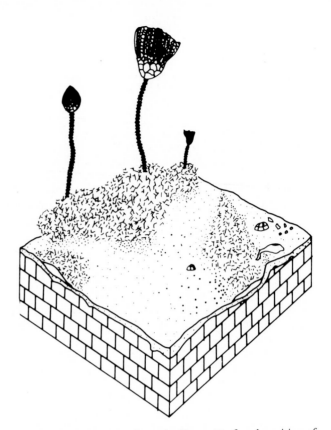

FIGURE 37. Wenlock age hardground surface (note evidence of solution and/or erosion in the form of pits and hollows) supporting epifauna of the crinoid *Eucalyptocrinites,* a chainlike auloporoid coral; and star-shaped pits representing juvenile crinoid roots. (From M. S. Halleck, 1973, *Lethaia, 6,* pp. 239–252, Figure 6. Copyright © 1973 by Universitetsforlaget.)

FIGURE 38. The area shown in Figure 37 after deposition of a small amount of mud (shale). Growths of algae have collected around the crinoids' basal portions and have filled in the surface irregularities. (From M. S. Halleck, 1973, *Lethaia, 6,* pp. 239–252, Figure 7. Copyright © 1973 by Universitetsforlaget.)

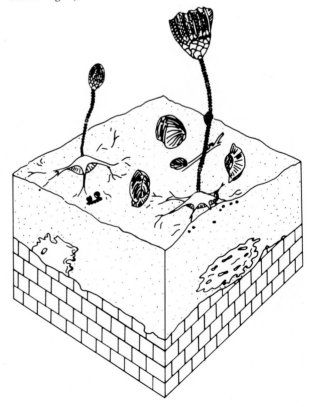

FIGURE 39. The same area shown in Figures 37 and 38 after additional mud (shale) as well as transported "clods" of debris have been deposited. Note the presence of a soft-bottom shelly community of brachiopods, and of additional crinoids employing shells as hard substrates. (From M. S. Halleck, 1973, *Lethaia, 6,* pp. 239–252, Figure 8. Copyright © 1973 by Universitetsforlaget.)

54

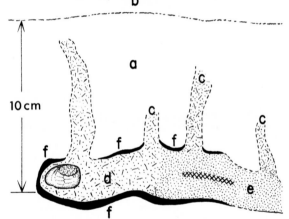

FIGURE 40. (Top left): *Brissopheustes danicus* in its burrow in indurated White Chalk. Skelding, Sterns Klint, before preparation. (Top right): During preparation (test removed). (Bottom): Possible echinoid burrow complex reconstruction in Cretaceous Chalk hardground. (*a*) Indurated White Chalk. (*b*) Bryozoan Limestone. (*c*) Respiratory funnel filling of chalk and bryozoans. (*d*) Anterior part of burrow. (*e*) Posterior part of burrow, partly silicified. (*f*) Limonite along walls of burrow. (From H. W. Rasmussen, 1971, *Lethaia, 4,* pp. 191–216, Figure 8. Copyright © 1971 by Universitetsforlaget.) The echinoid may have fallen into the burrow, rather than having made it (Weinberg Rasmussen, letter, 1977).

FIGURE 41. Local stratigraphic relations adjacent to Cretaceous hardground reconstruction shown in Figure 40. Note how the hardground surface with burrows cuts across more than one stratigraphic unit. (*a*) White Chalk with few bryozoans, and discontinuous flint concretions. (*b*) White Chalk rich in bryozoans. (*c*) Indurated White Chalk hardground with burrows. (*d*) Fish Clay. (*e*) *Cerithium* Limestone hardground with burrows. (*f*) Bryozoan Limestone deposited in mounds. (*g*) Brecciated surface of limestone. (*h*) Quaternary till.) (From H. W. Rasmussen, 1971, *Lethaia, 4,* pp. 191–216, Figure 1. Copyright © 1971 by Universitetsforlaget.)

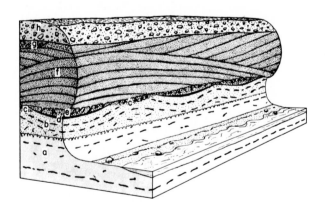

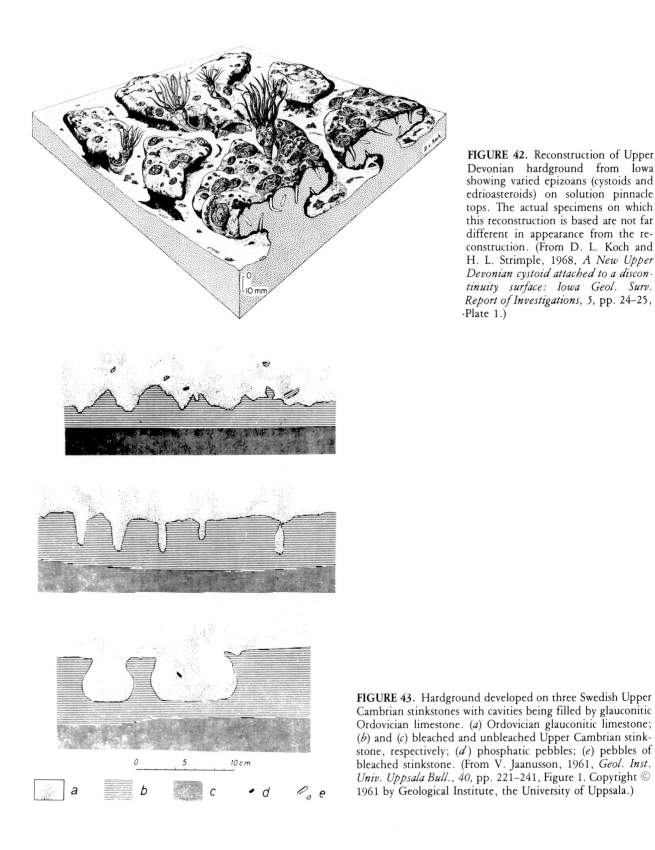

FIGURE 42. Reconstruction of Upper Devonian hardground from Iowa showing varied epizoans (cystoids and edrioasteroids) on solution pinnacle tops. The actual specimens on which this reconstruction is based are not far different in appearance from the reconstruction. (From D. L. Koch and H. L. Strimple, 1968, *A New Upper Devonian cystoid attached to a discontinuity surface: Iowa Geol. Surv. Report of Investigations,* 5, pp. 24–25, ·Plate 1.)

FIGURE 43. Hardground developed on three Swedish Upper Cambrian stinkstones with cavities being filled by glauconitic Ordovician limestone. (*a*) Ordovician glauconitic limestone; (*b*) and (*c*) bleached and unbleached Upper Cambrian stinkstone, respectively; (*d*) phosphatic pebbles; (*e*) pebbles of bleached stinkstone. (From V. Jaanusson, 1961, *Geol. Inst. Univ. Uppsala Bull.,* 40, pp. 221–241, Figure 1. Copyright © 1961 by Geological Institute, the University of Uppsala.)

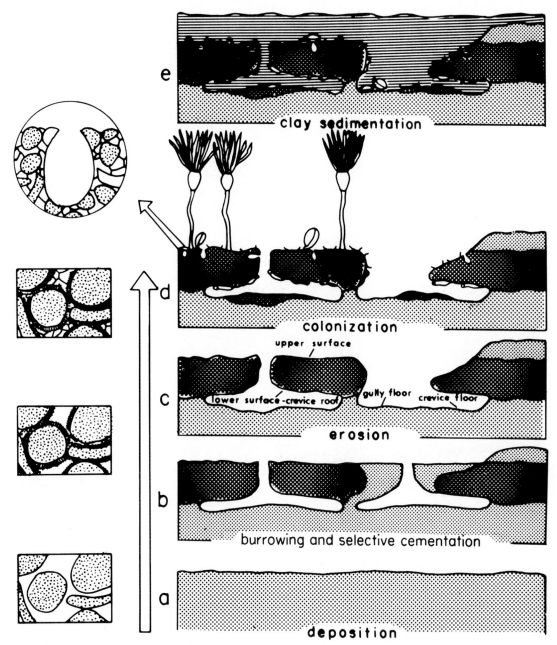

FIGURE 44. Diagrammatic history of cementation and colonization of the Jurassic Bradford hardground: (*a*) Deposition of lime-sand. (*b*) Crustaceans excavated burrow systems; discontinuous lithification of a thin layer of sediment began at or near the sediment–water interface. (*c*) Lithification continues; the bottom surfaces of these lithified layers were exposed by removal of underlying uncemented sediment. (*d*) The exposed hard surfaces were colonized by boring and encrusting animals. Periods of shell accumulation on the hardground alternated with periods of bioerosion, during which boring activity removed some encrusting shell material. The floors of the crevices started to lithify. (*e*) Shell material derived from the hardground accumulated within crevices. Eventually, clay deposition buried the hardground and its associated fauna. The rectangular insets on the left of the figure show the growth of cement around the grains; sediment passes from loose → lightly cemented → well cemented. Note that the borehole biota is different than that on the exposed, upper surface of the hardground. Torunski (1979) provides a good overview of the importance of limestone coasts and of the amount of fine-grained lime mud generated by bioerosion. (From T. J. Palmer and F. T. Fürsich, 1974, *Palaeontology, 17,* pp. 507–524, Text–figure 2. Copyright © 1974 by the Palaeontological Association.)

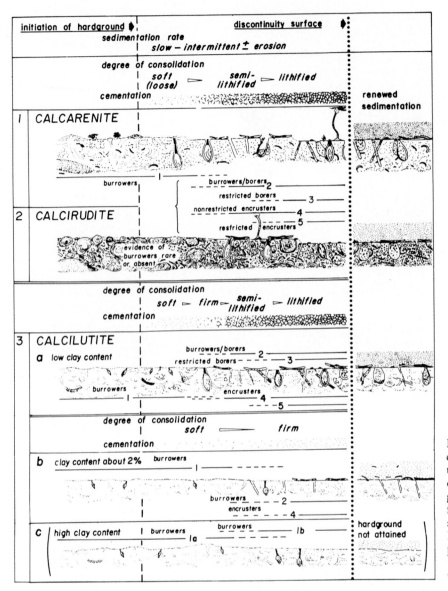

FIGURE 45. Diagram illustrating the correlation between different sediment types undergoing progressive consolidation and cementation into hardgrounds with biota, as well as the effects of differing rates of sedimentation and erosion. (From R. Goldring and J. Kazmierczak, 1974, *Palaeontology, 17,* pp. 949–962, Text–figure 2. Copyright © 1974 by Palaeontological Association.)

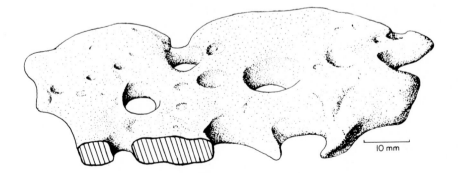

FIGURE 46. Strongly corroded bit of limestone hardground from the Swedish Ordovician. (From M. Lindström, 1963, *Sedimentology, 2,* pp. 243–292, Figure 12. Copyright © 1963 by Elsevier Publishing Company.)

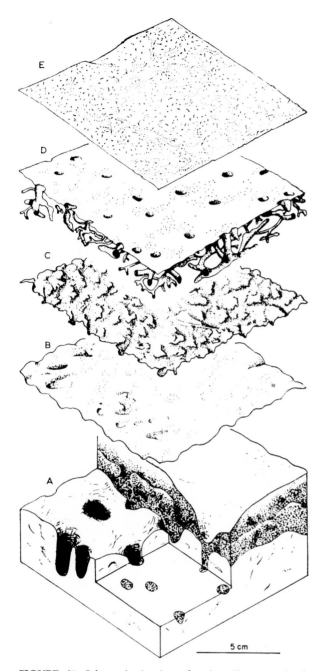

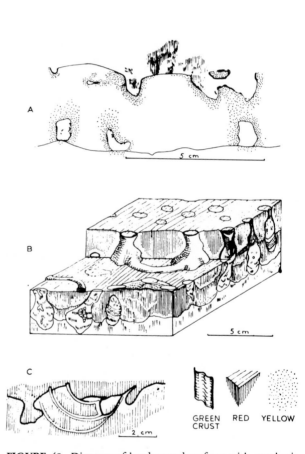

GREEN CRUST RED YELLOW

FIGURE 47. Schematic sketches of various limestone hardgrounds and hardground surfaces from the Swedish Ordovician: (*a*) is typical for the lowermost Ordovician limestone beds (glauconite is shown by black spots; phosphorite with vertical shading); (*b*) is from the top of the V-bed at Horns Udde; (*c*) is from the top of the X-bed at Horns Udde (this has no glauconite crust); (*d*) is typical for the beds (*l*) and (*h*) at Horns Udde; (*e*) is from the Billingen Stage at Kopings Klint (loc. 2). The surface is greenish (in an otherwise red limestone) and contains as ridges the exposed fillings of branching tubules. The surface is also studded with slightly protruding shell fragments. (From M. Lindström, 1963, *Sedimentology, 2,* pp. 243–292, Figure 9. Copyright © 1963 by Elsevier Publishing Company.)

FIGURE 48. Diagram of hardground surfaces with emphasis on the iron compound coloration present in examples taken from the Swedish Ordovician. (*a*) Section through the upper part and top of a bed. (*b*) Dissected stereogram of hardground development. Note in the foreground the trilobite carapace penetrated by circular borings. The bottom bed and the filling of the borings are marly. Filling usually is cavernous and may contain patches of crystalline calcite. (*c*) Nautiloid shell cut by a hardground surface. (From M. Lindström, 1963, *Sedimentology, 2,* pp. 243–292, Figure 11. Copyright © 1963 by Elsevier Publishing Company.)

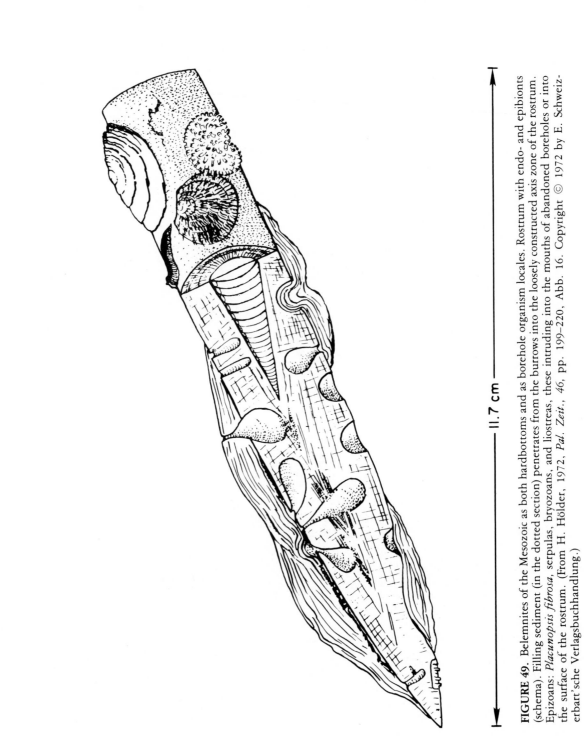

| 11.7 cm

FIGURE 49. Belemnites of the Mesozoic as both hardbottoms and as borehole organism locales. Rostrum with endo- and epibionts (schema). Filling sediment (in the dotted section) penetrates from the burrows into the loosely constructed axis zone of the rostrum. Epizoans: *Placunopsis fibrosa*, serpulas, bryozoans, and liostreas, these intruding into the mouths of abandoned boreholes or into the surface of the rostrum. (From H. Hölder, 1972, *Pal. Zeit.*, 46, pp. 199–220, Abb. 16. Copyright © 1972 by E. Schweiz-erbart'sche Verlagsbuchhandlung.)

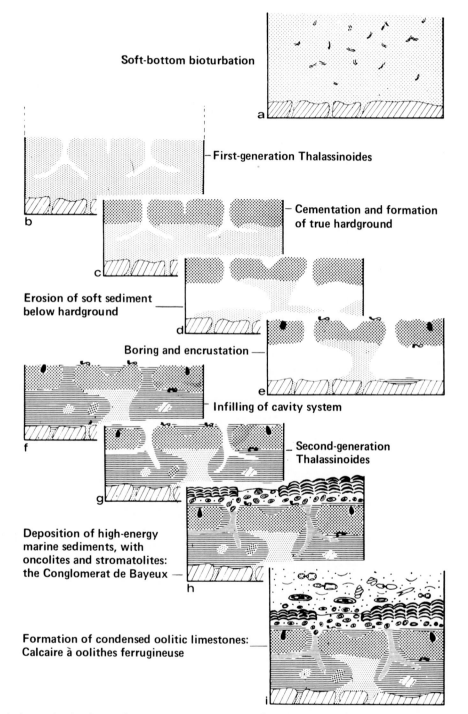

Soft-bottom bioturbation

First-generation Thalassinoides

Cementation and formation
of true hardground

Erosion of soft sediment
below hardground

Boring and encrustation

Infilling of cavity system

Second-generation
Thalassinoides

Deposition of high-energy
marine sediments, with
oncolites and stromatolites:
the Conglomerat de Bayeux

Formation of condensed oolitic limestones:
Calcaire à oolithes ferrugineuse

FIGURE 50. Typical complex hardground story. (From W. J. Kennedy, 1975, *In* R. W. Frey, Ed., *The Study of Trace Fossils*, Figure 17.9. Copyright © 1975 by Springer–Verlag New York, Inc.)

the surface, are all evidence of long exposure of a hard-ground surface, or at least of extremely slow rates of sedimentation (Glasby and Summerhayes, 1975). Boring, a short-term effect, can occur under conditions of either short-term or long-term exposure, and should not be confused with burrowing prior to lithification or solution pits and holes. Brecciation and shuffling of pieces of a formerly continuous hardground pavement can also occur under conditions of either long-term or short-term exposure of the bed in question.

Excellent examples of long-term exposure are provided by Jaanusson (1961), who favored a subaerial explanation of the solution phenomena before Shinn's most convincing studies became available, Lindström (1963), and the Terra Rossa of the Alpine Mesozoic. Goldring and Kazmierczak (1974) provide a review of the hardground situation giving particular attention to the effects of varying degrees of cementation on different materials. These authors emphasize the faunal changes (in their terminology, succession) occurring concurrently with changes in degree of cementation on different type materials.

Palmer and Palmer (1977) provide a synthesis of changes through time in the characteristics of hardground inhabitants. They suggest that the Paleozoic cover might have been sparser than it was during the post-Paleozoic, and that Paleozoic hardground epifauna may have had less current resistance than those of the post-Paleozoic.

Area

Jackson (1977) reports data that indicates that in some hard-bottom situations, the actual size of the area available for colonization is a determinant of faunal richness. Thus, it cannot be assumed that similar hard bottoms will always have the same diversity of species unless the total areas of two separate samples are equal.

THE REEF-COMPLEX ENVIRONMENT

The evolutionary implications of the reef complex have been considered by May (1973) and myself (Boucot, 1975), as well as many other workers, because of two particular reef attributes. First, it is generally assumed that the reef environment has been little changed through time, the physical factors remaining quite stable. This persistence and stability may have been important in the development of organisms specialized for a variety of microhabitats that do not exist in less persistent environments. Second, the structural complexity on the reef, from microscopic to massive scales, is of biological origin. Thus, the study of a reef complex involves thinking in terms of a true biological community, with a complicated system of organism interdependence and coevolution.

The term *reef environment* encompasses far more complexity, and misunderstanding, than do the terms *level-bottom, rocky-bottom,* or *hard-bottom environments.* There are almost as many definitions and usages of the term *reef* and *reef environment* as there are scientists interested in them (see Wilson, 1975, pp. 20–24 for a discussion of the problems). However, there appears to be some agreement in defining the reef environment as one in which calcium carbonate-secreting organisms play a major role in producing a reef framework and various types of sediments. The usage by some researchers of the term *reef* for any massive, organogenic carbonate structure of the past, regardless of its structure or form, is one of the major complications. The terms *bioherm*[1] and *buildup*[2] commonly, although not invariably, include the reef category, but the term *biostrome*[3] seldom, if ever, is properly applied to reef environments. Few biostromes appear to encompass as high a level of environmental heterogeneity as does the reef environment.

The overall solubility of calcium carbonate in cool and cold waters (Alexandersson, 1972; Wilson, 1975) restricts reef structures, as well as widespread limestones and dolomites of nonreef origin, to tropical and subtropical climatic regimes where secondary deposition and overgrowth of calcium carbonate, not solution, is the rule.

Studies of the reef environment are sharply divided into those of the Recent, in which emphasis has been placed on both the distribution of living organisms and their relations to topography (both biotically generated and biotically modified) and substrate character, and those of the geologic record, in which emphasis is on lithofacies. The studies of ancient reefs have, with few exceptions, devoted far too little attention to recurring associations of organisms. The geologist has found it far easier to treat the reef complex as an intricate fabric of rock types (whose biogenic origins are admitted although seldom pursued even to the family level) than as associations of genera and species. Thus, one often finds it difficult to bridge the gap between descriptions of ancient reefs and those of the present. Although Middle Triassic through Recent reefs display a reasonable element of taxonomic and community continuity, those of the Paleozoic (Figures 51–56) may be divided into at least three major groups with a low degree of either taxonomic or community continuity

[1]*Bioherm* is commonly defined by the geologist as a moundlike accumulation of organogenic debris. No position is taken as to whether or not it involves a biogenically cemented carbonate framework. It is a purely descriptive term for a moundlike body of organogenic material included in the stratigraphic record.

[2]*Buildup* is a term employed by the geologist to describe a carbonate moundlike body that may or may not be chiefly of organogenic origin, and may or may not have included a biogenically cemented carbonate framework.

[3]*Biostrome* is a term employed by the geologist for a tabular body of biogenic material. It may or may not have been biogenically cemented into a framework.

(Middle Ordovician through Late Ordovician; Late Wenlock through Frasne; Pennsylvanian through Permian).

The complexity of the reef environment derives in large measure from its intricate topography. Both the rocky-bottom and level-bottom environments are subject to conditions of varying turbulence, as are the reef, crevice, and cryptic environments. Differing slopes are also factors in the rocky-bottom environment, but the greater topographic complexity, intricate subdivision, and variety of the reef environment provide many more potential niches than other benthic environments. The reef environment, both past and present, is characterized by a wealth of turbulent, quiet, deep, shallow, steep-slope, rocky, crevice, cryptic, level-bottom environments (and all possible combinations of these). Adey (1978) has provided a number of insights into the complex of physical factors interacting to produce different reef characteristics and growth rates. The paleontologist is barely beginning the study of these factors. Even a brief perusal of modern or nineteenth-century descriptions of modern reefs should provide stimuli for the paleontologist.

Reef ecology is influenced by several physical parameters. Among these are (a) problems of macrotidal and microtidal ranges; (b) different ecology and aspect of the windward face as contrasted with the leeward face of the reef (in terms of prevailing winds; particularly those developed in the Trades); (c) monsoonal region reefs where annual wind reversal calls for different modifications; (d) the presence or absence of storms and oceanic swell; (e) various combinations and permutations of tide level, wind, storm, and swell; and (f) influx of large volumes of mud-laden river water. It is not clear whether it is the mud alone, or the mud in combination with the fresh water, or the fresh water that is incompatible with coral reefs in the same gulf or bay.

Despite its rich faunas and almost romantic attractiveness, the paleoecologist should be warned that the reef environment is poorly known, complex, and possibly different in more ways than are now imagined from the level-bottom environments; Conclusions based on reefal studies must be accepted with caution.

Reef Absence and Relative Success Factors

A major question in both zoogeography and paleontology has been why reef complexes are globally and temporally restricted in their distribution. The relatively greater solubility of calcium carbonate in cool water is thought to provide a mechanism for the restriction of modern reefs to warm regions. Alexandersson (1972, 1975) has reported solubility data for calcium carbonate that adequately supports this widely held view. Although carbonate solubility may represent a first-order control factor, the discontinuous distribution of reefs in the modern tropics clearly suggests the

involvement of other agents. A low carbonate solubility is adequate but not sufficient to allow successful reef formation.

Reefs forming along the margins of continents necessarily will come under the influence of interacting marine and terrestrial factors. Conspicuous among these is turbidity. Excessive turbidity (Dana, 1975; Manton and Stephenson, 1931; Pichon, 1971; Pillai, 1971; Wilson, 1975) appears to reduce or prohibit reef development. Both in the present and in the past, reefs commonly occur or are preserved in areas lacking excessive turbidity. Among the reasons for this may be decreased light levels, clogging of the feeding mechanisms of filter-feeders, or smothering of recently settled larvae or adult organisms. The high correlation between turbidity and turbulence implies that meteorological and sedimentological phenomena may act in conjunction to favor or disfavor reef success.

Regions adjacent to riverine or other low-salinity water influx are subject to salinity fluctuations that seem to preclude both the establishment and success of carbonate reefs in the present (Pillai, 1971; Whitehouse, 1973). Riverine influx during the past probably had an important effect upon reef formation as well. Thus, evidence of river discharge in the paleoenvironment can be an important clue in the explanation of reef absence in otherwise favorable environments if a strong paleogeographic and lithofacies argument can be made for such discharge. In the case of both turbidity and river discharge, which also affects turbidity, local current patterns can complicate interpretation. Shallow current patterns determined by meteorologic and topographic factors partially determine the local distribution of other factors.

Corollary to freshwater influx control is the problem of regression and transgression over continental platforms. Riverine influx will be a function of, among other things, drainage basin area. Thus, a trend toward higher levels of freshwater influx during times of high regression as contrasted with high transgression can be expected. Regression and transgression are also important in the creation of turbidity through provision of suitable materials.

The foregoing discussion is in reference to continental reefs. Currently, oceanic reefs are recognized as extending back into the Cretaceous; pre-Cretaceous reefs are considered to have been continental. Although positive evidence is lacking, we believe that oceanic reefs were present during the pre-Cretaceous. When they occur, oceanic reefs are affected by temperature and massive ocean chemistry changes. Although fluctuations in the CCD might constitute such massive chemical change, major fluctuations in the CCD do not appear to coincide with the presence or absence of reefs.

There is also the question of reef biota genesis following global reef extinction. We know that the time interval required for reef biota genesis was dif-

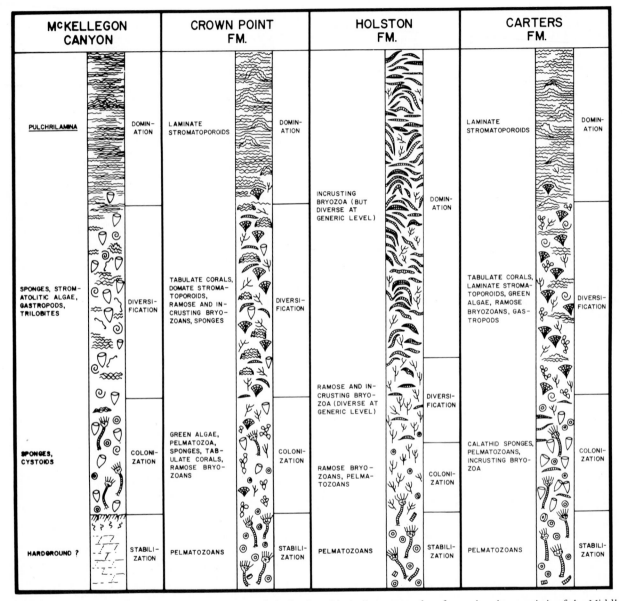

McKELLEGON CANYON			CROWN POINT FM.			HOLSTON FM.			CARTERS FM.		
PULCHRILAMINA		DOMIN-ATION	LAMINATE STROMATOPOROIDS		DOMIN-ATION				LAMINATE STROMATOPOROIDS		DOMIN-ATION
						INCRUSTING BRYOZOA (BUT DIVERSE AT GENERIC LEVEL)		DOMIN-ATION			
SPONGES, STROM-ATOLITIC ALGAE, GASTROPODS, TRILOBITES		DIVERSI-FICATION	TABULATE CORALS, DOMATE STROMA-TOPOROIDS, RAMOSE AND IN-CRUSTING BRYO-ZOANS, SPONGES		DIVERSI-FICATION				TABULATE CORALS, LAMINATE STROMA-TOPOROIDS, GREEN ALGAE, RAMOSE BRYOZOANS, GAS-TROPODS		DIVERSI-FICATION
						RAMOSE AND IN-CRUSTING BRYO-ZOA (DIVERSE AT GENERIC LEVEL)		DIVERSI-FICATION			
SPONGES, CYSTOIDS		COLONI-ZATION	GREEN ALGAE, PELMATOZOA, SPONGES, TAB-ULATE CORALS, RAMOSE BRYO-ZOANS		COLONI-ZATION	RAMOSE BRYO-ZOANS, PELMA-TOZOANS		COLONI-ZATION	CALATHID SPONGES, PELMATOZOANS, INCRUSTING BRYO-ZOA		COLONI-ZATION
HARDGROUND ?		STABILI-ZATION	PELMATOZOANS		STABILI-ZATION	PELMATOZOANS		STABILI-ZATION	PELMATOZOANS		STABILI-ZATION

FIGURE 51. Generalized diagram showing the upward shallowing, biotically generated reef complex characteristic of the Middle and Upper Ordovician type reefs (McKellegon, Crown Point, Holston and Carters). Note the similarities and differences characteristic of each time interval. The Upper Carboniferous-Permian reef complexes (unfigured here) provide still another story of this general type. (From K. R. Walker and L. P. Alberstadt, 1975, *Paleobiology, 1,* pp. 238–257, Figure 2. Copyright © 1975 by the Paleontological Society.)

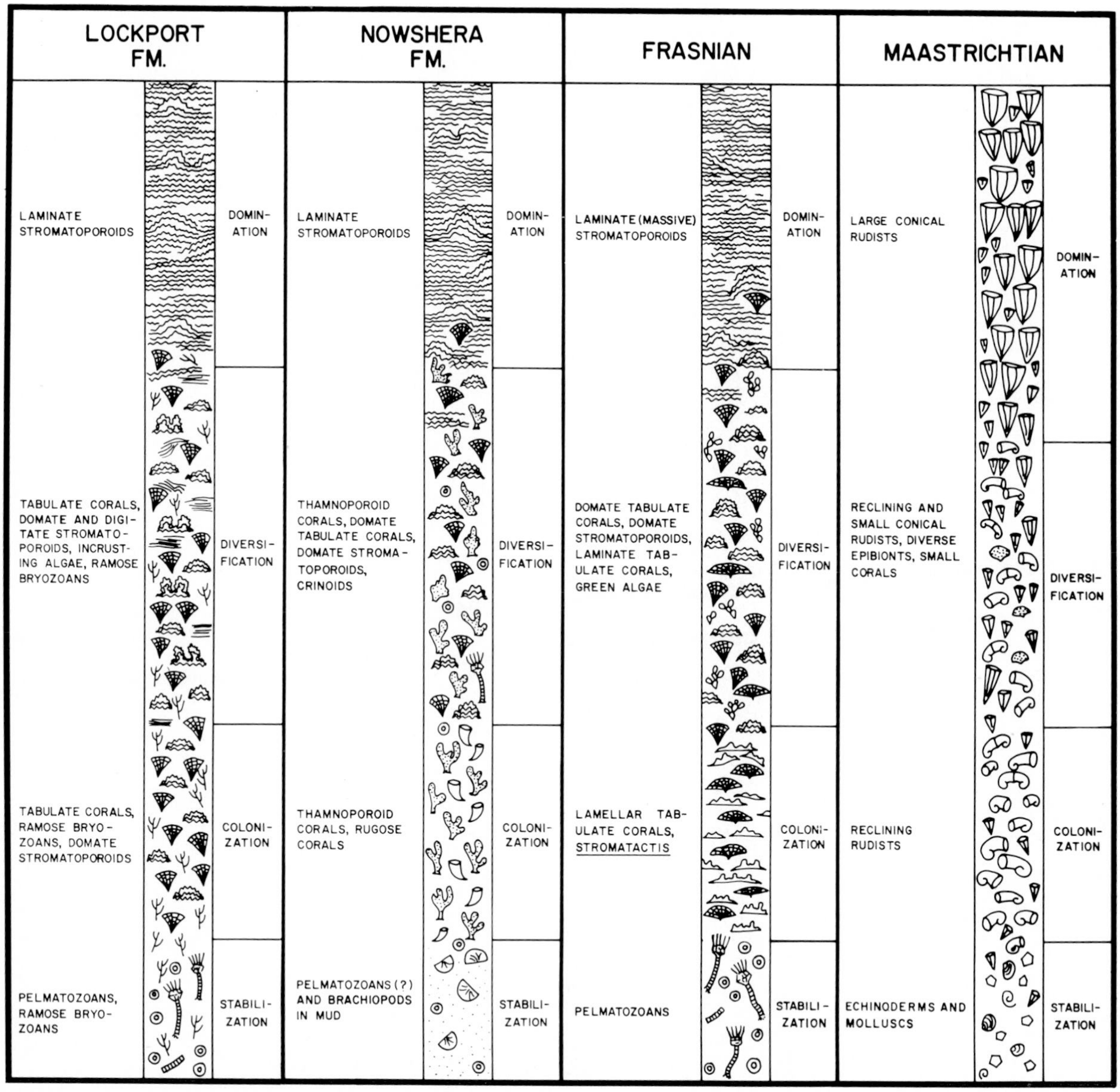

LOCKPORT FM.		NOWSHERA FM.		FRASNIAN		MAASTRICHTIAN	
LAMINATE STROMATOPOROIDS	DOMIN-ATION	LAMINATE STROMATOPOROIDS	DOMIN-ATION	LAMINATE (MASSIVE) STROMATOPOROIDS	DOMIN-ATION	LARGE CONICAL RUDISTS	DOMIN-ATION
TABULATE CORALS, DOMATE AND DIGI-TATE STROMATO-POROIDS, INCRUST-ING ALGAE, RAMOSE BRYOZOANS	DIVERSI-FICATION	THAMNOPOROID CORALS, DOMATE TABULATE CORALS, DOMATE STROMA-TOPOROIDS, CRINOIDS	DIVERSI-FICATION	DOMATE TABULATE CORALS, DOMATE STROMATOPOROIDS, LAMINATE TAB-ULATE CORALS, GREEN ALGAE	DIVERSI-FICATION	RECLINING AND SMALL CONICAL RUDISTS, DIVERSE EPIBIONTS, SMALL CORALS	DIVERSI-FICATION
TABULATE CORALS, RAMOSE BRYO-ZOANS, DOMATE STROMATOPOROIDS	COLONI-ZATION	THAMNOPOROID CORALS, RUGOSE CORALS	COLONI-ZATION	LAMELLAR TAB-ULATE CORALS, STROMATACTIS	COLONI-ZATION	RECLINING RUDISTS	COLONI-ZATION
PELMATOZOANS, RAMOSE BRYO-ZOANS	STABILI-ZATION	PELMATOZOANS (?) AND BRACHIOPODS IN MUD	STABILI-ZATION	PELMATOZOANS	STABILI-ZATION	ECHINODERMS AND MOLLUSCS	STABILI-ZATION

FIGURE 52. Generalized diagram showing The Upper Silurian–Devonian type reefs (Lockport, Nowshera, and Frasnian) and of the Jurassic–Cretaceous nonreef rudistid complex (Maastrichtian). (From K. R. Walker and L. P. Alberstadt, 1975, *Paleobiology, 1,* pp. 238–257, Figure 2. Copyright © 1975 by the Paleontological Society.)

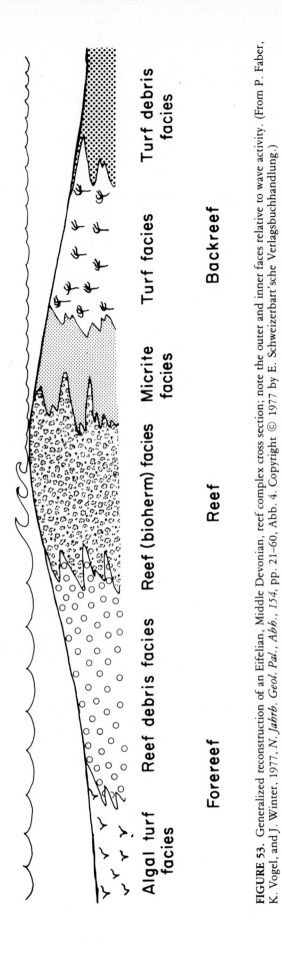

FIGURE 53. Generalized reconstruction of an Eifelian, Middle Devonian, reef complex cross section; note the outer and inner faces relative to wave activity. (From P. Faber, K. Vogel, and J. Winter, 1977, *N. Jahrb. Geol. Pal., Abh., 154*, pp. 21–60, Abb. 4. Copyright © 1977 by E. Schweizerbart'sche Verlagsbuchhandlung.)

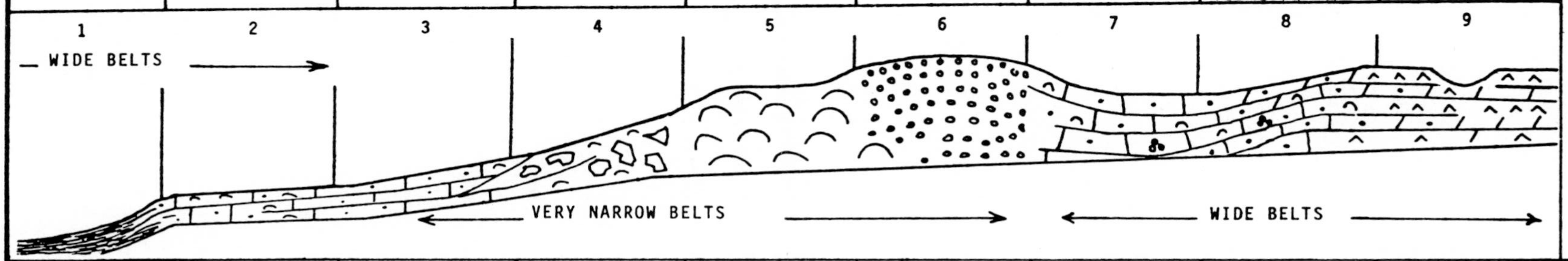

FACIES PROFILE		2ND ORDER SEDIMENTARY BODIES	STANDARD MICROFACIES
BASIN	1		1 SPICULITE 2 MICROBIOCLASTIC CALCISILT 3 PELAGIC MICRITE RADIOLARITE SHALE
OPEN SEA SHELF	2		2 MICROBIOCLASTIC CALCISILT 8 WHOLE SHELLS IN MICRITE 9 BIOCLASTIC WACKESTONE 10 COATED GRAINS IN MICRITE
DEEP SHELF MARGIN	3	DEBRIS FLOWS & TURBIDITES IN FINE LAMINATE STRATA. MOUNDS ON TOE OF SLOPE.	2 MICROBIOCLASTIC CALCISILT 3 PELAGIC MICRITE 4 BIOCLASTIC-LITHOCLASTIC MICROBRECCIA
FORESLOPE	4	GIANT TALUS BLOCKS. INFILLED LARGE CAVITIES. DOWNSLOPE MOUNDS.	4 BIOCLASTIC LITHOCLASTIC MICROBRECCIA. LITHOCLASTIC CONGLOMERATE. 5 BIOCLASTIC GRAINSTONE-PACKSTONE. FLOATSTONE. 6 REEF RUDSTONE.
ORGANIC BUILD UP	5	DOWNSLOPE MOUNDS. REEF KNOLLS. BOUNDSTONE PATCHES. FRINGING AND BARRIER FRAMEWORK REEF. SPUR AND GROOVE.	7 BOUNDSTONE 11 COATED, WORN, BIOCLASTIC GRAINSTONE. 12 COQUINA (SHELL HASH)
WINNOWED EDGE SANDS	6	ISLANDS. DUNES. BARRIER BARS. PASSES AND CHANNELS.	11 COATED, WORN BIOCLASTIC GRAINSTONE 12 COQUINA (SHELL HASH) 13 ONKOIDAL BIOCLASTIC GRAINSTONE 14 LAG BRECCIA 15 OOLITE
SHELF LAGOON OPEN CIRCULATION	7	TIDAL DELTAS. LAGOONAL PONDS. TYPICAL SHELF MOUNDS. COLUMNAR ALGAL MATS. CHANNELS AND TIDAL BARS OF LIME SAND.	8 WHOLE SHELLS IN MICRITE 9 BIOCLASTIC WACKESTONE 10 COATED GRAINS IN MICRITE 16 PELSPARITE 17 GRAPESTONE ONKOIDS IN MICRITE 18 FORAM, DASYCLADACEAN GRAINSTONE
RESTRICTED CIRCULATION SHELF AND TIDAL FLATS	8	TIDAL FLATS. CHANNELS. NATURAL LEVEES. PONDS. ALGAL MAT BELTS.	16, 17, 18 19 FENESTRAL PELOIDAL LAMINATE MICRITE 24 RUDSTONE IN CHANNELS 21 SPONGIOSTROME MICRITE 23 NON LAMINATE PURE MICRITE 22 ONKOIDAL MICRITE
EVAPORITES ON SABKHAS-SALINAS	9	ANHYDRITE DOMES. TEPEE STRUCTURES. LAMINATED CRUSTS OF GYPSUM. SALINAS (EVAPORATIVE PONDS). SABKHAS (EVAPORATIVE FLATS).	20 STROMATOLITIC MICRITE. 23 NON LAMINATE, PURE MICRITE. NODULAR-PEARL ENTEROLITHIC ANHYDRITE. SELENITE BLADES IN MICRITE.

Facies profile belt widths: WIDE BELTS — VERY NARROW BELTS — WIDE BELTS

FIGURE 54. Wilson's (1975) Standard facies belts for the carbonate lithofacies. Note the position of the organic buildup to the other facies, and keep in mind that the "organic buildup" includes the reef facies as well as several other categories of nonreef materials. (From J. L. Wilson, 1975, *Carbonate Facies in Geologic History*, Figure XII–1. Copyright © 1975 by Springer-Verlag New York, Inc.)

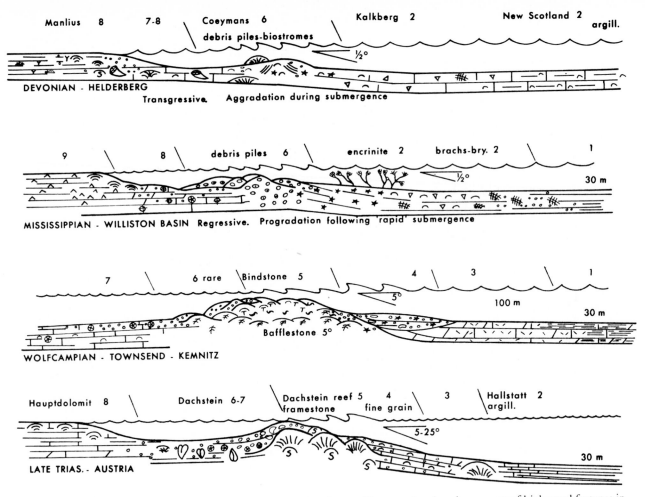

FIGURE 55. Compare the generalized subdivisions of Figure 54 to the specific examples given here: nonreef biohermal features in the Devonian and Mississippian examples and reef type biohermal features in the Permian and Triassic examples. (From J. L. Wilson, 1975, *Carbonate Facies in Geologic History*, Figure XII–2. Copyright © 1975 by Springer–Verlag New York, Inc.)

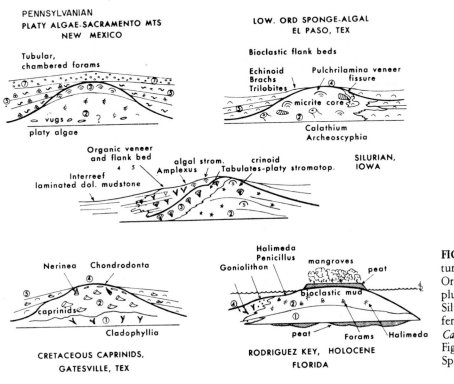

FIGURE 56. Nonreef biohermal features from the Pennsylvanian, Lower Ordovician, Cretaceous and Holocene plus a reef biohermal feature from the Silurian; note the similarities and differences. (From J. L. Wilson, 1975, *Carbonate Facies in Geologic History*, Figure XII–4. Copyright © 1975 by Springer–Verlag New York, Inc.)

ferent during the Phanerozoic, but the factors govern-
ing the phenomenon have not been delineated. The
possibility that oceanic reefs existed in the pre-
Cretaceous infers the possibility that reef taxa existed in
the oceanic environment far from continental influence
during the past, just as they do at present. The lack of
taxic continuity of the different Phanerozoic reef biotas
permits the conclusion that there has never been an
oceanic reservoir from which reef taxa were ultimately
derived following times of reef extinction affecting the
continental areas. Circumstantial evidence favoring this
conclusion is the evident ''weakness'' of island
endemics today, when they are faced with competition
from similar-type continental taxa in terms of herbivorous competition or predation.

Post-Paleozoic Reefs

Most of our knowledge of reefs concerns those of the
post-Paleozoic, particularly those of the Recent. The
following section provides a few generalizations per-
taining to the Recent. Ladd (1970), supplies an apt
introduction:

> Most students of existing reefs think of that wave-
> resisting structure, whatever its shape, as a *reef com-
> plex.* This term includes the entire reef mass: the off-
> reef areas; the wave-resisting margin; the reef flat; the
> beaches; the lagoon, if present, and its patch reefs; and
> any sediment underlying the area to which reef
> builders have contributed. A wave-resisting wall com-
> posed of the skeletons of sedentary shallow water reef
> organisms is an essential though quantitatively unim-
> portant part of all reefs. The actual existence of such a
> wall has been suggested by some of the Marshall Islands
> drilling (Ladd and Schlanger, 1960, p. 901). The wall
> permits the existence of a sheltered lagoon in which a
> great variety of reef organisms can thrive, forming
> knolls or patch reefs and adding their skeletons to the
> debris eroded from the wall.
>
> Anyone who has ever walked over, sailed over, or
> dived on a reef edge or in a reef lagoon realizes that
> only a small part of the area is carpeted with living
> organisms. Reef-drilling has demonstrated that such
> conditions also prevailed in the past. It appears that
> more than 90 percent of a three-dimensional reef com-
> plex is composed of sediments. On atolls and on open-
> sea shoals all sediments are reef-derived carbonates; on
> volcanic islands and on reefs near continental areas,
> non-carbonate materials are added. . . The term reef
> has been defined and redefined; reefs have been
> classified and subdivided. The general subject has been
> expounded in all forms of scientific writing from the
> scientist's presidential address to the special journal ar-
> ticle. As major reef types we still have the atoll, the bar-
> rier reef and the fringing reef, but there are now more
> minor reef types than ever before [pp. 1273–1274].

Cloud (1959b) adds to Ladd's view:

> Geologically the reef complex is an aggregate of
> calcium carbonate-secreting and frame-building

organisms, associated biota, and mainly biogenic
sediments. Ecologically it is an essentially steady-state
oasis of organic productivity featured by high
population-density, intense calcium metabolism, and
complex nutrient chains, and generally surrounded by
waters of relatively low mineral nutrient and plankton
content. Any reef complex at any given time is the
resultant of its nutrient chains and their disintegration
products (and of the past history of the reef) [p. 387].

and from Newell (1971): ''They create and maintain
favorable substrate, food, and shelter by a kind of community homeostasis. Their organisms are superlatively
coadapted [p. 2].''

The modern reef complex may be viewed as both a
lithofacies complex and a biofacies complex. Considered alone, the lithofacies approach would emphasize the hard structural features of the reef, and the
biofacies approach would place emphasis upon the
identity of the organisms forming and associated with
these lithofacies. In both of these interrelated views,
facies may be correlated with topographic features and
climatic factors. Although some confusion arises from
the detailed terminology of many independent
workers, a broad appreciation of reef zonation can be
obtained from a generalized atoll profile (see Figures
57 and 58 from Wells, 1957, Figure 2; and Yonge,
1963a, Figure 2; Figure 59). Wilson (1975) has provided an overview of the reef-type cross section and its
seaward and landward or lagoonward facies (Figures
54–56) with examples taken from the geologic record.
Employing submersible observations, the profile view
may be extended subtidally below the depth observ-
able by the scuba diver and in far greater detail than is
available from a dredge sample (Ginsburg and James,
1974; Lang, 1974). The limitations of a profile for
characterizing the wealth of environmental variety of a
reef complex must be kept in mind. The map (Figure
63) of the Low Isles (Stephenson *et al.,* 1931) shows
some appreciation of this variety in planar view. Additional cross-sections and illustrations of shore zonation
show more of the variants possible in Recent reef complexes (Cribb, 1973; Morton and Challis, 1969; Stoddart, 1973; Figures 64–66). The excellent reviews of
modern reef development provided by Ladd (1970),
Stoddart (1969), and Wells (1957) should be consulted
for more detailed accounts.

There are a fair number of lithofacies analyses of
modern reefs reviewed in the general references already
mentioned. In these, the lithic complexities are
realistically shown to be correlated with the degree of
turbulence (Figures 57–66) and the availability of
various biotic building materials under the differing
physical conditions. Although there has been much
surveying of the biota of reef complexes, such
community analyses have not been carried out as
systematically as the easier lithologic studies.

A certain number of studies have, however, been
devoted to the generalized problems of reef develop-

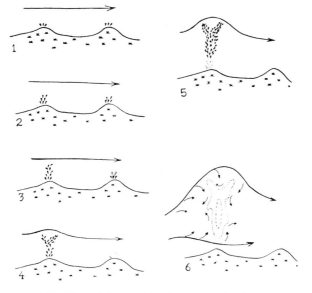

FIGURE 71. The origin of Gulf Coast oyster banks growing in a longshore current outward from the shoreline region. Note how the "upstream" oysters are thought to monopolize available nutrients, and how eventually the growing oyster bank becomes detached from the shoreline region and itself modifies the local current pattern. The numbers 1–6 represent progressive growth stages of the bank, beginning with nearshore colonization. (From J. Hedgpeth, 1953, *Pub. Inst. Mar. Sci.*, Univ. Texas, Port Aransas, *3*, pp. 111–224, Figure 32, modified from Graves, 1905.)

Discontinuous Shelly Substrates. When shelly debris is so sparsely scattered over the sediment surface that no continuous shell pavement is formed, the condition is termed a *discontinuous shelly substrate*. Some investigators (Buchanan, 1958) have viewed the discontinuous substrate as scattered bits of rocky bottom in an otherwise muddy and level environment. The presence on many benthic shells of fouling organisms different from those of the adjacent substrates, both in life and after death, point to the importance of such scattered shell debris in structuring bottom communities.

Many present examples can be found to parallel those of the past. The epibiota of large scallop shells have been discussed in some detail (Allen, 1953; Wells, Wells, and Gray, 1964; see Table 3.4 and Figures 72–74 in modern cases).The work of Waller (1969, Figure 75) and Zakharov (1966a,b, Figure 78) illustrates shells of Cenozoic and Mesozoic pectinids that suggest the possibility of similar epibios flourishing in a like manner during the past (compare these with an oyster epibiota, Boekschoten, 1966, Table 1; see our Table 3.3). Takemura and Okutani (1955) describe the complex epibiota found on pearl oysters *Pinctada* (also see Herdman, 1906). Scanland (1979) describes the epibiota of two species of *Arca*.

Comparable in some ways to the Cenozoic pectinid epibiota is the inoceramid epibiota of the later

Mesozoic (see Kauffman, 1975, for a well-illustrated example, and Frey, 1972).

Both the pectinid and inoceramid epibiotae (see Bottjer *et al.*, 1978, for an inoceramid example) may be considered separate communities when the fauna of a stratum is being analyzed. Failure to do so will seriously cloud comparisons with equivalent level-bottom strata lacking the epibios attached to such shelly substrates.

Kauffman and Sohl (1974) discuss the well-known epibionts present on the rudistid bivalves of the Jurassic and Cretaceous (Figure 76).

Heckel (1974, Figure 77) has used an illustration that provides some insight into the environmental tolerances and boundaries of many of the nonreef environmental complexes of the past. These complex relations suggest that similar situations have existed during the past.

Thomsen (1977) presents data showing the substrate selectivity of some species of encrusting bryozoans for specific bryozoan substrates—this is an instance of highly developed shelly substrate selectivity.

Stachowitsch (1977) describes the epibios carried about by some hermit crabs, which act as a movable hard substrate enabling their epibios to avoid burial by the sediment rain.

Gundrum (1979) describes the epizoan fauna fossilized on and in a Pennsylvanian sponge-thicket community complex. Her work is an excellent example of what can be done with epizoans from the fossil record.

Waterlogged Wood. Allied to the concept of discontinuous shelly substrate is the possibility that waterlogged wood can be a benthic substrate. Seilacher *et al.* (1968) illustrate a Jurassic example in which oysters and crinoids occur attached to a log of wood that Kauffman (in press) has shown rested on the bottom rather than floating on the surface. Involved here is the problem of estimating how heavy a load of fouling organisms a piece of wood can support in a floating position. The specific gravity and volume of calcium carbonate shells attached to a log provides at least a minimum estimate, even in the absence of information about potential soft-bodied fouling organisms. Wood-boring organisms are well known in the fossil record, although it is commonly difficult to determine whether or not the boring was initiated and completed while the wood was floating or while it was on the bottom (see Chapter 3, pp. 96–99).

Thickets and Forests

Kelp forests are well-known in the temperate photic zones but absent in the tropics (see Dayton, 1975a,b; Dexter, 1944, 1968; Duffus, 1969; Ebling *et al.*, 1948; Kain, 1962; Kitching and Ebling, 1967; Kitching, Macan, and Gilson, 1934; North, 1971). They are

TABLE 3.3
Adriatic Sea Oyster Epibionts[a]

Depth	Sample	Ostrea edulis	Pycnodonte cochlear	Epilithic algae	Endolithic algae	Plan- orbulina sp.	Placo- psilina	Cliona	Bryozoa ctenosto- mata	Other Bryozoa
8	92	×								
8	122	×		×					×	
12	93	×			×					
19	48	×								
20	80	×								
22	110	×		×	×	×		×		
28	103							×		×
30	131			×						×
32	141			×	×			×		×
35	135									×
36	136			×				×		
40	170			×						
50	246		×			×		×	×	
60	73		×		×					×
70	54		×						×	×
75	53		×		×	×			×	×
77	52		×		×					×
100	13		×			×		×		
105	240		×			×		×	×	×
112	280		×		×			×	×	×
123	252		×							×
124	259		×	×	×		×	×		×
132	339		×		×					
136	311		×	×	×		×	×		×
137	336		×		×		×			
138	266		×		×					×
140	257		×		×		×	×		×
141	351		×		×		×	×		×
152	264		×				×			
155	28		×		×		×			
165	288		×		×		×	×		×
175	262		×	×	×		×	×		×
179	340		×		×		×			×
185	37		×		×					
927	308	No ostreids		×	×		×	×		×

[a]From G. J. Boekschoten, 1966, *Palaeogeogr., Palaeoclimatol., Palaeoecol., 2*, pp. 333–379, Table 1. Copyright © 1966 by Elsevier Publishing Company.

recognized as special environments occupied by a varied epifauna (Table 3.5; Moore, 1973; Ryland, 1962; Sloane, Ebling, Kitching and Lilly, 1957). This biogenic complex is of interest to the paleoecologist as an example of a specialized community and as a paleoclimatological indicator. Unfortunately, the kelp fronds are seldom preserved, and the fieldworker may make the mistake of considering the preserved kelp epifauna part of the bottom fauna. Whatley and Wall (1975) point out that there are consistent relations in some cases between seaweed floras and associated ostracode faunas. These algal–ostracode relations are similar to those existing between some fish faunas and associated algal thickets where the fish live; implications for the fossil record after decomposition of the

soft-bodied algae are obvious. The biotic diversity (and communities) of the kelp forest will produce an above-average taxic diversity after condensation into the fossil record. Stauffer's (1937) comparison (Table 3.6) of shallow marine-area invertebrate data from an eel grass-based community before and after the removal of the eel grass is most instructive in terms of the much greater taxic diversity for the fauna living directly on and in the mud before their removal. North (1971) has emphasized that the kelp epizoans are very different from the animals present on the bottom between the plants. Voigt (1956) and Plaziat (1970) suggest how it is possible, by carefully studying the attachment face of fossil epibionts and xenomorphs, to understand the nature of their host plant or animal.

TABLE 3.4
Frequency of Occurrence of Sea Scallop Epifauna on 100 Specimens[a]

Epifauna	Frequency	Epifauna	Frequency
Balanus amphitrite	100	*Sarcodictyon* sp.	10
Pomatoceros caeruleus	100	*Pontonia margarita* (internal)	9
Sabellaria floridensis	99	*Phascolosoma verrilli*	8
Nereid polychaetes	99	Other encrusting Bryozoa	7
Amphipods	99	*Lineus socialis*	7
Polydora websteri (blisters)	99	*Ervilia concentrica*	4
Nereiphylla fragilis	81	*Hiatella artica*	4
Balanus calidus	70	*Didemnum candidum*	3
Odostomia seminuda	70	*Arbacia punctulata*	3
Oerstedia dorsalis	68	*Murex fulvescens*	2
Schizoporella unicornis	55	*Leptogorgia setacea*	2
Ceratonereis tridentata (external)	53	*Pinnotheres maculatus*	2
Hydroids	48	*Ostrea equestris*	2
Nemerteans	37	*Polyandrocarpa tincta*	2
Crepidula fornicata	35	*Odostomia dianthophila*	2
Eupomatus spp.	34	*Bugula turrita*	2
Aetea anguina	31	Isopods	2
Mitrella lunata	26	*Leptogorgia virgulata*	1
Ceratonereis tridentata (blisters)	25	*Micropanope nuttingi*	1
Cantharus multangulus	21	Holothurian juvenile	1
Barbatia spp.	19	*Pteria colymbus*	1
Anachis spp.	14	*Musculus lateralis*	1
Chione grus	13	*Sabella melanostigma*	1
Anomia simplex	13	*Odostomia teres*	1
Anthopodium rubens	12	*Chama macerophylla*	1
		Haliclona molitba	1
		Anemone	1

[a]From H. W. Wells, M. J. Wells, and I. E. Gray, 1964, *Bull. Mar. Sci. Gulf and Caribbean, 14,* pp. 561–593, Table 2. Copyright © 1964 by Rosenstiel School of Marine and Atmospheric Sciences, University of Miami.

The specialization of the kelp epifauna is so varied that at least three separate faunal zones can be recognized. There is a different biota associated with the bottom holdfasts, the intermediate region, and the upper portions of the plants. Dell and Fleming (1973) point out that the occurrence of a particular species of *Chlamys,* which today lives only on the holdfasts and fronds of a species of the laminarian *Macrocystis* in quiet, cool–temperate waters, is good evidence that in the Miocene a particular area of Antarctica was ice-free (kelps would be removed from the bottom by moving ice), quiet, and warm enough to permit a shallow-water, cool–temperate species to exist.

Brasier (1975) provides an excellent review of the seagrass biota and communities both past and present, including data from the past (especially that of Voigt, 1956, 1973). Ogden *et al.* (1973) describe how tropical reef organisms (both fish and sea urchins) keep the shallow-water sea grasses mowed down in the "Randall Halo" adjacent to the reef complex base, although predators prevent them from venturing too far from the reef. Think of the biofacies complexities made possible by these interactions.

Both the sea grasses and the kelps are important food sources in the photic zone. Sloane *et al.* (1957) have found that varying degrees of turbulence, in an area of turbulence, are associated with differing abundances of kelp epizoans as well as presence or absence phenomena for some taxa; these relations guarantee that laminarian, and probably seagrass communities, will form a complicated community complex.

When the biogenic substrate is a calcareous alga, then the paleontologist may be able to study both the epifauna and the substrate organisms. Toomey (1976) has illustrated the calcareous phylloid algal thickets of the Pennsylvanian and shown how they are responsible for sediment-baffling sufficient to generate algal mounds. They also offer enough environmental complexity to provide shelter for a number of epibiont and associated taxa within the confines of the algal thicket.

The widespread and abundant pelmatozoan limestones of the past, as well as similar level-bottom aggregations of coral material and bryozoan material, dictate that thickets must have been important factors in modifying environments. Too little attention has been directed toward these problems to date.

One must consider the consequences of pelmatozoan thickets, as well as of coral and bryozoan thickets (Bretsky and Bretsky, 1975). Such thickets will act either as restrictive factors in lowering shelly taxic diversity or

FIGURE 72. Diagram of a North Carolina calico scallop community. Included are *Balanus amphitrite*, *B. calidus*, *Pomatoceros caeruleus* and *Sabellaria floridensis*. The sabellarid would probably leave no trace in the fossil record; the barnacle plates would probably disarticulate and drop off into the surrounding level bottom area, but the calcareous worm tube would remain in good condition. (From. H. W. Wells, M. J. Wells, and I. E. Gray, 1964, *Bull. Mar. Sci. Gulf and Caribbean*, 14, pp. 561–593, Figure 4. Copyright © 1964 by Rosenstiel School of Marine and Atmospheric Sciences, University of Miami.)

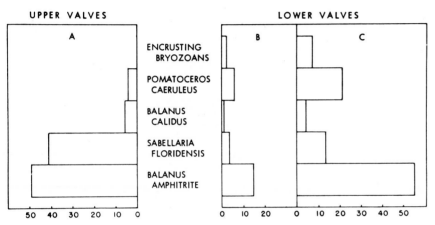

UPPER VALVES   LOWER VALVES

A

ENCRUSTING
BRYOZOANS

POMATOCEROS
CAERULEUS

BALANUS
CALIDUS

SABELLARIA
FLORIDENSIS

BALANUS
AMPHITRITE

B   C

50 40 30 20 10 0   0 10 20   0 10 20 30 40 50

FIGURE 73. Comparison of total area occupied by dominant species on upper and lower calico scallop shells by the various epizoans of the epizoan community; the width of the bars indicates the percentage of the total area occupied by the encrusting organisms. (From H. W. Wells, M. J. Wells, and I. E. Gray, 1964, *Bull. Mar. Sci. Gulf and Caribbean*, 14, pp. 561–593, Figure 3. Copyright © 1964 by Rosenstiel School of Marine and Atmospheric Sciences, University of Miami.)

have the reverse effect. It is important to note taxic-diversity data in terms of associated possibilities for coral (Asgaard, 1968), crinoid and cystoid (Lane, 1973; Koch and Strimple, 1968; Watkins and Hurst, 1977) thickets as compared to nearby environments lacking abundant pelmatozoan, coral, or bryozoan-type debris. Asgaard (1968) points out that the brachiopod genera associated today with thickets of ahermatypic corals on the deeper continental shelf and upper bathyal zone are about the same as those found fossil in association with coral thickets of the older Tertiary. This is a good example of a thicket association and of community continuity through the Cenozoic. Coates and Kauffman (1973) describe a Cretaceous deeper-water ahermatypic thicket lacking a varied set of associated invertebrates.

Squires (1964) and Vella (1964) have produced papers on New Zealand Neogene ahermatypic coral thickets presenting a model for the formation of nonreef biostromal and bioherman production processes (see Figure 79). They emphasize the differing sedimentologic processes involved, whether in deep or shallow water, and the very different animal communities present in the thicket and adjacent nonthicket environments (Table 3.7).

Neumann *et al.* (1977) describe a modern, relatively deep biohermal situation in which organogenic carbonate debris has been cemented subtidally to form a structure that simulates a reef in many regards, recalling the Lower Carboniferous Waulsortian "reefs." Such reefs probably represent thicket accumulations.

Sabellarid Worm Reefs

Extant sabellarid worms build massive reefs from the midtide level to about 10 m below sea level (Kirtley, personal communication, 1975). They cement sand grains to form a tube, sorting the grains according to size and shape (Multer and Milliman, 1967, provide a useful summary of sabellarian sorting and building). Such reefs of massed tubes serve as rocky bottoms in many areas, with almost no known animals strictly endemic to the sabellarian reefs, as compared with many known animals endemic to rocky bottoms in the same region. Kirtley (personal communication, 1975) reports that such reefs occur in both temperate and tropical regions, and that in colder waters they are subtidal. If the pipe rocks and *Skolithus* quartzites of the past are a product of sabellarian activities, a point as yet

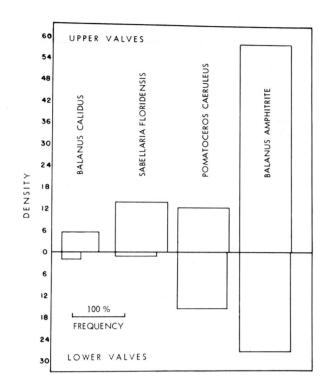

FIGURE 74. Diagram showing differences in faunal density and frequency of sea scallop epizoans on the upper and lower valves. (From H. W. Wells, M. J. Wells, and I. E. Gray, 1964, *Bull. Mar. Sci. Gulf and Caribbean, 14,* pp. 561–593, Figure 2. Copyright © 1964 by Rosenstiel School of Marine and Atmospheric Sciences, University of Miami.)

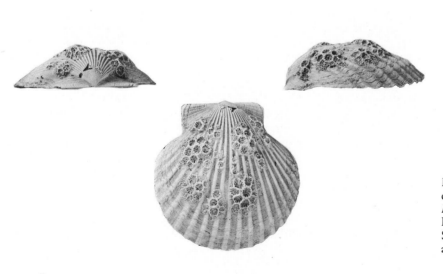

FIGURE 75. Miocene pecten epizoan corals. (From T. R. Waller, 1969, *Paleontol. Soc. Mem. 3,* Plate 6, Figures 5, 7, 9. Copyright © 1969 by Society of Economic Paleontologists and Mineralogists.)

TABLE 3.5
Distribution of Epizoan Mollusks on Green Algae, *Caulerpa* mat [a,b]

Mollusc	Caulerpa prolifera	C. racemosa	Caulerpa mat	Codium bursa	C. (tomentosum?)	Halimeda tuna
Cantharidus exasperatus	1520	1240	120	40	30	2040
Rissoa costulata	1400	—	940	—	10	50
Bittium reticulatum	2060	20	400	50	—	170
Tricolia pullus	60	20	710	—	—	20
Pusia tricolor	20	20	40	—	—	160
Cerithium vulgatum	40	—	10	40	—	50
Columbella rustica	—	—	10	—	—	50
Trophon muricatus	20	—	10	—	—	10
Arca lactea	—	—	—	30	—	—
Marginella miliaria	—	—	20	—	—	
Eulima alba	—	—	10	—	—	—

[a]From J. H. Duffus, 1969, *Proc. Malacol. Soc. London, 38,* pp. 343–349, Table 2. Copyright © 1969 by the Malacological Society of London.
[b]Numbers of mollusks are expressed per kilogram dry weight.

FIGURE 76. Distribution of borings and epibionts on adult Upper Cretaceous *Titanosarcolites*. Drawings are a composite from several specimens (× 1/6). (*b*) Average distribution and density of borings and epibionts on the flange (upper) side of paired valves. Note concentration of borings on protected areas around beaks and within coil. Epizoans are more common around margins than on the elevated crest of valves. (*a*), (*c*) Flange (upper) and tubular (lower) surfaces, respectively, showing distribution and density of borings; *Cliona* is average for many shells, other borings represent total information for about 40 specimens. Note preference of *Lithophaga* and *Cliona* borings for protected areas within coil, under shell, and especially around beaks. *Lithophaga* is densest under beaks. There are few borings on the tubular side which apparently contacts the sediment during life. Boring bryozoans prefer most exposed, elevated surface of upper (flanged) shell surface. (*d*), (*g*) Distribution of calcareous algae, corals and hydrozoans on the flanged upper surface, and tubular lower surface, respectively, (total information from about 40 specimens). Note great restriction of epibionts over smooth tubular surface where it is in contact with sediment during life; strong preference for calcareous algae and hydrozoans for exposed flanged surface, and of corals for more protected, shaded areas of down-facing tubular surface where it is not in contact with the substrate. On lower tubular shell surface, algae are zoned strongly around the shell margins where light is maximum for this environment. (*e*), (*f*) Views of upper flanged and lower tubular surfaces, respectively, showing distribution of epizoan bivalves, worms and foraminifera (total information from about 40 specimens). Note stronger preference among oysters and several types of rudists for exposed, well lighted upper surface, especially lateral to commissure around the outer edge of coil (current facing?), and for serpulid worms and rarer cemented bivalves (*Spondylus, Plicatula*) for protected, shaded undersurface of valves. This rudist story is reminiscent of a Sung Dynasty celadon bowl in my possession. Half of the bowl is somewhat discolored from the normal gray–green. The normal half is partially encrusted by calcareous worm tubes and a few hermatypic calices. It is clear that the bowl was dropped long ago into shallow harbor water, and became encrusted with a few epizoans. Its lower half, which was buried in the mud, was altered by the action of the mud. The bowl was "fished" out of the mire and preserved with its burden of epizoans. The small number of epizoans and their juvenile condition indicates that the bowl was in the seawater for not more than part of a season. (From E. G. Kauffman and N. F. Sohl, 1974, *Verhandl. Naturf. Ges. Basel, 84,* pp. 399–467, Text–figure 9. Copyright © 1974 by Naturforschers Gessellschaft Basel.)

ment discussed previously is a prime example of a localized environment that may have a different light level than the larger environment around it. Such localized environments may exist from the intertidal to the lower limits of the photic zone. However, they are not small extensions of deeper, subphotic faunas, but are occupied by a specialized cryptic fauna. Remember that the infauna has different light levels than the overlying bottom. This observation has introduced some confusion into the literature. Reese (1966) comments on the fact that most sea urchins do not like too much light; the remark does not pertain to the burrowing urchins, but to the epifaunal types.

There is a tendency to think of light gradients as correlating with vertical depths or turbidity, but the paleoecologist working with benthic material is most often interested in reconstructing the slope of the paleoenvironment. Kitching, Macan, and Gilson (1934) discuss the correlation between the slope and taxa with various algae giving way ultimately to sponges when light drops off severely under rock overhangs. Riedl (1966) discusses limestone caves in the Mediterranean region in terms of the high correlation between light and taxa. Jackson, Goreau, and Hartman (1971) described the correlation between light, or absence of it, and the inhabitants of tropical reef, cryptic environments. The rarity of rock slopes and overhangs in the geologic record makes this class of environments of little concern for most geologists, but the abundance of potential cryptic environments in reeflike structures of the past merits serious attention.

Measuring light is difficult enough for the paleoeco-

logist, but the question of seasonal and nonseasonal light levels must also be considered. Light supply under both polar and nonpolar regimes as well as in the intermediate regions is critical to many questions of both paleoecology and biogeography. The problem of permanence of the earth's obliquity in time is involved; this question will be solved either empirically by the geologist or from physical principles by the student of mechanics.

The empirical approach attempts to characterize the properties of polar biotas as contrasted with equatorial biotas, and then by a search for past parallels.

In general, a lower taxic diversity exists in polar regions than in equatorial regions. The overall cooler temperature regimes of the the polar regions, partly a consequence of the lower incident angle of solar radiation in the polar areas, tends to result in absence of carbonate-based reef structures and absence or rarity of limestones, as well as nontropical type weathering phenomena and an absence of marine evaporites. The nonseasonal rainfall regime of certain equatorial and wet–temperate regions (the rainforests), and certain temperate regions finds no counterpart in the polar regions in terms of the presence or absence of tree rings. The overall lower taxic diversity of the nontropical regions is clearly a complex matter, but the fact that highly seasonal regions require that successful taxa possess greater environmental breadth is certainly involved.

Light is also important in bodies of fresh water, as emphasized in Boycott's (1936) discussion of freshwater mollusks.

SIGNIFICANCE OF COLOR PATTERNS

Color patterns such as dots, stripes, concentric banding, and zig–zags are very common in many groups of marine animals but are seldom preserved in fossil material. It is rare to find color-patterned fossils even once in an entire career—such a find normally results in a short paper announcing the discovery. What is the possible ecological significance of this color patterning, and can the paleoecologist make use of the rare color-patterned fossils to do anything more than confirm that past animals had some points in common with the present ones? As has been discussed in the section on light, color patterns may have some importance in triggering behavioral response to visual cues. However, many colored organisms are strictly infaunal, blind to complex patterns, and virtually unseen.

Color patterns today are not randomly distributed among the biota; they are most frequent within the photic zone. Brightly colored patterns characterize the tropical–subtropical zone. Dull colors are more characteristic of the temperate–polar shells, and of the subphotic zone (Nicol, 1967, discusses some aspects of

these questions). Freshwater shells are not commonly patterned. The color-patterned marine shells known from the fossil record, rare as they are (Hoare, 1978, summarizes most of the published occurrences) fit easily into benthic assemblage 2 and 3 positions (i.e., the photic zone). In addition, a variety of factors, including the abundance of limestone, indicate that most of them were collected from tropical–subtropical regions. Cool or cold biogeographic units such as the Gondwana realm of the Upper Paleozoic have yielded few color-patterned shells. Thus, color patterns on fossils provide some circumstantial confirmation of temperatures, and of light penetration with depths of the past. The agreement of these data with independent measures of the same variables is encouraging.

Color patterns may, of course, provide some idea of an animal's life habits. Bivalved shells with color banding on both valves may have lived in such a position that both valves were exposed to light (i.e., individual valves are not commonly color patterned if they habitually rest on the bottom). Counter-shading in cephalopods may be interpreted in terms of predator–prey capabilities or, alternatively, in terms of the unpatterned portion of the shell resting on or closely approximating the bottom.

The generation of different colors and color patterns is not a simple matter. Spight (1976) has presented evidence and an account of previous work indicating that diet plays a role in determining which color or absence of any color appears in the shell of colored forms. In addition, there is evidence that color patterns in the same species may be different, and the presence of different colors may vary from environment to environment. Both genetic factors are partially sorted out by the selective agencies operating in the environment, and interact with the nature of the diet to determine both colors and their patterns. Spight also points up the importance of eroding agencies such as solution (possibly assisted by boring algae) and abrasion in removing the colored shell material. Olsen (1968) outlines how sensitive color banding may be to annual successional changes in plant foods—with resultant bands reflecting the annual change in food-type—obviously a type of banding unrelated to natural selection, but also a type of banding that would not be expected in subphotic occurrences as compared to nearshore and shallow-subtidal sites. Both Spight (1976) and Turner (1958) present experimental evidence that carnivores may obtain coloring material for their shells from their prey; Moore (1936) has presented similar evidence.

Oberling (1968) reviews the geometric problems involved in generating the multiplicity of color patterns by the mantle margin. The mantle margin may be thought of as a genetically controlled jaquard loom.

Color patterns of one sort or another have been noted in many of the major groups of fossil invertebrates. As would be expected for a variety of sampling reasons, color patterns amongst invertebrate fossils are more commonly encountered in Cenozoic and Mesozoic materials than in those of the Paleozoic. Mollusca are well represented, with color patterns known from the fossil record (including Bivalvia, Gastropoda, and Cephalopoda). Brachiopods, trilobites, eurypterids, insects, and some pelmatozoans are also recorded with well-preserved color patterns.

The number of color patterns found in the fossil record can be considerably enhanced if routine examination of fossils by means of ultraviolet light is employed. Rolfe (1965, pp. 353–355) summarizes some of the older accounts of color patterns and banding on fossil shells revealed by means of ultraviolet light, and Boni (1938), Nuttall (1969), and Neuffer (1971) provide additional data. Spjeldnaes (personal communication, 1978) has obtained a large number of younger fossil mollusks, chiefly Cenozoic in age, which display varied dramatic patterning visible in ultraviolet light.

Wilson (1975) has rendered a number of fluorescent, color-patterned Cenozoic gastropods; Vokes and Vokes (1968) have illustrated additional color-patterned, fluorescent Cenozoic mollusks; and Plas (1969) has summarized earlier work indicating that the fluorescence of fossil-patterned shells is enhanced by treatment in sodium hypochlorite solution, and that the fresh shells of some modern forms do not fluoresce, whereas abraded and weathered portions of the same modern shells do fluoresce. Plas's (1969) studies indicate that the cause of the color-pattern fluorescence probably has something to do, at least in some cases, with alteration and chemical changes of the original, primary pigments.

TEMPERATURE

Temperature is probably the most important variable affecting both plants and animals of the present. Despite much discussion of other physical variables, few would question the value of temperature (see Figure 17, p. 26). The significance of temperature is commonly evaluated observationally by noting the local and/or regional presence or absence of various taxa as a correlate of temperature. Experimentally, temperature is evaluated in terms of the physiological reactions and behavior of organisms to conditions of varying temperature. Kinne (1970, pp. 321–616) provides an excellent summary of the effects of temperature in the marine environment (see Gunter, 1957, for an older summary). Temperature affects organisms in a variety of ways, such as inducing death at varying rates (Table 3.9) and limiting reproduction, larval success, and successful competition. The total influence of temperature on a particular species is too complex for gross simplification—even without considering the problem of acclimatization or interactions and reactions with other variables.

Hutchins (1947) has emphasized how (Figures 99 and 100) minimum temperature for adult survival sets a winter poleward boundary; minimum temperature for reproduction sets a summer poleward boundary; maximum temperature for reproduction sets a winter equatorward boundary; and maximum temperature for adult survival sets a summer equatorward boundary—with the reverse situations present in the southern hemisphere. Hutchins makes it clear that the diverse seasonal and surface current temperature regimes on the north–south coastlines, particularly in the temperate sectors, have a profound effect in biotically changing the various coastal temperate strips in terms of combinations of expected species. One cannot expect a perfect latitudinal, global series of beltlike

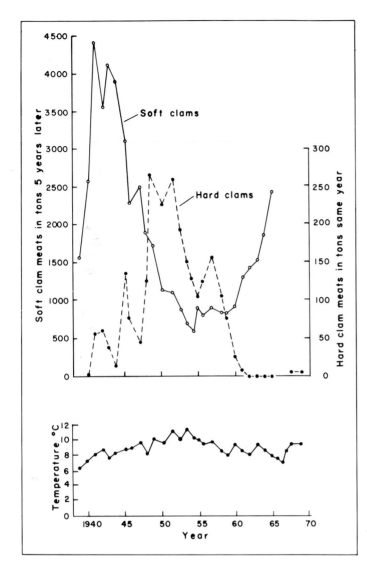

FIGURE 96. Correlation between mean temperature in Gulf of Maine (data from Boothbay Harbor, Maine) and landings of warm versus cool clam species. (From R. L. Dow, 1972, *Jour. Conseil Internat. L'Explor. de la Mer, 34,* pp. 532–534, Figure 1. Copyright © 1972 by International Council for the Exploration of the Sea.)

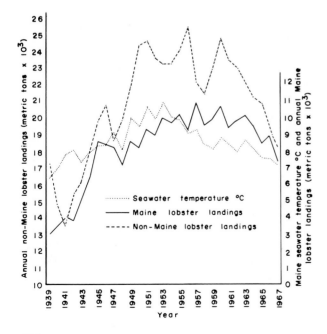

FIGURE 97. Correlation between lobster landings in Gulf of Maine region and mean annual seasurface temperature (Boothbay Harbor). (From R. L. Dow, 1969, *Science,* pp. 1060–1063, Figure 3. Copyright © 1969 by American Association for the Advancement of Science.)

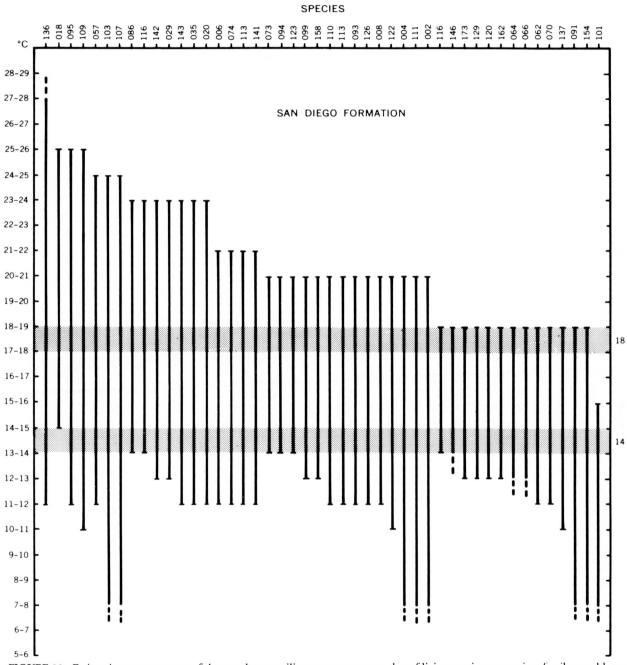

FIGURE 98. Estimating temperatures of the past by compiling temperature overlap of living species present in a fossil assemblage; ostracodes from the Upper Pliocene of California are shown in this example. (From P. C. Valentine, 1976, *U.S. Geol. Surv. Prof. Paper 916,* Figure 12.)

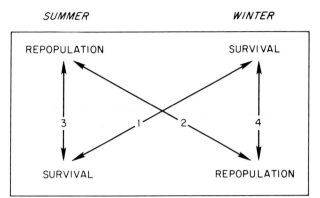

FIGURE 99. Diagram indicating 4 basic types of zonation related to summer and winter temperature limits of adult existence and for effective reproduction. (From L. W. Hutchins, 1947, *Ecological Monographs, 17,* pp. 325–335, Figure 1.)

habiting the tropical and subtropical regions will also have lived in similar temperature regions. This assumes a certain degree of environmental or ecological conservatism on the part of the organisms. The conservatism of community evolution in the shelf-depth regions of the Phanerozoic provides some confidence in the assumption (Boucot, 1978). Additionally, the global biogeographic picture for the Middle Ordovician to Recent biota suggests a distributional picture that is consistent with the temperature extrapolations. However, the circumstantial nature of the evidence requires tentative conclusions.

The few examples in the Phanerozoic of continental glaciation (Ashgill and Permo–Carboniferous) are associated with global taxic diversity gradients, and so on, that are in general agreement with what we know of the present and Neogene.

Aarseth and others (1975) present a good example, with Holocene materials, of the recognition of warm- and cold-water faunas in a single stratigraphic section by comparison with modern representatives. Gustavson (1976) provides a similar, additional example of this approach (Figure 101). This type of logical extrapolation may be carried back through time. However, the farther back one goes, the less reliable the method becomes. Comparisons of identical species of the past with those of the present is, of course, very reliable but, for practical purposes, can seldom be extended below the Pliocene. Comparisons of identical genera with differing species although less reliable, appear to give fair results, although Arkell (1956) presents a Jurassic association that is enough to chill the blood of any paleoecologist given to unthinking taxic extrapolations. Addicott (1969, 1970) provides a good example (Figure 102) of a type of logic that employs different species of the same genera for purposes of determining north–south climatic fluctuations during the Cenozoic of Western North America. Pre-Mesozoic extrapolations are restricted, however, to comparisons with the same

superfamily or higher levels, and to structures that are thought to have been functionally similar—thus, a high level of caution is indicated.

Some organismal distributions may even be used to calibrate "thermometers" (Figures 103–106) for use with fossil representatives of the same group, as has been done by Wells (1967a) for the ahermatypic corals, by Valentine (1976) for ostracodes (Figure 98), and by Strauch (1968, 1971) with a modern bivalve species that has a lengthy fossil record into the earlier Cenozoic. Lutz and Jablonski (1978b) provide a "thermometer" of sorts that employs change in bivalve protoconchs size for a single species from locality to locality. Rosen (1977, Figure 104) provides information, based on numbers of hermatypic genera, that may be used to deduce temperature if depth is known, and vice versa; his points (see legend, Figure 104) are well taken. In Wells' case, advantage is taken of the known, overlapping temperature ranges of a large number of genera that cooccur in varying combinations correlated with temperature, and in Strauch's case, advantage is taken of a phenotypic gradient correlating highly with temperature. Mitchell (1975) has found a high correlation within a number of bivalve and brachiopod species between number of radially disposed costae and temperature, with evidence that the number is partly a function of daylight duration (latitude); this is another phenotypic example. The setting up of such "thermometers" is a most valuable exercise, even if the absolute calibration breaks down as we go back in time due to the disappearance of extant species and genera.

The correlation of some biogeographic boundaries with temperature gradients (such as shown in Figures 102 and 107; also involved in the interpretation of the data in Figures 96 and 97 and Table 3.8) provides still more evidence of the prime importance of temperature as a distributional factor. Hayden and Dolan (1976) show the high correlation between regional temperature and fauna; their data is comprehensive and convincing.

Dell and Fleming (1973) mention that many shelf-depth polar bivalves show solution pits of a circular form in the umbonal region where the protective periostracum has been penetrated at discrete points. They note that such solution pits are seen today from the Arctic down to Puget Sound and in the Southern Ocean as far north as Kerguelen. These solution pits are another means of measuring temperature in a relative manner.

Philip and Foster (1971) suggest that the presence of marsupiate echinoids in some abundance and taxic diversity is suggestive of cool to cold conditions at present, and also back into the Tertiary.

DuBar and Taylor (1962) provide a good example of estimating temperatures represented by a Tertiary fauna obtained by comparing the overlapping temperature ranges of modern species of those genera repre-

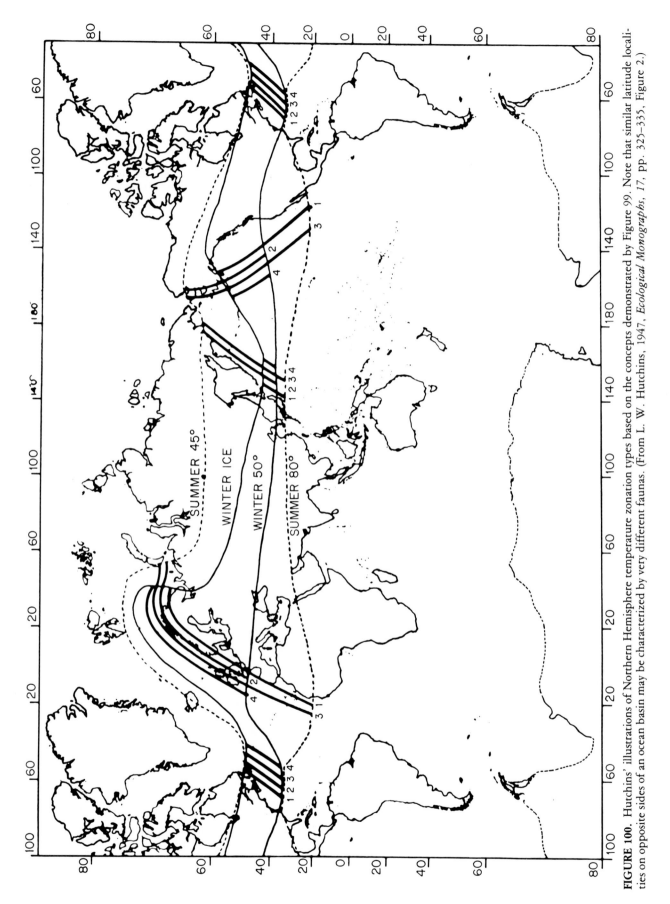

FIGURE 100. Hutchins' illustrations of Northern Hemisphere temperature zonation types based on the concepts demonstrated by Figure 99. Note that similar latitude localities on opposite sides of an ocean basin may be characterized by very different faunas. (From L. W. Hutchins, 1947, *Ecological Monographs*, *17*, pp. 325–335, Figure 2.)

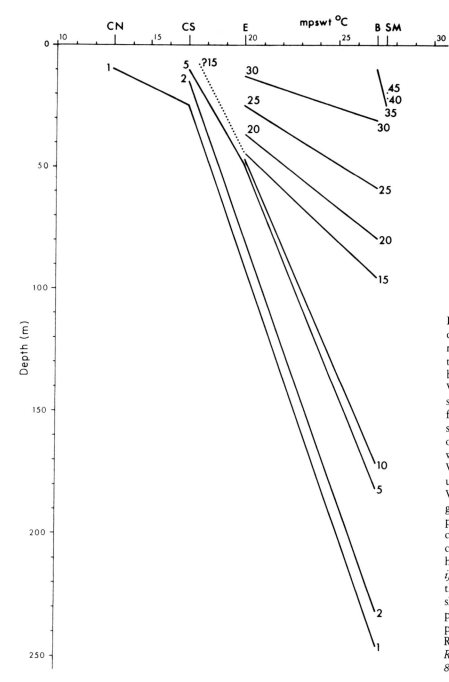

FIGURE 104. A decided, regional change in the generic diversity of hermatypic corals can be seen as one leaves the areas of their greatest abundance in both the Old World and the New World. This decrease in generic diversity is greatest from areas of warm surface waters to those of cool water; similar decreases in generic diversity occur with depth. This figure suggests, with data from the Old World (New World diversities are generally lower unit for unit than those of the Old World) that any given number of genera is a function of *both* temperature and depth. Therefore, conclusions about temperatures of the past cannot be inferred from number of hermatypic genera alone. For example, *if* other evidence shows conclusively that a particular fauna represents very shallow water conditions, then a temperature may be deduced with some precision; otherwise not. (From B. R. Rosen, 1977, *Mémoires du Bureau de Recherches Géologiques et Mineres, 89,* pp. 507–517, Figure 2.)

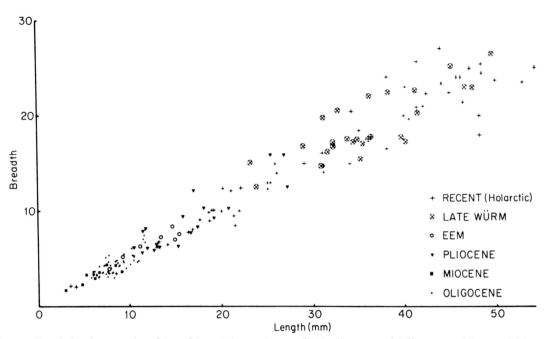

FIGURE 105. Correlation between breadth and length in specimens of *Hiatella arctica* of different age. These variables correlate highly with temperature when compared with the data plotted on Figure 106. (From F. Strauch, 1968, *Palaeogeogr., Palaeoclimatol., Palaeoecol., 8,* pp. 213–233, Figure 3. Copyright © 1968 by Elsevier Scientific Publishing Company.)

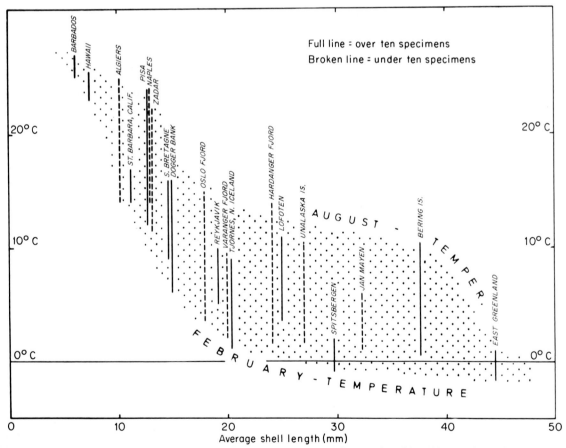

FIGURE 106. Plot of average shell length against yearly temperature span for a series of localities yielding adult specimens of *Hiatella arctica.* Note the high correlation between size and temperature. (From F. Strauch, 1968, *Palaeogeogr., Palaeoclimatol., Palaeoecol., 8,* pp. 213–233, Figure 2. Copyright © 1968 by Elsevier Scientific Publishing Company.)

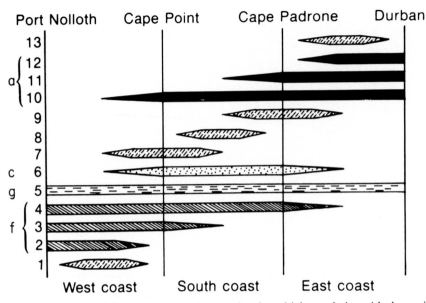

FIGURE 107. Geographic ranges of selected groups of intertidal taxa showing a high correlation with change in temperature. Thirteen ways in which organisms may be distributed around the South African Coast. The coast is divided into three sections; distances are not drawn to scale. Category 12, for example, includes those warm water species that do not extend as far south as Cape Padrone; category 11, those that reach beyond Cape Padrone to the south coast; and category 10, those that stretch beyond Cape Point—the other categories can be interpreted similarly. Categories 1, 7–9, and 13 are various local components, each composed of species restricted to particular localities; the cold-water component (*f*) includes categories 2–3; the ubiquitous component (*g*) is coextensive with category 5; the south coast (or warm–temperate) component (*c*) is coextensive with category 6; and the warm water component (*a*) includes categories 10–12. (From T. A. Stephenson and A. Stephenson, *Life Between Tidemarks on Rocky Shores*, Figure 3.1. Copyright © 1972 by W. H. Freeman and Company.)

sented in the Tertiary fauna. This assumes, of course, little change in overall temperature tolerances for the genera involved.

In principle, the changing ratio of planktotrophic and nonplanktotrophic protoconchs found from locality to locality and stratum to stratum provides a useful means of studying temperature gradients in the fossil record (see Chapter 4, p. 273).

The overall abundance of endemic species suggests that most poikilotherms will be stenothermal, and the compilation of experimental studies (Moore, 1940) as well as the vertical zonation of most poikilotherms tends to confirm this assumption.

SALINITY

As with light and temperature, we have no simple, direct means of measuring salinities of the past. There are isotopic methods, however, (see Allen *et al.*, 1973; Allen and Keith, 1965; Dodd and Stanton, 1975; Keith and Parker, 1965; Lloyd, 1972; Tan and Hudson, 1974; Figures 108–110, Table 3.10), for estimating by means of oxygen and carbon isotopes whether fresh water, brackish, or fully marine waters are most likely (Williams, 1976). It is encouraging that the isotopic data, when applied to fossiliferous materials as far back as the Jurassic, yields results consistent with those deduced independently from study of associated plants, animals, and geologic criteria. Once into the fully marine environment, there is no technique available for measuring the small but significant differences common to that environment. The small amount of isotopic work carried out to date on salinity measurement testifies that both biotic and physical criteria provide similar results, with the former giving greater speed at far less cost.

Hypersaline conditions are evidenced in the marine environment by the presence of evaporites. Meyerhoff

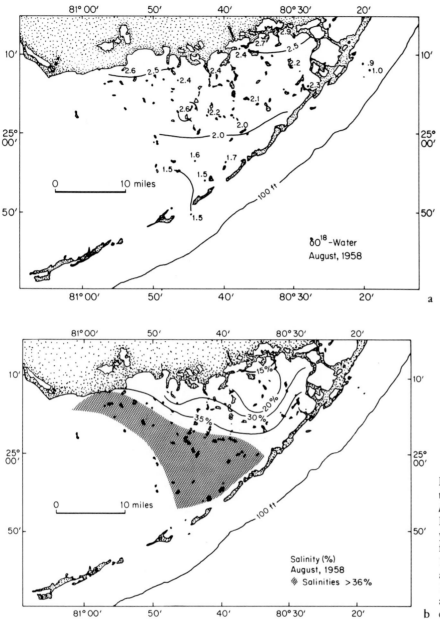

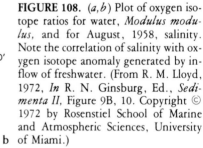

FIGURE 108. (a,b) Plot of oxygen isotope ratios for water, *Modulus modulus,* and for August, 1958, salinity. Note the correlation of salinity with oxygen isotope anomaly generated by inflow of freshwater. (From R. M. Lloyd, 1972, *In* R. N. Ginsburg, Ed., *Sedimenta II,* Figure 9B, 10. Copyright © 1972 by Rosenstiel School of Marine and Atmospheric Sciences, University of Miami.)

(1970) has compiled much of the available evidence relevant to the occurrence of Phanerozoic evaporites. Marine evaporites have been important intermittently, but we have no reason to believe that major evaporite deposition in the shallow marine environment has been continuous since the Cambrian. Evaporitic deposition is a function of both evaporation in excess of rainfall, and of circulation, where the oceanic reservoir cannot make up for the excess lost through evaporation. Therefore, the mere existence of evaporites is somewhat ambiguous in terms of strict causation by rainfall and/or circulation. In addition to the presence of marine-type evaporites and of associated structures, the organisms themselves are considered (Hedgpeth, 1957a). Hypersaline faunas are remarkable for their taxic impoverishment (Bosellini and Hardie, 1973, page 15; Hecker, Ossipova and Belskaya, 1963, Figure 114; Hedgpeth, 1957a, page 717; Figures 111 and 112; Hudson, 1963, Figure 113). Students of all intervals containing widespread hypersaline deposits have no difficulty in providing faunal lists showing the typically reduced numbers of species. Arthropods are prominent among the species present; the ostracodes are numerous. Bivalves and certain gastropods are also available, but echinoderms, brachiopods, corals, and many other invertebrate groups are either absent or extremely rare.

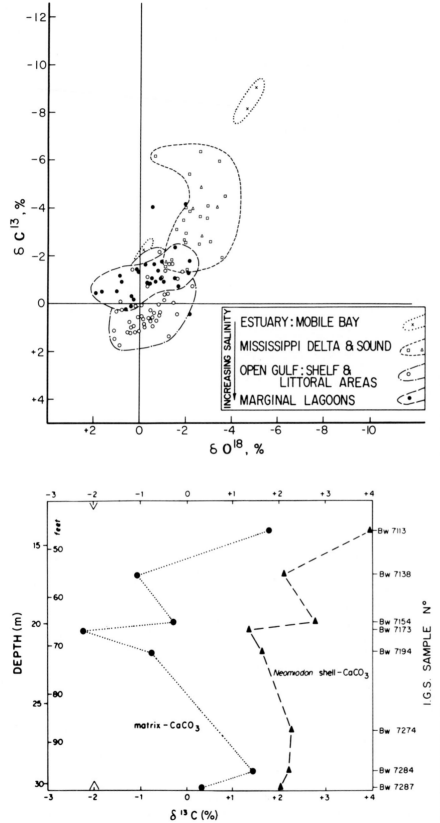

FIGURE 109. Plot of carbon and oxygen isotope data indicating the changes obtained in going from the fully marine into the brackish water environment. (From M. L. Keith and R. H. Parker, 1965, *Marine Geology, 3,* pp. 115–129, Figure 2. Copyright © 1965 by Elsevier Scientific Publishing Company.)

FIGURE 110. Carbon isotopic data indicating variations from the fully marine into the brackish environments in the English Jurassic–Cretaceous boundary area. (From P. Allen, M. L. Keith, F. C. Tan, and P. Deines, 1973, *Palaeontology, 16,* pp. 607–621. Copyright © 1973 by the Palaeontological Association.)

TABLE 3.10
Carbon Isotope Data Obtained from Fully Marine to Brackish Water Environments[a]

Environment	Mean $\delta^{13}C(^o/_{oo})$		Approximate observed mean salinity ($^o/_{oo}$)
	All shell samples	Gastropods only	
Estuary	−6.51	—	8
Delta	−3.62	−2.49	14
Sound	−3.19	—	18
Marginal Bays	−1.03	—	—
Aransas	−1.26	−1.66	34
Laguna Madre	−1.14	—	47
Copano & St. Charles	− .70	—	34
Littoral zone (open gulf)	− .53	− .45	36
Shelf areas (open gulf)	+ .53	+ .17	36

[a] From M. L. Keith and R. H. Parker, 1965, *Marine Geology, 3,* pp. 115–129, Table IV. Copyright © 1965 by Elsevier Scientific Publishing Company.

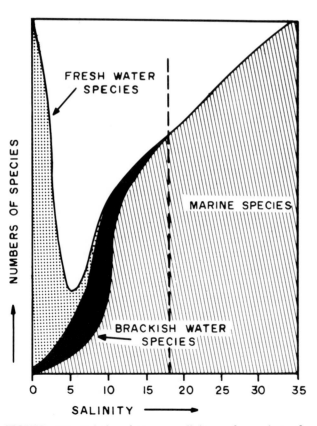

FIGURE 111. Relation between salinity and number of species. (From J. Hedgpeth, 1957a, *Geol. Soc. Amer. Mem., 67,* Figure 18A, after Remane, 1934.)

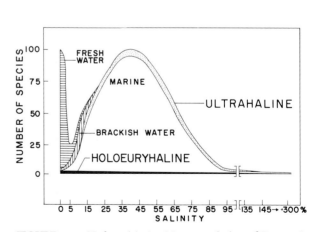

FIGURE 112. Hedgpeth's (1967) extrapolation of Remane's curve showing the relation between number of species and salinities. (From J. Hedgpeth 1967, *In* G. H. Lauff, Ed., *Estuaries, AAAS Pub. No. 83,* Figure 5. Copyright © 1967 by American Association for the Advancement of Science.)

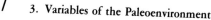
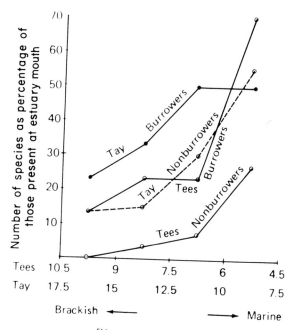

FIGURE 115. Diagram indicating that marine infaunal organisms penetrate further into the brackish environment than do epifaunal organisms (i.e., salinity is more stable in the sediment than above in the water.) (From D. C. Rhoads, 1975, *In* R. W. Frey, Ed., *The Study of Trace Fossils*, Figure 9.3. Copyright © 1975 by Springer–Verlag New York Inc.)

number of the truly brackish-water species probably reflects the long-term absence of conditions of isolation and a small population induced by the shifting, transitory nature of the brackish-water locales through time. Observations in the Zuyder Zee during its flooding and dike construction showed that the marine fauna was gradually replaced by freshwater fauna with no evidence for the development of an endemic brackish fauna during the brief brackish interval (van Benthem Jutting, 1965). This shifting, evanescent problem is reminiscent of the fact that, on land, ephemeral ponds have a very cosmopolitan fauna (Jackson, 1974) rather than a series of endemics. The cosmopolitan nature of many of the brackish-estuarine species is commented on over and over again. Old lakes, many of which are deep, and tropical reef areas of high environmental complexity harbor many endemics. Thus it is possible to infer that these environments have been available for a very long time in these locales. In the Cretaceous to Recent, mangrove and saltmarsh areas may be recognized, in some instances with the aid of pollen evidence, but for the pre-Cretaceous, the situation is far more difficult.

Zenkevitch (1963) has clarified the fact that an extensive shelf or platform covered with hyposaline waters can exist if an adjacent continent has a river system, like that of northern Siberia and Russia, adequate to provide a steady influx of fresh water. The Arctic situation is complicated, of course, by the sum-

mer melting ice problem. In such environs today, there is a low taxic diversity; only extremely euryhaline species are able to stand the salinity fluctuations. Filatova (1957), for example, reported only 119 species in 47 genera for the entire, vast shelf area off the Arctic coast of the Soviet Union, in the shallow-water regions where hyposalinity is a problem. There is no reason to think that extensive river systems of the past, flowing off large land masses, may not have had the potential for producing similar hyposaline environments with appropriate taxic consequences. That the effect of a freshwater influx adjacent to a river mouth is no Siberian anomaly in promoting a concentration of euryhalines to the exclusion of stenohalines is pointed out by Fleming's (1951) observation of a euryhaline concentration adjacent to river mouths in the Fiords of Western Southland, New Zealand.

Kinne (1971a) emphasizes not only the importance of changes in salinity, but that the relative abundance of the various ionic species can be critical in both the brackish and hypersaline environments; also that the number of truly endemic brackish and hypersaline species is small. The morphologic changes affecting euryhaline taxa within the brackish and hypersaline environments should, in principle, be of use to the paleoecologist trying to decipher the environmental history.

Segerstrale (1957) provides an excellent example of what may be expected in biotic terms across a salinity gradient from normal ocean water into fresh environment in terms of the Baltic. However, the complex Neogene history of the Baltic, with the comings and goings of different geographies and climates, has complicated things far more than may have been the case with most bodies of water that show a transition from normal sea water to fresh water. In any event, the decrease in the number of "normal" marine creatures as the environment becomes fresher, and their size changes (Figures 116 and 117) as one follows this gradient should be kept in mind. The size-gradient data, however, must be employed thoughtfully; Bateson's (1889) classic paper presents a good case for the decrease in shell size and thickness as one goes into **both** the brackish and the hypersaline environments from the normal marine. Changes in shell size and shell thickness in a given species can be employed as a type of salinity meter for both the present and the past. (Caution is clearly in order when one constructs any sort of phenotypic gradient "instrument" for measuring salinity, temperature, or anything else, since organisms can react in a similar manner, such as change in size, to more than one physical variable; these variables and responses can be sorted out if care is exercised).

The tendency for species in brackish and hypersaline waters to develop thinner shells and to reach smaller size-limits with age is also emphasized by Boltovskoy and Wright (1976), dealing with foraminifera (Table 3.12).

Barthel and Boettcher (1978) describe a Jurassic ex-

FIGURE 116. Size reduction in Baltic Sea mollusks paralleling the decreasing salinity. (From S. G. Segerstrale, 1957, *In* J. Hedgpeth, Ed., *Geol. Soc. Amer. Mem.*, *67*, Figures 1–2.)

FIGURE 117. Normal salinity seaweed (Bergen, Norway, at surface) *Callophyllis laciniata* and dwarf form from the Baltic (Kristineberg at 25 meters depth). (From F. Gessner and W. Schramm, 1971, *In* O. Kinne, Ed., *Marine Ecology,* vol. 1. part 2, Figure 4–62. Copyright © 1971 by John Wiley & Sons.)

ample from the midst of the old Nubian Sandstone in southwestern Egypt. Surrounded by barren, presumably nonmarine strata, is a shaly lens that includes a very shallow marine, possibly estuarine, fauna, grading into typical brackish and freshwater environments. Transported land plant debris is also present. The abundance of lingulids is notable, as is the abundance of an isopod that was once confused with a Cambrian trilobite, leading to a miscorrelation of the Jurassic with the Cambrian. The faunal changes and mixtures present make this Egyptian example very useful in terms of what to expect in the transition from the very shallow marine into the nonmarine.

As we approach the present, the discrimination between fresh, brackish, and normal marine faunas becomes progressively easier because of the increasing percentage of living genera, genera that we may assume for the most part to have lived in similar-salinity waters and environments. Keen (1977) presents a very good example from the English Eocene ostracodes (see Figure 119; Table 3.13).

Ostracodes have proved particularly useful in laying out salinity gradients from the fresh to the fully marine (Cronin, 1977; Keen, 1977) because of the abundance of the group as fossils. Cronin's (1977) example (Figure 118) is particularly instructive. Cronin outlines how analyzing the present temperature and salinity correlations of the same species enables one to evaluate more realistically the multifactorial control exercised over the presence and absence of taxa.

The monsoonal climatic regimes may exert a profound effect on the shallow-water, nearshore organisms (Paul, 1942) in terms of the season of freshwater influx.

It has long been recognized that both hypersaline and hyposaline conditions act to decrease taxic diversity. It is also clear that conditions of alternating hypersalinity, normal salinity, and hyposalinity, such as those encountered in some coastal lagoons, and situations where restricted circulation with the sea plus heavy rain, riverine input, or dry spells, can result in very low taxic diversity. The Laguna Madre off the Texas coast (Gunter, 1957) is an excellent modern example and Nicol (1965) attributes some Permian, molluscan dominated faunas to the same type of environment (see Table 3.14).

Wells (1961) reports that some of the typical brackish-water species, such as the brackish-water oysters, do well in this environment not so much because they cannot tolerate normal salinity conditions but because their competitors and predators cannot tolerate brackish conditions. This point should be kept in mind when attempting to construct overly rigid schemes categorizing some species as ''brackish'' and some as ''normal.''

Furst et al. (1978) have suggested that the trace element boron in siliceous sponges may vary with salinity, providing a salinometer for the past.

FOG AND UPWELLING

An example of how easy it is to overlook environmentally critical variables is provided by fog. How many of us would immediately think of it? In the intertidal region, desiccation is a key factor in determining the vertical distribution and occurrence of both intertidal and shallow-subtidal organisms. Glynn (1965) presents data that convincingly shows that fogbound coasts have a distinctive intertidal community structure that differs significantly from that of nonfoggy coasts. Stephenson and Stephenson (1972, p. 228) emphasize how fog can help to keep evaporation and some types of radiation down to the point where intertidal biotas are richer than would normally be the case, particularly if coupled with a tidal regime having high tides coinciding with the noonday sun.

Foggy coastal regions tend to occur today as a result of coastal upwelling. Similar situations may have characterized the past. Upwelling, of course, is related to a peculiar interplay of wind and currents that brings nutrient-rich, colder-bottom waters to the surface, resulting in great increases in surface productivity. It is

no accident that some of the largest fisheries today are located in areas of coastal upwelling. Part of the sedimentary phosphate deposits, as well as areas of high diversity, may have been located in areas of ancient upwelling.

Burnett (1977) summarized much of the data relating to the accumulation of phosphate-rich deposits in the vicinity of upwelling regions. He emphasized their occurrence along the western sides of continents. A more profound knowledge of the older paleogeographies than is currently available is needed before all our uncertainties concerning the proper usage and interpretation of sedimentary phosphate deposits as indices of local upwelling are dispelled.

Upwelling, in the purely oceanic environment, is an important phenomenon in modern seas. It has been deduced on purely paleontologic evidence by extrapolating some of the planktonic characteristics of oceanic waters (see Valencia, 1977, for a Pleistocene example).

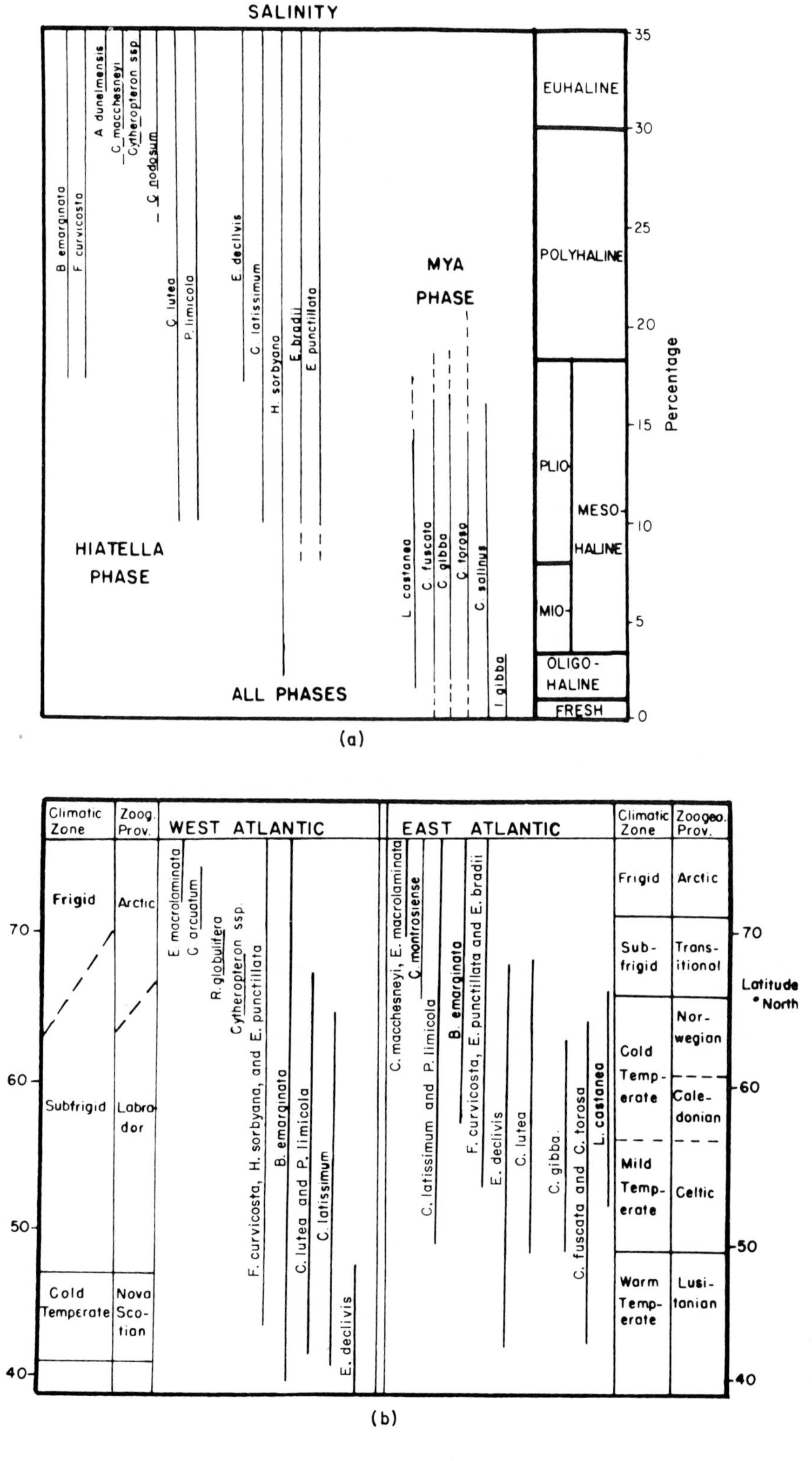

shelf-
viron
living
fauna
fauna
migh
into
sion-
latior
sity
could
fer,
and
trilol
trast
bival
elsev
logic
ance
oligc
Cam

SED

V
upo
othe
deca
mat
gists
mar
bon
ime
mal
sam
assu
reco
hav
ten
the
exp
the
is t
tua
ana
in
def
me

Th

a
bec
con
sec
ecc

FIGURE 118. Cronin's (1977) plot of salinity ranges presently tolerated by a group of Pleistocene ostracodes in the Champlain Sea of eastern North America (*a*), and the modern temperature ranges of many of the same species (*b*). Note the coincident changes in both temperature and salinity tolerances of the Pleistocene forms. (From T. M. Cronin, 1977, *Quaternary Research*, 7, pp. 238–253, Figures 5–6. Copyright © 1977 by Academic Press/ University of Washington.)

133

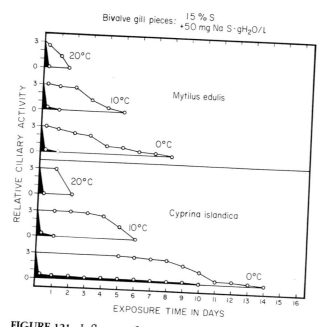

Bivalve gill pieces: 15 % S
+50 mg Na S·gH₂O/L

FIGURE 121. Influence of temperature on ciliary activity and survival time of isolated gill tissues of two lamellibranch species from the Baltic Sea (15°OOS) in O_2 deficient seawater, and of water to which sulphide (50 mg Na_2Sx9H_2)/1 has been added. (From H. Theede et al., 1969, Marine Biology, 2, pp. 325–337, Figure 6. Copyright © 1969 by Springer–Verlag New York, Inc.)

and a "normal" epifauna. Do such anomalous occurrences reflect the fact that fouled sediment conditions occurring below "normal" waters support a "normal" epifauna?

Rhoads and Morse (1971) provide diagrams making clear how the progressive restriction of oxygen level correlates with marked reduction in taxic diversity (Figures 122–125). Variables other than oxygen are changing as well, but it is certain that the decrease in oxygen to the virtually anaerobic point is chiefly responsible for the observed effect in their examples.

The emphasis on high free carbon must be understood in the context of coalification (Gray and Boucot, 1975). Carbonaceous materials undergo progressive increase in carbon content accompanied by loss of water, carbon dioxide, and methane, which exit the system as stable products. Thus, one may expect a progressive change in original carbon content with metamorphism (induced by thermochemical, piezochemical, and, rarely, radiochemical processes).

The presence of marine beds rich in ferric pigments, especially those produced on land under conditions of lateritic weathering (red beds in the generic sense), also provides evidence favoring an oxidizing environment, although the presence of beds rich in ferrous iron does not necessarily reveal anything about oxygen tenor above the sediment–water interface during deposition, nor about postdepositional reduction processes occur-

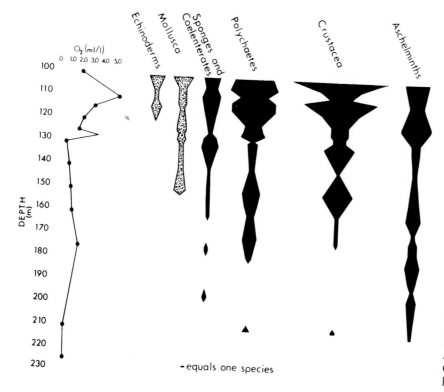

—equals one species

FIGURE 122. Species abundance of major benthic invertebrate groups in the Black Sea related to depth and dissolved oxygen (heavily calcified taxa represented by dotted pattern). Note the obvious correlation between oxygen level and taxic presence. (From D. C. Rhoads and J. W. Morse, 1971, Lethaia, 4, pp. 413–428, Figure 1. Copyright © 1971 by Universitetforlaget.)

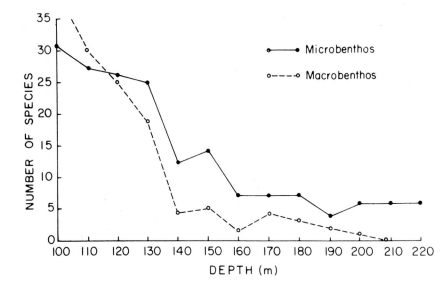

FIGURE 123. Species abundance of microbenthos and macrobenthos related to depth (and dissolved oxygen; see Figure 122 for oxygen variation with depth) in the Black Sea. Another way of viewing the high correlation between oxygen level and taxic abundance. (From D. C. Rhoads and J. W. Morse, 1971, *Lethaia, 4,* pp. 413–428, Figure 2. Copyright © 1971 by Universitetsforlaget.)

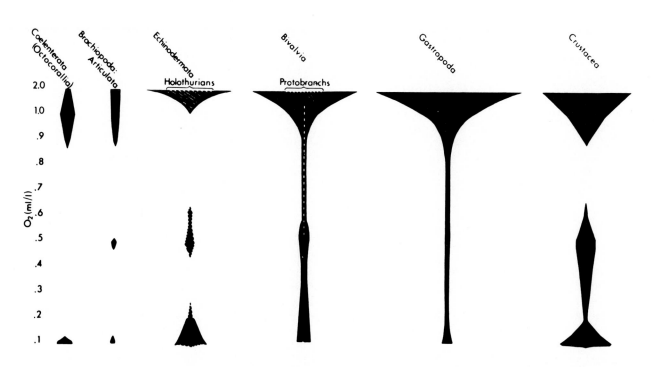

FIGURE 124. Relation between species abundance and dissolved oxygen in the Gulf of California; poorly calcified echinodermata (holothuroidea) and mollusca (protobranchia) are represented by diagonal line pattern. Here again, note the high correlation between taxic abundance and available oxygen. (From D. C. Rhoads and J. W. Morse, 1971, *Lethaia, 4,* pp. 413–428, Figure 3. Copyright © 1971 by Universitetsforlaget.)

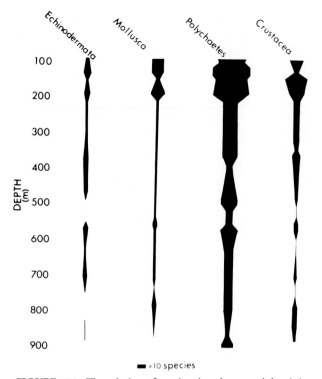

FIGURE 125. The relation of species abundance and depth in the San Pedro Basin; dissolved oxygen above 200 m is greater than 1.0 ml/liter, decreasing to .5 ml/liter between 200–500 m basin floor water contains .2 ml/liter dissolved oxygen. Here again, note the high correlation between taxic abundance and oxygen level. (From D. C. Rhoads and J. W. Morse, 1971, *Lethaia, 4,* pp. 413–428, Figure 4. Copyright © 1971 by Universitetsforlaget.)

ring after the sediment had passed below the zone of infaunal life. For example, the Silurian Bouleaux Formation of southern Gaspé consists in part of red, hematite-bearing limestone, and contains a rich, "normal" fauna of shelly marine invertebrates similar to that of the gray-colored Henryhouse Formation of Oklahoma. In both cases, it is reasonable to deduce a moderately quiet, but not still-water environment due to the articulated condition of many of the bivalved shells, as well as to the presence of algal oncolites in the well-oxygenated Bouleaux.

Basinal analysis that pays attention to all available biotic and physical evidence is probably the best tool for evaluating the overall conditions. But the paleoecologist has no proper tool with which to evaluate the differing oxygen levels affecting benthic infauna and epifauna of the past. The best that can be done is to segregate obvious end-member situations of Black Sea, anoxic, and red bed kinds from the "normal." In a general way, the absence of organic microfossils in unmetamorphosed fine-grained beds is consistent with oxidizing conditions, although certainly insufficient as proof of their existence (Gray and Boucot, 1975).

There is a general correlation between finely laminated marine sediments, and an absence of bioturbating organisms in areas where rates of sedimentation are not unusually high, low-oxygen or even anoxic, as well as relatively still-water conditions exist (Jerzmanska and Kotlarczyk, 1976; and Soutar and Isaacs, 1974, provide excellent examples). These laminated sediments sometimes contain unusually well-preserved fossils; the Solenhofens, Holzmadens, and Burgess Shales of the record.

Kauffman (in press) provides an extended treatment for the paleoecology of the Holzmaden beds. The precise level of the oxygenated–anaerobic horizon may change with time to positions within the water column or near or at the sediment–water interface, as well as down into the sediment. These changes may be ascertained in large part if special attention is paid to the nature of the preserved fauna. It is emphasized that laminated sediment, although indicative of a low level of bioturbation, does not always indicate anoxic conditions. These are most important papers because they provide guides for gathering the necessary classes of evidence for arriving at a well-documented, convincing, conclusion.

PRESSURE

There is a tendency for many ecological studies to subordinate the effects of hydrostatic pressure to those of other variables; perhaps on the misleading assumption that pressure is important only to organisms that contain compressable gas bladders or bubbles. However, Siebenaller and Somero's (1978) discussion of enzyme-moderated pressure problems in marine fishes indicates that reasonable biochemical pressure guages are conceptually possible. However, hydrostatic pressure has profound effects on many physiological systems due to the kinetics and stereospecificity of physiological biochemistry (see Kinne, 1971a) and must not be disregarded. The present conspicuous vertical zonation of the biota from the subphotic–photic boundary to the bottom of the hadal region necessitates that large pressure differences not be disregarded. They are an important factor in determining distributions. We would like to point out that even in the relatively shallow depths (continental shelf–upper bathyal) of most fossil deposits (except for the Jurassic and younger deep-sea deposits), hydrostatic pressure may play a very important role in determining depth distribution.

Pressure in the submarine environment can be

estimated through a combination of physical and biotic factors. First, depth must be judged in terms of the intertidal region, the upper and lower limits of the photic zone, and the shelf-margin region (see Boucot, 1975). The presence of abundant unimploded shelled cephalopods provides a lower limit (probably somewhere in the upper reaches of the bathyal zone). The overall reduction in amount of calcareous skeletal material in both fishes and cephalopods, beginning somewhere in the midbathyal region and extending down (as a means of weight loss helping to conserve energy that would otherwise be wasted in maintaining depth position), and its replacement by more gelatinous materials (or even by fatty or oily materials) should be noted where possible (Figure 126).

For the depth zonation of the photic zone, Golubic et al. (1975; see Figure 133, p. 152) have suggested that the microscopic and characteristic boreholes left behind by certain algal groups might be of service. This possibility should be explored.

Most of the benthic invertebrates showing distinct vertical zonation that might be due to pressure effects also possess larval dispersal stages. Flügel (1972) summarizes a large amount of data providing evidence that adult benthic organisms are not too sensitive to pressure differentials found on the continental shelf.

However, Morgan (1972) reviews the data of pressure sensitivity and emphasizes the reaction of the very shallow-water organisms. He points out that pressure sensitivity is coordinated with reactions to light and gravity to produce the resultant feedback through which the organisms satisfactorily maintain position.

It is obvious that if the planktic larvae of benthic organisms are pressure sensitive (as indicated by experimental evidence of the type reviewed by Morgan, 1972), we have a mechanism for controlling their depth distribution. Larvae, which will resist transport below a certain depth, will certainly tend to metamorphose on bottoms above or below that critical depth. Plankton depth regulation, by means of pressure receptors of uncertain type (Digby, 1972), is in agreement with this deduction. However, pressure cannot be assumed to be the key variable in all benthic fossil distributions until other variables, such as substrate and turbulence, have been explored and found wanting as an explanation for distributions. It will be particularly difficult to isolate the effects of light from that of pressure in controlling photic-zone organisms.

Rice (1964) asserts that depth maintenance of adult plankters or of planktic larvae by light alone is impractical because of the vagaries of cloud cover, as well as the day–night and seasonal variations and that many

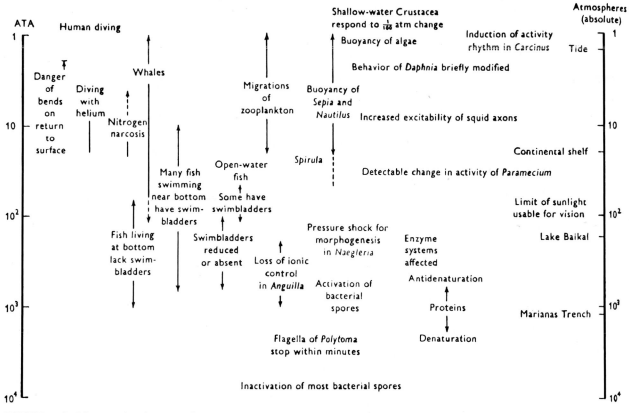

FIGURE 126. Diagram showing part of the depth spectrum of effects ascribed to hydrostatic pressure. (From J. A. Kitching, 1972, *Symp. Soc. Exper. Biol., 26,* pp. 473–482, Figure 1. Copyright © 1972 by The Society of Experimental Biology.)

depth-distribution regularities of plankton can best be explained in terms of pure depth sensitivity, despite the absence of air bladders or other depth gauges. He also points out that prior to 1951 there was no appreciation of the depth sensitivity of organisms lacking large gas spaces such as the bladders of fishes.

Naylor and Atkinson (1972) comment as follows on the significance of pressure in influencing the rhythmic behavior of marine animals: "Depending upon the sign of the response, pressure sensitivity is seen as enabling planktonic organisms and hyperbenthic explorers to undergo flood transport to return to estuaries or inshore settling sites, or ebb transport to avoid stranding between tidemarks" (p. 413).

Bayne (1963) performed crucial experiments with early growth stages of *Mytilus edulis* larvae. It was clear that an increase of even one-half an atmosphere hydrostatic pressure produced an upward swimming response.

Knight–Jones and Quasim (1955) have reported on the number of marine planktonic organisms that are sensitive to small changes in hydrostatic pressure and tend to swim upwards when subjected to increases (however, they did not find this response in all the species tested; there is a possibility that some are more reactive at certain growth stages than at others). The changes to which larval stages of some benthic organisms are sensitive are well within the range (measurements in meters) in which an important role may be played in determining in what depth of water the larvae actually settle and metamorphose into adulthood. A pressure-sensitivity threshold of .01 atmosphere (atm) has been observed for some larvae (Figure 127).

Figures 129–132 present data that suggests correlations of both benthic and planktic abundances with depth. The interpretation of such data should include an appraisal of other variables before ascribing the effects to depth alone.

Basan (1978) reviews many of the depth-correlated occurrences of trace fossil assemblages. Trace fossil assemblages have a potential similar to that of shelly benthos for working out relative shoreline proximity and relative depth. Conclusions derived from both shelly and trace fossil data should be synthesized to provide an overall consistent picture. Golubic *et al.* (1975) discuss (Figure 133) the depth-measuring potential of microboring algae—both modern and fossil.

Figure 128 (from Buchanan, 1958) illustrates the difficulty of directly correlating taxa with depth, (i.e., pressure). It is a reasonable deduction that the combined depth–substrate correlations are largely due to the substrate factor. In order to be reasonably certain that pressure is the prime variable in controlling taxic changes across a depth transect, one must be certain that other factors are held constant, but such a certainty rarely exists. Depth as a significant variable should not automatically be discounted merely because

correlations with other cooccurring variables have been established.

Boltovskoy and Wright (1976) provide insight into how the depth-correlated distribution of foraminifera (Figure 134) may be employed for depth-discrimination purposes. Similar data may be developed for other shelly groups.

DEPTH

As previously mentioned, the conspicuous change in the fauna and flora occurring in progressively deeper water was one of the earliest observations in marine biology. Unfortunately, many physical variables change with depth, preventing the attribution of any changes in fauna to simple hydrostatic pressure, which changes linearly with depth, about one atomosphere for every 10 m depth. For example, light alters in spectral composition and intensity along any depth transect. Temperature decreases at various rates with increasing depth, and even substrates show a marked change from coarse- to fine-grained. Therefore, it is very difficult to determine from simple biotic change just which factors are causative and which are correlative. Detailed faunal studies over widely differing natural conditions may eventually illumine the relative importance of involved variables. In the Arctic and Antarctic regions, where there is very little temperature variation with depth changes, there is still marked vertical zonation of the benthic fauna. Similar zonation is found in low latitude oceans where there is marked thermal change. It may be permissable to conclude that temperature is not the most important variable in determining vertical zonation of some fauna. How is it decided whether nutrient supply to the bottom, pressure, chemical con-

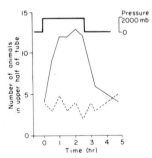

FIGURE 127. Graph illustrating the number out of 25 zoea and megalopa larvae of *Portunus* and *Carcinus* swimming in the upper half of the pressure vessel when subjected to pressure changes of 2000 mb (continuous line). The broken line indicates the number of larvae in the upper half of a control tube kept at atmospheric pressure. The pressure regime in the experimental situation is indicated by the heavy line above the graph. (From E. Morgan, 1972, *Proc. Roy. Soc. Edinburgh* (B), 73, pp. 287–299, Text–figure 1. Copyright © 1972 by The Royal Society of Edinburgh, after Hardy and Bainbridge, 1951.)

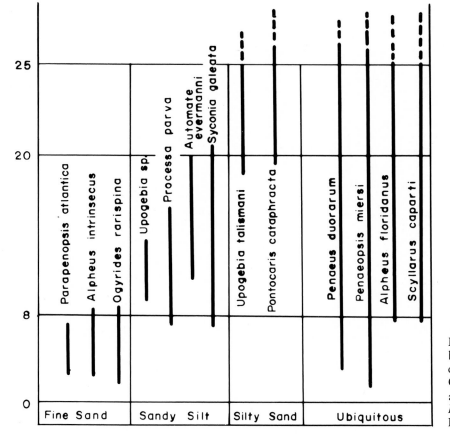

FIGURE 128. Depth correlated distribution of important species of crustacea from 0–25 fathoms off Accra, Ghana, with substrate type indicated as well. (From J. B. Buchanan, 1958, *Proc. Zool. Soc. London, 1,* pp. 1–56, Figure 9.)

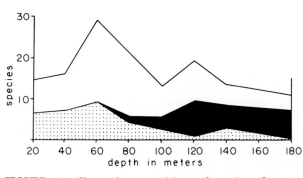

FIGURE 129. Change in composition and number of species of the benthos with depth on the St. Ann and Guinea shelf sectors; black–deep species, stipple–shallow species, white–ubiquitous species. (From A. R. Longhurst, 1958, *Fishery Publications, 11,* Figure 7, with permission of the Controller of Her Britannic Majesty's Stationery Office.)

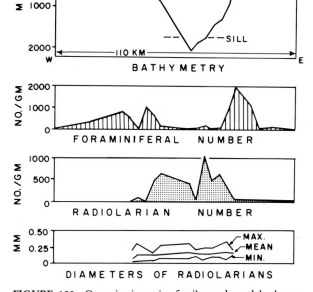

FIGURE 130. Quantitative microfossil trends and bathymetry across Guaymas Basin, Gulf of California. (From J. C. Ingle, Jr., 1972, *Proc. Pacific Coast Miocene Biostrat. Symp.* 47th Ann. Pacific SEPM Conv., Figure 5. Copyright © 1972 by The Society of Economic Paleontologists and Mineralogists.)

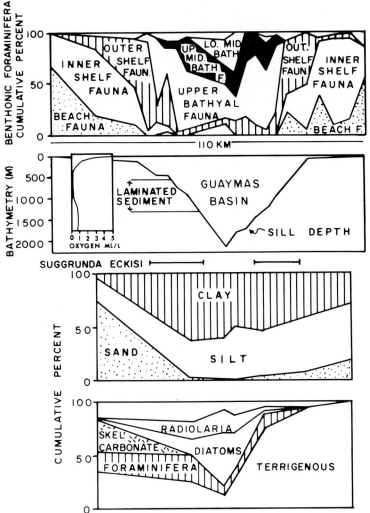

FIGURE 131. Relation between depth, substrate, and fossil types along an east–west profile across Guaymas Basin, Gulf of California. (From J. C. Ingle, Jr., 1972, *Proc. Pacific Coast Miocene Biostrat. Symp.*, 47th Ann. Pacific SEPM Conv., Figure 6. Copyright © 1972 by the Society of Economic Paleontologists and Mineralogists.)

ditions, sediment composition or other factors determine such vertical zonation? It is obvious that the environmental significance of depth-correlated changes is not a simple process; nevertheless, environmental depth is one of the most important physical parameters that can be determined from the fossil material, its environment of deposition, and associated geological information.

ABSOLUTE DEPTH

Shabica and I (Boucot [1975]; Shabica and Boucot [1976a,b]) have reviewed much of the data useful for determining absolute depth in the marine environment of the past. Of prime importance is the location of the intertidal zone and the establishment of faunal criteria for recognizing the lower intertidal faunas (see Chapter 2, pages 5–9). Second, the recognition of the lower limits of active reef growth and of active plant growth (as determined by the presence of calcareous algae, or even

of boring algae) provides a useful clue to the lower limit of the photic–phytal zone. Ginsburg *et al.* (1974) provide a vivid account of the lower limits of active reef growth at one subtropical location (about 65 m) based on actual observation. However, it has long been recognized that the lower limits of the photic–phytal zone vary in accord with the turbidity of the water, both from season to season and from place to place, and with the transparency of the water. Therefore, the lower limit of the photic–phytal zone is an approximation rather than an absolute number. It is important to recognize that available evidence strongly favors the view that active reef growth today does not reach above mean low water except in a few spots where spray and pools permit limited growth of reef-type taxa in special situations of only local significance.

The next "isobath" of practical value for the paleontologist is the shelf margin region, which can be recognized by a combination of physical and biotic criteria such as incoming of the flysch facies; incoming of geosynclinal, bedded cherts; dropping out of a rich,

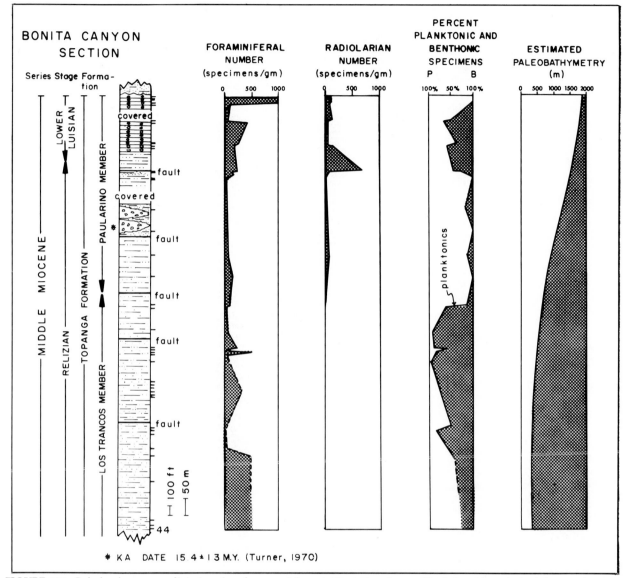

FIGURE 132. Relation between radiolarian abundance and foraminiferal abundance of both benthic and planktic types in terms of the depth transect. (From J. C. Ingle, Jr., 1972, *Proc. Pacific Coast Miocene Biostrat. Symp.,* 47th Ann. Pacific SEPM Conv., Figure 7. Copyright © 1972 by the Society of Economic Paleontologists and Mineralogists.)

shelly, level-bottom fauna with a high percentage of filter and suspension feeders, smaller size of the shelly benthic forms as contrasted with those of the nearer shore and shallower portions of the continental shelf. Gage (1977) mentions that some of the abyssal macrobenthos is represented by forms much smaller than those seen on the shelf; his conclusion supports the earlier data reviewed by Shabica and I (1976a,b), indicating that shelf margin shells tend to be much smaller than similar taxa found in shallower waters. In addition to the incoming of the flysch facies in the shelf margin, it is worth noting that dark, commonly carbonaceous shales and bedded cherts (lydites), together with phosphorites, and nodules high in

manganese carbonate, are known in many places near this shelf margin position.

At present the shelf margin averages a few hundred meters, but in the Antarctic the shelf margin occurs at about 800 m (see Shabica and Boucot, 1976a,b, for a brief discussion). Since many shelf margin-type faunas of the past are known from nonpolar regions characterized by sediments rich in calcium carbonate, it is probably reasonable to conclude that such margins represent the shallower limit rather than the deeper limit.

At present we have not developed any reliable criteria for discriminating between bathyal deposits and abyssal deposits of the past. Much of the problem has to do with the virtual nonpreservation of abyssal ben-

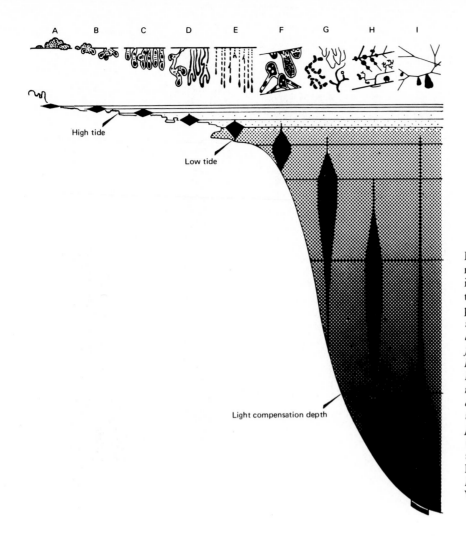

FIGURE 133. Depth distribution of modern microborers in carbonate; this is a potential depth measuring tool for the photic zone. (*a*) Coccoid cyanophytes; (*b*) *Hormathonema luteobrunneum, H. violaceonigrum.* (*c*) *Hormathonema paulocellulare;* (*d*) *Solentia foveolarum, Kyrtuthrix dalmatica;* (*e*) *Hyella tenuior;* (*f*) Chlorophytes *Eugomontia sacculata, Codiolum polyrhizum;* (*g*) *Hyella caespitosa, Plectonema terebrans, Mastigocoleus testarum;* (*h*) *Conchocelis*—stages of *Porphyra,* and *Ostreobium* spp. I, fungi. (From S. O. Golubic, R. D. Perkins, and K. J. Lucas, 1975, Figure 12.2. In R. W. Frey, Ed., *The Study of Trace Fossils.* Copyright © 1975 by Springer–Verlag New York, Inc.)

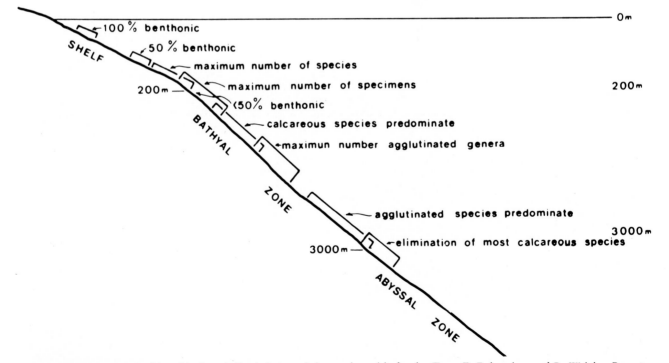

FIGURE 134. Summary of benthic foraminiferal characteristics varying with depth. (From E. Boltovskoy and R. Wright, *Recent Foraminifera* Figure 55, Copyright © 1976 by Dr. W. Junk b.v.)

thic fossils (with the conspicuous exception of certain ostracodes; see Benson, 1975; and foraminifera). An example of the paleontologic data necessary for discriminating between bathyal and abyssal depths is provided by the Paleogene fish-scale data of David (1946a,b), where bathyal and abyssal fish-scale assemblages could be discriminated. Conceptually, the vertical change in calcium carbonate solubility in the oceans could be very useful in determining the depth of deposition of abyssal and bathyal sediments. Carbonate tests are preserved above the CCD; below the CCD they are dissolved. Unfortunately, the temporal and geographic persistence of the CCD has neither been mapped from fossil evidence nor been predicted by chemical model. Until this very basic work is completed, the CCD cannot be accepted uncritically as an indicator of paleobathymetry.

An important aspect of absolute-depth determination is the use of reef height. Klovan (1974; see Figures 14 and 15; Figure 135) provides one view of how data from the geologic record of reef-type structures may be employed to estimate absolute depth. In principle, one studies a reef cross-section from the lowest successional, quiet-water, initial stage (see Walker and Alberstadt, 1975, Figures 51 and 52 for a number of successional examples in diagrammatic form) up to the rough-water, shallow-subtidal environment, and measures the total height of the vertical succession. The assumption is made that there has been no change in either relative sea level position or in sediment influx about the reef base significant enough to perturb the successional history. If the reef represents a relatively brief moment in geologic time, where these assumptions have a better chance of holding true, one can rely on the measured

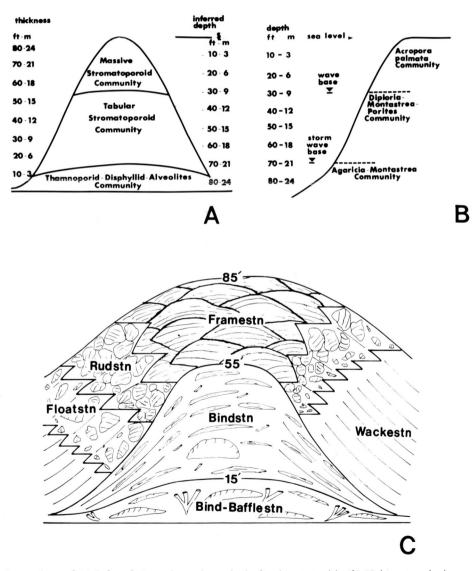

FIGURE 135. Comparison of (*a*) inferred Devonian paleoecologic depth zones with (*b*) Holocene ecologic zones (after Logan, 1969); (*c*) illustrates facies distribution within typical Late Devonian bioherm. (From J. E. Klovan, 1974, *Amer. Assoc. Pet. Geol. Bull.*, *58*, pp. 787–799, Figure 14. Copyright © 1974 by the American Association of Petroleum Geologists.)

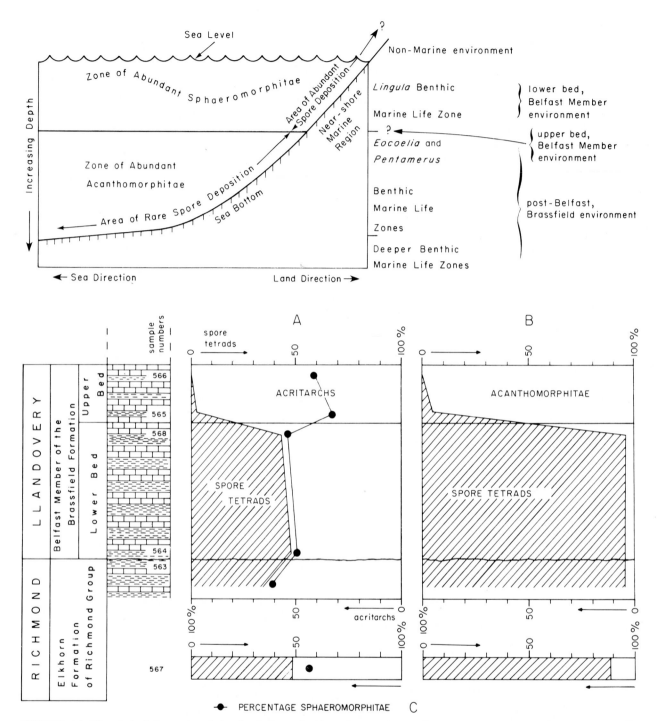

FIGURE 136. Use of planktic taxa in relative depth determination. (From J. Gray and A. J. Boucot, 1972, *Geol. Soc. Amer. Bull.*, *83*, pp. 1299–1314, Figures 3–4. Copyright © 1972 by the Geological Society of America.)

(Top) Distribution of planktonic acritarch communities and invertebrate benthic life zones according to our hypothesis of depth control at Ohio Brush Creek during Late Ordovician–Early Silurian time. Abundant spore tetrads at Ohio Brush Creek appear to be confined to nearshore sites in the zone of abundant Sphaeromorphitae. The *Lingula* benthic life zone is not represented by megafossils at Ohio Brush Creek although the environment represented by abundant spore tetrads is no deeper than that of the *Eocoelia* benthic life zone represented in the lower part of the overlying Brassfield Formation.

(Bottom) A composite spore tetrad–acritarch diagram for the Ohio Brush Creek section, Ohio. The percentage frequencies are based on a combined total of between 100 and 125 spore tetrads and acritarchs. Chitinozoans and scolecodonts (including

reef height as an absolute depth. However, the effects of intensive stylolitic solution and compaction phenomena must be considered as they may have diminished the original reef height. Most absolute-depth measurements are conceptually influenced by current thinking about the depths of maximum, active reef growth, and facies relations. Therefore, one must accept such measurements critically. It is encouraging to note, however, that most of the absolute-depth measurements calculated on such assumptions provide reasonable numbers in terms of modern reefs. It is clear that reef growth taking place in a subsiding area will provide reef heights far in excess of true absolute heights, and that elevation of an area during reef growth should result in erosion and shortening.

Gray and I (1972) have shown how planktic taxa may also be used for relative-depth determination (Figure 136).

DuBar and Taylor (1962) provide an example of depth estimation for a Tertiary fauna obtained by measuring the overlapping depth ranges of modern species of those genera represented in the Tertiary fauna, assuming that there has been no overall change in the depth range of the genera involved.

TURBULENCE

Turbulence, as an important factor in the distribution of living marine plants and animals, is noted by all (see Kinne, 1971b, for data and references) concerned with the intertidal and shallow subtidal environments (Figures 137 and 138).

Possible relations between turbulence and animal distributions in deeper regions have not been explored (turbulence is commonly noted and studied through the photic zone, but not much farther down), but this absence of study does not necessarily indicate lack of importance. The presence, even as far down as the abyssal–hadal boundary of hard substrate filter and suspension-feeding organisms would suggest (unless downward movement of nutrient particles supplies all their needs) that differing levels of turbulence must be important in all submarine positions. Table 3.17 tabulates some of the characters useful for indicating level of turbulence.

For the geologist, the high correlation between substrate grain size and fauna indicates that rheology was important in the past. Rough water morphologies for colonial organisms (the nonbranching form as opposed to the quiet-water branching form) are notable in this regard. Thomsen (1977) describes the thicker-walled bryozoan colony form in a high current-velocity situation as contrasted with thinner-walled form of the same species in a lower current-velocity situation, and (1976) deals as well with the relevant sedimentary and bryo-

zoan parameters. The more generally massive shells present in rough-water environments are also notable (the Russian doll beak end-deposits of certain brachiopods, for example). There is a general tendency for heavy shells to stabilize themselves on unconsolidated substrates (Rosewater, 1965), and this fact has been taken advantage of for purposes of paleoecologic interpretation (see Seilacher, 1968, for an example from the Devonian; and Surlyk, 1972, for one from the Cretaceous). Eagar (1977; 1978) carefully documents the different shell morphologies encountered within the same species in a fast-water environment as contrasted with a quiet-water environment, and despite the fact that he is dealing with freshwater bivalves, the moral about employing phenotypic gradients as aids in sorting out environmental variables should be noted. The difficulties of using shell morphologies alone for estimating quiet as compared to rough water are well-illustrated (as a warning) by Harger's studies (1968, 1969, 1970, 1971, 1972) of two species of *Mytilus* in which one species (*M. edulis*) does well in quiet water and the other (*M. californianus*) does well in rough water. Would any sensible paleoecologist place *M. edulis* shells in a different rheological environment than *M. californianus?* (I think not!) However, the studies summarized in Faber, Vogel, and Winter (1977) show conclusively that correlations between shell form and grain size, even when dealing with extinct higher taxa, do

---

recognizable fragments of both) were tabulated above the combined totals for spore tetrads and acritarchs in preparing Table 1, but are excluded from the diagrams for the following reasons: chitinozoans because of their scarcity; scolecodonts because their dispersal and accumulation is probably largely unrelated to that of the water or wind dispersed spore tetrads and the water dispersed acritarchs. Panel (*a*) shows the ratio of spore tetrads to acritarchs; the basis for calculation is the total spore tetrads and acritarchs from each sample. The separate curve indicating relative abundance for Sphaeromorphitae acritarchs (also included in the total) should be read on the acritarch scale from the right. Panel (*b*) shows the ratio of spore tetrads to Acanthomorphitae. The basis for the calculation is the total spore tetrads and Acanthomorphitae acritarchs, and (*c*) records counts for a single sample from the same horizon as sample number 564 in the stratigraphic column, but approximately 100 ft along the outcrop from 564. For (*c*), the notation is the same as that of the diagrams immediately above each of the separate bars.

The columnar section is a stylized representation of the calcareous shales of the Elkhorn formation and the Belfast member of the Brassfield. Shales and limestones do not alternate in this section.

TABLE 3.17
Criteria for Recognizing the Quiet- to Rough-Water Spectrum

Very quiet (still)	Quiet (very slowly moving)	Moderately agitated (moving)	Very agitated (turbulent)
Anoxic	Moderately well oxygenated	Well oxygenated	Well oxygenated
Black, no bioturbation	Gray, bioturbated	Gray, bioturbated	Gray-to-red if tropical–subtropical bioturbated
	Slabby oncolites, pelmatozoans	Round oncolites, pelmatozoans	
Clays	Silts, sands, muds	Silts, sands	Sands, gravels
Very scattered shells, low diversity	Scattered shells, low to high diversity	Clumps and scattered shells, medium to high diversity	Blankets of shells, low diversity
Micrite		*Sparite*	
None to few shelly specimens	Many shelly specimens	Many shelly specimens	Many shelly specimens
Few species			Few species
Articulated shells	Articulated shells	Articulated to disarticulated shells	Disarticulated to broken shells
Camerate : Inadunate Ratios			
← *(low–high)* →			
Life position common	Life position common	Life position rare	Life position very rare. Unattached shells or loosely attached
Some modified for soft muds	Some modified for soft muds	"Normal" varied shells	Massive shells, Russian dolls
Laminated, no cross beds	Ripples–small, small cross beds	Ripples–medium, cross beds–small and moderate	Ripples–large. Big cross beds

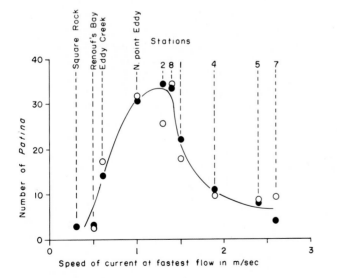

FIGURE 137. Population density of *Patina pellucida* on *Saccorhiza polyschides* (whole plant) in relation to the speed of current at fastest flow (in meters per second). Solid dots are mean number of *Patina/Saccorhiza* plant; open circles number of *Patina/*3 lb. of *Saccorhiza*. (With permission from J. A. Kitching and F. J. Ebling, 1967, *In* J. H. Cragg, Ed., *Advances in Ecological Research,* Figure 18. Copyright © 1967 by Academic Press Inc. [London] Ltd.)

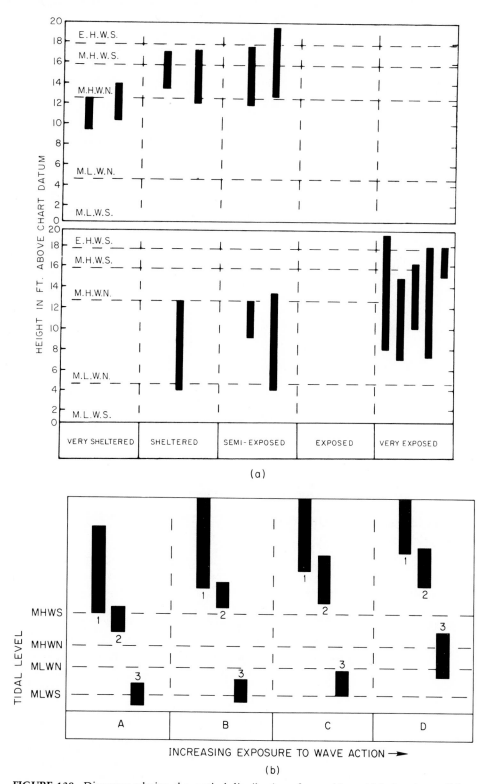

FIGURE 138. Diagrams relating the vertical distribution of several intertidal algae in conditions of differing wave action. (*a*) Diagram showing the vertical distribution of the brown alga *Pelvetia canaliculata* and the lichen *Lichina pygmaea* (below) in relation to exposure to wave action. (Based on Evans, 1947). (*b*) Diagram showing the vertical distribution of *Verrucaria*, Myxophycae and *Gigartina* in relation to exposure to wave action. 1 *Verrucaria*, 2. Myxophycae, 3. *Gigartina* (Based on Lewis, 1964). (From R. C. Newell, 1970, *Biology of Intertidal Animals,* Figures 1.5 and 1.6 after Evans, 1947, and Lewis, 1964. Copyright © 1970 by Elsevier/North Holland Biomedical Press.)

provide real possibilities for employing shell form as an index of current strength (see Figures 139 and 140). It must always be kept in mind, however, that grain size correlates with depth as well as with current velocity. Therefore, grain-size correlations do not invariably indicate that current velocity is a more important factor than others that are also correlated with depth.

Alexander (1975) has provided a synthesis of a variety of physical and biological factors that led him to conclude that Late Ordovician strophomenoid brachiopods from Indiana possessed both size- and shape-parameters sensitive to the degree of turbulence characteristic of their respective quiet- and rough-water environments. The phenotypic gradient observed by Alexander does appear to correlate very well with fac-

tors suggestive of quiet and turbulent conditions, and flume experiments with models of his brachiopods support his conclusions.

Batham (1958) points out that stable intertidal stones commonly have a fauna beneath them, whereas the converse is true for unstable stones. Such a character is of potential paleoecologic value in terms of conglomeratic rocks or boulders such as those discussed by Surlyk and Christensen (1974). The well-developed attached fauna on gneissic boulders (Figure 141) in the Danish Cretaceous are good evidence for boulder stability, and the ancillary evidence of rounded oncolites (''algal biscuits'') indicates enough turbulence to provide adequate food for the variety of attached filter-and suspension-feeding organisms. Anyone who has ever

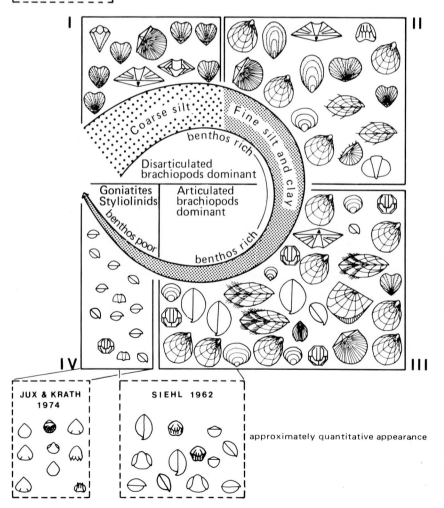

FIGURE 139. Distribution of Devonian brachiopods in four hydrodynamic categories. The correlation is between grain size of entombing sediment and the morphology of some Middle Paleozoic brachiopods. Note that the styliolinid–goniatite facies occurs seaward of the brachiopods, and that disarticulation is more common shoreward. Note also that many other parameters potentially correlate with the ones shown in this figure; thus determination of causality is not simple. (From P. Faber, K. Vogel, and J. Winter, 1977, *N. Jahrb. Geol. Pal. Abh., 154,* pp. 21–60, Figure 11. Copyright © 1977 by E. Schweizerbart'sche Verlagsbuchhandlung.)

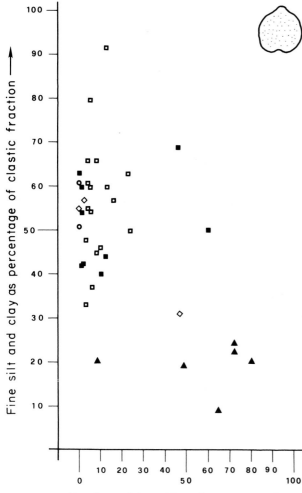

FIGURE 140. Correlation between the presence of punctae in a brachiopod and the nature of the entombing sediment. Note that Mitchell (1975) has shown something similar on a regional scale for number of plications and inferred temperature. Foster (1974) suggests a similar correlation between density of punctae and temperature for living brachiopods. It is possible, therefore, that density of punctation responds to more than one physical variable in the environment. (From P. Faber, K. Vogel, and J. Winter, 1977, *N. Jahrb. Geol. Pal., Abh., 154,* pp. 21–60, Figure 8. Copyright © 1977 by E. Schweizerbart'sche Verlagsbuchhandlung.)

FIGURE 141. Reconstruction of Cretaceous epizoans on PreCambrian boulders in Scania. Note the distributions on upper and lower boulder faces, as well as on the associated level bottoms. (From F. Surlyk and W. K. Christensen, 1974, *Geology, 2,* pp. 529–534, Figure 10.)

listened to the clinkety–klink–klink music of ball-milling cobbles on a steeply sloping, wave-pounded beach needs no explanation for the absence of attached biota (Willow Creek, south of Monterey, California, 1955).

Loesch's (1953, pp. 208–209) little jeep "experiment" is dramatic evidence of how the ubiquitous surf clam *Donax* is adapted to the pounding of the surf on sandy, intertidal, level bottoms. When one considers the *Donax* behavioral complex, is it not surprising that the intertidal, sandy, turbulent substrates are low in taxic diversity. Mori (1938) has suggested that the responses of *Donax* may reflect a relation between the thixotropic–dilatant change in the sandy substrate as the sands become wetted down by the advancing and retreating waves rather than of the actual pounding. Is it possible that Loesch's jeep, in rolling across damp sand, actually injected enough water lateral of the tires to have the desired effect on the thixotropic state of the sand? Figure 142 presents actual *Donax* densities. Tiffany (1971) appears to have answered the question by implicating both "acoustic shock" and water saturation, (i.e., thixotropy), acting together.

During the summer the shoreline at Mustang Island consists of a series of cusps and intercusps. It is in the intercusps that the majority of the *Donax* are found. There are on the average about sixty cusps per mile, varying with the time of year. A concentration was generally found in each intercusp. The number of concentrations of *Donax* per mile of beach at different times of the year is given in Table 4.

When *Donax* are only slightly buried, they can be located by their excurrent and incurrent siphon tube holes in the sand. These two holes are about half as far apart as the length of the clam.

Donax are among the few animals that have become adapted to the wave impact area where the land meets the sea. As the tide rises and falls *D. variabilis* moves up and down the beach slope synchronous with the tide. As the clams have no independent means of locomotion other than the ability to dig in or out of sand, they hitch a ride by using the movement of the water in a wave. Within the short period of one wave the entire population, both lower and upper limits, may change its location by as much as ten yards, or it may not move at all. One change in position from inshore to offshore by one clam was observed to be 24 yards. After riding a wave the clam buries in 1.5 to 3 seconds with ten to twenty probing movements. The movements are slower in winter. The size of the clam has no effect on the time or on the number of movements required to bury.

On a rising tide *D. variabilis* appears to be influenced by the vibrations of the breaking waves transmitted through the sand. When the clams are at the height of a rising period they can be made to rise by running a jeep about 20 yards or nearer the population. As the jeep runs along the beach parallel with the water's edge, the clams rise just about even with the jeep. The picture in Figure 5, which shows a heavy con-

centration of *Donax* lying on the beach, was made after inducing the clams to rise to the false stimulus of a jeep [Loesch, 1953, pp. 208–209].

Concentrations of *Donax* per Mile[a]

Date	Stations					Average
	1	2	3	4	5	
June 27, 1951	—	—	—	—	—	55
August 3, 1951	110	73	71	75	51	80
September 3, 1951	—	—	71	70	50	64
September 29, 1951	14	14	15	14	14	14
November 17, 1951	Not concentrated, scattered					
December 16, 1951	Not concentrated, scattered					
June 2, 1952	Not concentrated, in bands 1–1.5 yds wide					
June 27, 1952	71	65	61	57	52	59

[a]From Loesch, 1953, p. 209

In any event, the high correlation observed by ecologists from the intertidal-down between biota and rheic conditions, the similar correlation observed by the geologist among fossil biota, grain size, and sedimentary structures correlating with rheic conditions, leaves little doubt about the significance of rheic factors in the distribution of plants and animals. However, care must be exercised because, as always, from place to place more than one important factor may vary with rheic factors to produce a dual or even false correlation. Table 3.18 and Figures 290 and 291, pp. 335 and 336, provide data on a mixed, quiet-water—rough-water assemblage in which quiet-water fauna grew on and about a group of rough-water shells **after** conditions changed from rough to quiet.

When trying to work out a means for inferring turbulence, one should note the turbulence limits favored by various groups. For example, most echinoderms favor a moderately turbulent environment; rock-dwelling intertidal barnacles favor a very turbulent environment.

Loya (1972) suggests that the branching form of many colonial organisms, or at least of hermatypic corals, may be a cleaning modification making life on quiet-water, muddy substrates practical since more massive forms cannot thrive unless they develop special modifications for cleaning. Of course, the high correlation between branching forms and muddy matrices as contrasted with massive forms and arenitic matrices has long been noted for many groups of colonial organisms such as Paleozoic stromatoporoids, tetracorals, and stony bryozoans.

Lane (1971) summarizes data indicating that the ratio of camerate to inadunate crinoids provides a measure of turbulence with a predominance of camerates indicating somewhat rough water and a predominance of inadunates indicating relatively quiet, although not still, conditions. The dominance relations are best expressed on muddy level bottoms and in reef-associated biofacies.

FIGURE 142. *Donax* burrows from above, in side view, and side view by X-ray. Note the high population density. (From R. W. Frey, Ed., *The Study of Trace Fossils*, Figure 2.10. Copyright © 1975 by Springer–Verlag New York, Inc.)

In the various carbonate facies of biogenic origin, Wilson (1975) summarizes data indicating the existence of high correlations between abundant micritic matrix and relatively quiet-water conditions as contrasted with the rougher-water sparite matrix material.

Ettensohn (1976) suggests that the stemless crinoids of the Paleozoic lived in somewhat turbulent environments, as did the Jurassic–Recent comatulids.

Although these crinoids are not common as fossils, they may prove of some use when they are recognized.

Kier (personal communication, 1977) points out that regular echinoids are rarely seen today on muddy substrates, probably because their food supplies (including kelps) seldom occur associated with muddy substrates.

It is important to note, in studying the importance of turbulence, that there is a "normal" level, at which taxic diversity is relatively high. Departures from the "normal" in either direction are associated with sharp drops in taxic diversity. The very quiet-water environments are commonly anoxic or low in oxygen and host few species. The increasingly turbulent environments impose more and more rigorous requirements, and also correlate with decreased diversity at all levels.

The subtidal shoreface environment is commonly defined as the region of wave activity. A normal level of daily wave activity should be viewed as an environmental variable with a biota modified to exist in it. Abnormal wave activity, such as that encountered in unusually severe storms, may be regarded as a catastrophic event rather than as a normal factor in the environment.

ONCOLITES AND PHOTIC ZONE-TURBULENCE MEASUREMENT

The term *oncolite* (see Ginsburg, 1970; rhodolite [Bosellini and Ginsburg, 1971]) is restricted by Bosellini and Ginsburg (1971) to spherical and subspherical bodies with or without a nonalgal core or nucleus of finely laminated calcium carbonate detrital grains. Bossellini and Ginsburg employ the term *rhodolite* for similar balls formed from coralline algae. Their restricted term *oncolite* involves the activities of bluegreen algae. The presence of laminae on all sides of the oncolites and rhodolites provides strong evidence for a certain level of turbulence-induced movement. The actual frequency of the turbulence appears to correlate highly with oncolite shape. Oncolites of a platelike or slabby form with growth occurring on only one side have clearly grown in relatively quiet waters intermittently turbulent enough to flip the plate over so as to

permit intermittent growth on both sides (Figure 143–145). Spherical and subspherical oncolites are indicative of frequent enough turbulence to keep the algal growth from becoming irregular in form (Rodriguez and Gutschick, 1975).

As might be expected, the bulk of the oncolites and rhodolites of the past are associated with carbonate-rich marine sediments, obviously photic zone, inferred to have developed under tropical and subtropical regimes. However, Alexandersson (1974) has directly observed subtidal oncolites growing today in the north temperate waters of the Skagerrak, so that one must employ caution when using oncolites for climatic interpretation. Bosence (1976) and Farrow *et al.* (1978) describe British examples.

The rate at which spherical oncolites move, in at least one instance, is indicated by Herdman's (1906, p. 112) illustration of an example from the Indian pearl banks occurring with attached pearl oysters (*Pinctada*), at least a year old.

Functionally related to algal oncolites are spherical-coral colonies of the present and past (see Kissling, 1973, for Recent and Silurian examples), which indicate a reasonable level of turbulence, as are the spherical bryozoan colonies figured by Müller (1966).

The most recent, useful synthesis dealing with oncolites is that of Bosellini and Ginsburg (1971). They suggest restricting the term *oncolite* to algal nodules formed by plants that act to cement sediment grains, omitting those formed by lime-secreting algae. They present data on the growth rate of certain types of oncolite, suggesting that many of them are tens of years old, which is certainly useful in estimating the sedimentary influx into an area. However, it must be remembered that the rolling about that occurs during growth of oncolites is consistent with a fair sediment influx without consequent burial.

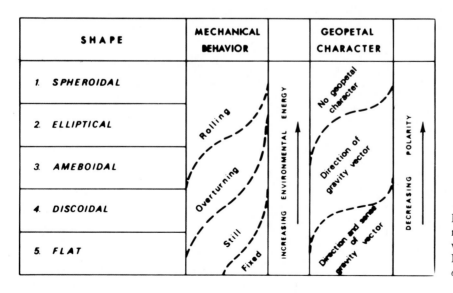

FIGURE 143. Diagram indicating the relation between oncolite form and environment. (From A. Bosellini and R. N. Ginsburg, 1971, *Journal of Geology, 79*, pp. 669–682, Figure 12.)

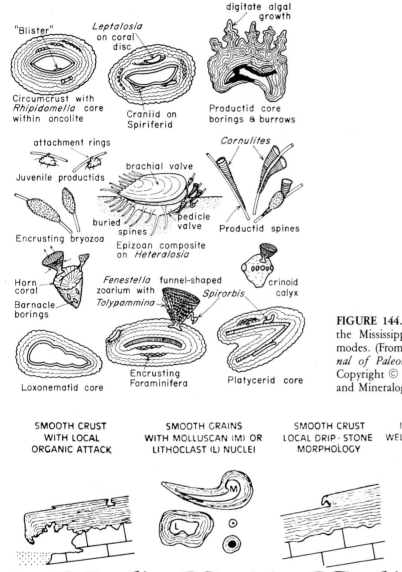

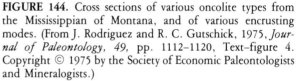

FIGURE 144. Cross sections of various oncolite types from the Mississippian of Montana, and of various encrusting modes. (From J. Rodriguez and R. C. Gutschick, 1975, *Journal of Paleontology*, 49, pp. 1112–1120, Text–figure 4. Copyright © 1975 by the Society of Economic Paleontologists and Mineralogists.)

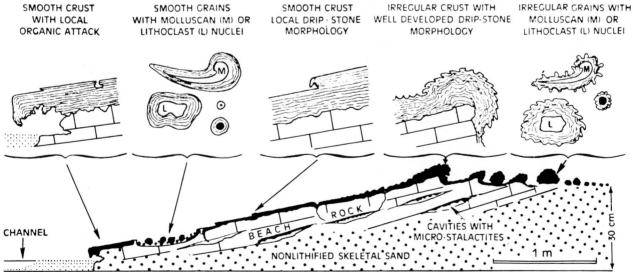

FIGURE 145. Various encrusting mode cross sections from the supratidal region of the Persian Gulf area. These encrustations have been interpreted as both inorganic precipitates and as the results of algal activity. (From B. H. Purser and J. P. Loreau, 1973, *In* B. H. Purser, Ed., *The Persian Gulf, Holocene Carbonate Sedimentation and Diagenesis in a Shallow Epicontinental Sea*, Figure 11. Copyright © 1973 by Springer–Verlag New York, Inc.)

Review of the references cited here indicates that oncolites and rhodolites form in both the intertidal and photic–phytal zone subtidal environment. Purser and Loreau (1973) provide excellent descriptions of a variety of intertidal oncolites, which in their opinion, may be inorganic (Figure 145).

Similar in environmental significance to oncolites are certain coral "balls" such as the favositids of the Lower Paleozoic, which also indicate a certain amount of rolling during the growth of the spherical to subspherical colony with the implication that a certain level of turbulence is involved. Wilson (1975) employs the term *onkoid* to refer to all ball-shaped or disclike aggregations, no matter what their biologic affinities, that represent a certain amount of rolling, flipping, and turning during growth.

Bosence (1976) provides a useful amount of actual data on the currents needed to move coralline algal rhodolites and details the relation of growth form to velocity environment more thoroughly than any that has been previously available. Bosence points out that branching density, as well as size and form, is involved in determining the velocity needed to provide movement (Figures 146 and 147; Table 3.19).

LAMINATION

Odom (1967) exhibits a high correlation between organic content of fine-grained shales and the excellent lamination (Figure 148) that accompanies fissility. Grain size, fissility, and high organic content all correlate with quiet-water conditions of the type that commonly accompany the black shale facies. Byers (1974) points out that these fissile black shales are not subject to bioturbation. The lack of bioturbation also correlates with low oxygen levels and unusually quiet water conditions. Odom (1967) has pointed out that such fissile shales also display a high degree of clay–mineral orientation. This high level of clay–mineral orientation is, of course, largely responsible for the fissile nature of the shales and may itself be assumed to result from quiet-water settling of clay–mineral particles or their formation parallel to the sediment–water interface.

SPACING

The spacing of marine benthos correlates to a large degree with conditions of turbulence. Under very turbulent conditions, there is a tendency for blankets of

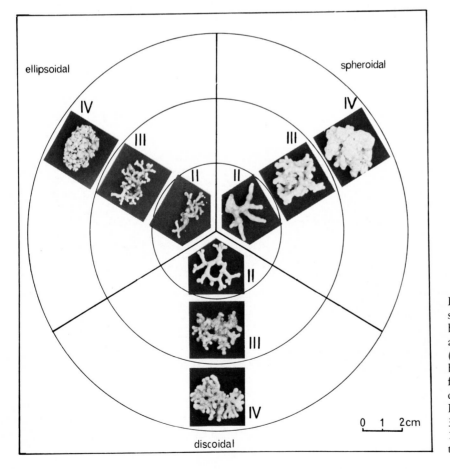

FIGURE 146. Bosence's (1976) classification of oncolite form in which both shape (ellipsoidal, spheroidal, and discoidal) and branching density (open, frequent, and densely branched) are taken into account. Both factors are involved in accounting for competent velocity. (From D. W. J. Bosence, 1976, *Palaeontology, 19,* pp. 365–396, Text–figure 10. Copyright © 1976 by the Palaeontological Association.)

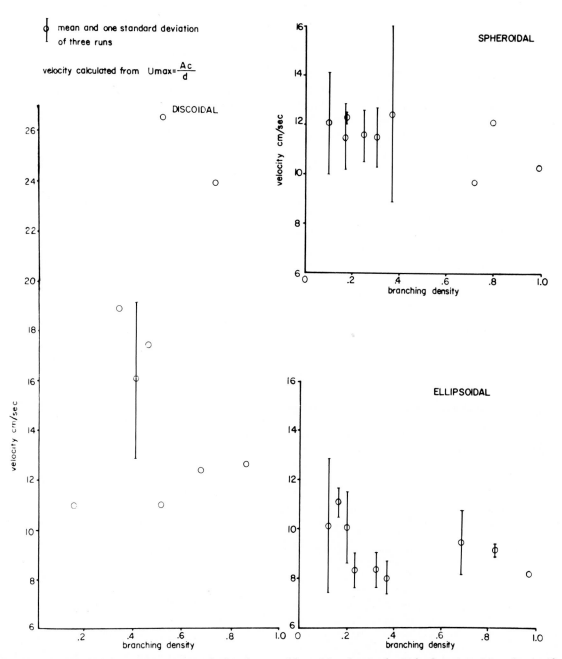

FIGURE 147. Relationship between competent velocity, shape and branching density for *Lithothamnium*. Note the significantly different results correlated with both shape and branching density. Presumably this type of result is predicted for oncolites made up from many different type materials. (From D. W. J. Bosence, 1976, *Palaeontology, 19*, pp. 265–296, Text–figure 12. Copyright © 1976 by the Palaeontological Association.)

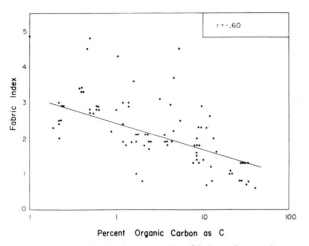

FIGURE 148. Relation between clay fabric and organic carbon content. Note how more laminated (i.e., ordered) sediments have a higher carbon content. (From I. E. Odom, 1967, *Journal of Sedimentary Petrology, 37,* pp. 610–623, Figure 7. Copyright © 1967 by the Society of Economic Paleontologists and Mineralogists.)

shells to form with very high dominance of a single species, such as are encountered today with species such as *Mytilus californianus*, some balanid barnacles, as well as, in the past, with for example, the rough water pentamerinid species of the Silurian such as *Pentamerus oblongus* and *Kirkidium knighti*. In conditions of relatively quiet water, there are more scattered shells with low dominance of any one species. Under some quieter water conditions, clumps or nests with high dominance tend to develop. As conditions become very quiet, and even poorly oxygenated, dominance of a single species tends to increase but shells become more scattered. Spacing can thus be used to provide additional clues about rheic conditions, especially when combined with other information.

TABLE 3.18
Abundance of Taxa Associated in a Mixed Quiet Water and Rough Water Community Association from Locality, USNM 18067, Gaspé, Quebec, in Beds of C₆-Low Wenlock Age[a]

Taxa	Quantity
"Leptostrophia" sp.	
Skenidioides? sp.	5
"Sphaerirhynchus" sp.	3
small rostrospiroid	13
Leptaenisca sp.	1
Mesopholidostrophia sp.	31
eatonioid	6
Protomegastrophia sp.	37
"Ancillotoechia" sp.	15
dalmanellid	
orthotetacid	11
multiplicate *Howellella*	14 brachial
smooth *Howellella*	67 pedicle
Eospirifer sp.	1 whole
	40 brachial
	132 pedicle
Atrypa "reticularis"	36
Plectodonta sp.	23
Isorthis sp.	70
Dalejina sp.	47 brachial
gypidulid	34 brachial
	149 pedicle
trilobites	2
Drummockina sp.	3
syringoporoid	1
platyceratid	6
oriostomatid	
Leptaena "rhomboidalis"	3
Costistricklandia gaspensis	236 pedicle
	192 brachial
	1 whole
Rhabdocyclus	23

[a] See Figures 290–291. Collection made available through courtesy of P. A. Bourque, Laval University, Quebec.

STORMS

Storms that disturb the bottom sediment have the capability of seriously disrupting the structure and content of benthic shelly communities. Davidson-Arnott and Greenwood (1976) comment: "The transport of sediment, whether in water, air, or some other fluid, is usually accompanied by some deformation of the sediment–water interface." Clifton (1976) comments:

Most coasts are subject to intense but relatively infrequent storms or abnormally large swell. The shoreward sequence of structures produced under these conditions may resemble the normal sequence but generally reflects larger orbital velocities, and possibly velocity asymmetry. The large waves will rework sediment most deeply in shallow water; the thickness of the reworked

layer should decrease in progressively deeper water until at some depth at which waves normally have no effect on the bottom, only the sediment surface is disturbed [p. 144].

Clifton's comment suggests that there will be one realm where storm- and swell-generated waves will be very important in modifying expected community structure and content, but that below a certain depth the effects of bioturbation will predominate. There will, naturally, be an intermediate region where the effects of wave activity and bioturbation will alternate. The depth to which significant bioturbation can extend is important in this context. "Supershrimp" of the type discussed later are probably unusual, but the activity of nearshore Callianassid crustaceans with their *Ophiomorpha*-form burrows commonly extend down as far as a meter.

With reference to the previous statements, it is important to note the difference between quiet-water areas affected by rare storms that tend to retard the successional history, and relatively rough-water regions where rare storms merely intensify the effects of the day-to-day environment.

SLOPE

The rate of environmental change experienced by any area of the benthic environment subject to a vertically fluctuating physical variable will be determined by the rate of vertical change and the slope of the bottom or other substrate. Thus, slope is a significant variable in the modern environment (see Figures 57 and 58, pp. 70–71), especially in the rocky bottom and the reef complex. Little attention has been paid to the effects of slope on environments of the past as a factor in the distribution of organisms (however, see Ingels, 1963; Figure 149). Here are some examples of the consideration of slope as a potential factor in biotic distributions of the present. Connell (1970) found that barnacle predators could climb an intertidal slope to their prey if it were vertical, but the time involved in ascending a sloping surface prevented them from reaching the prey in time to escape their own desiccation. The complexities of the slope situation are suggested by the studies of Brander, McLeod, and Humphreys (1971) who found a rocky substrate, subtidal, tropical pass in which the sides had a much richer fauna than the bottom; on further inspection it was found that the bottom was swept by far more rapid, rock abrading currents than was the case for the sides, so that slope was one factor insofar as it led to differences in current velocities.

Many ecologists have noted the correlation between biotic distribution and slope from one area to another; some have seen a connection with light (see Chapter 3, pp. 108–110) in those cases involving overhangs (particularly where algae disappear and sponges become dominant), but few have done enough work to insure causal relation rather than a chance occurrence. Until the completion of adequate experimental work, the role of slope in determining biotic distributions will remain uncertain.

WIND

Any suggestion that wind might be an ecologically significant variable for marine creatures, other than as an agent involved in the generation of submarine turbulence, might be disregarded, although it is a recognized factor of great importance in the terrestrial environment. However, Courtney (1972) points out how the desiccating effect of wind on certain substrates appears to partially control the distribution of certain intertidal gastropods. The drying and desiccation permits loosening of the snail from its substrate. The role of desiccation in the intertidal region has long been recognized as an important variable, and wind, temperature, and humidity are crucial in determining the rate at which desiccating processes progress.

Wind, as the key factor in generating shallow water turbulence, is of vital importance. The windward side of most Pacific reefs with their algal ridge environment, contrasts markedly with their leeward, quiet side (see Figures 57 and 58, pp. 70 and 71). Ingels (1963) has described a reef complex from the Silurian of the Chicago area that is interpreted as having had a windward and leeward set of biotic and lithic complexes (Figure 149). Regions affected by regular storm tracks have more complex characteristics.

Most of us commonly associate wind with waves, but the wind-generated oceanic swells (chiefly emanating from Southern Ocean storms today) are important in generating shallow-water turbulence unassociated with local winds or storms. Davies (1964) has provided an excellent overview of global swell, storm, wind, and tidal patterns covering these phenomena (Figures 150–153).

The monsoonal winds and accompanying rains presumably have a regional effect on the biota unlike that seen in nonmonsoonal regions, but adequate comparisons of the two types of regions have not yet been carried out.

Farrow (1974) relates how *Cardium* in the intertidal region is transported from one biotope into another by wind. He even illustrates how an attached epiphyte may be transported along with the bivalve. This is presumably a minor mechanism, but should be considered.

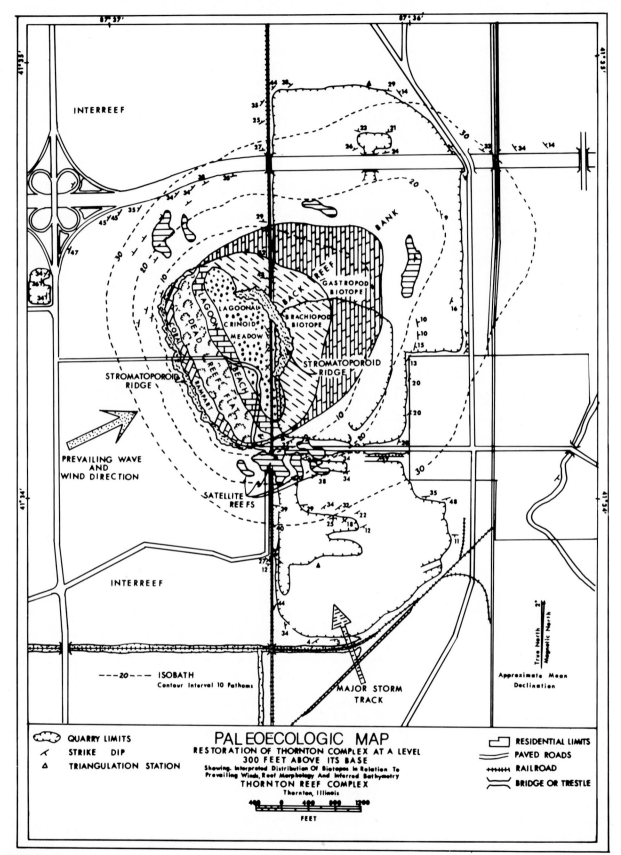

FIGURE 149. Interpretative community complexes related to both prevailing wind and wave directions as well as major storm track for the Upper Silurian, near Chicago, Illinois, in a reef complex by Ingels (1963). (From J. J. C. Ingels, 1963, *Amer. Assoc. Pet. Geol. Bull.*, 47, pp. 405–440, Figure 16. Copyright © 1963 by the American Association of Petroleum Geologists.)

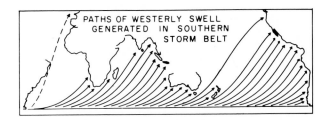

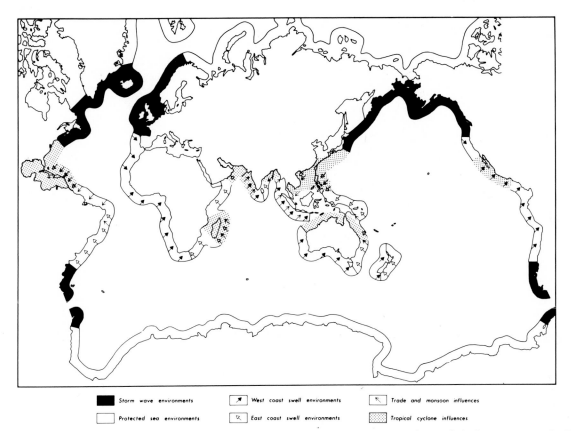

FIGURE 150. Median paths taken by swell generated in westerly gales at 55°S (Mercator projection). (From J. L. Davies, 1964, *Ann. Geomorph. Sonderheft, 8(N.F.)*, pp. 127–142, Figure 3. Copyright © 1964 by Gebruder Borntraeger.)

■ Storm wave environments	⬈ West coast swell environments	⬉ Trade and monsoon influences
☐ Protected sea environments	⬋ East coast swell environments	▨ Tropical cyclone influences

FIGURE 151. Major types of wave environment. (From J. L. Davies, 1964, *Ann. Geomorph. Sonderheft, 8*, pp. 127–142, Figure 4. Copyright © 1964 by Gebruder Borntraeger.)

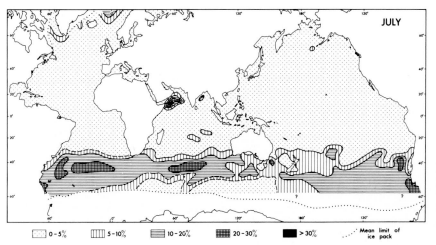

0-5%   5-10%   10-20%   20-30%   >30%   Mean limit of ice pack

FIGURE 152. Percentage frequency of gale force winds (Beaufort Scale Force 8 and over) for July. (From J. L. Davies, 1964, *Ann. Geomorph. Sonderheft, 8*, pp. 127–142, Figure 1B. Copyright © 1964 by Gebruder Borntraeger.)

169

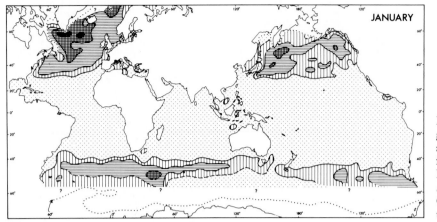

FIGURE 153. Percentage frequency of gale force winds (Beaufort Scale Force 8 and over) for January. See Figure 152 for explanation of symbols. (From J. L. Davies, 1964, *Ann Geomorph. Sonderheft, 8,* pp. 127–142, Figure 1A. Copyright © 1964 by Gebruder Borntraeger.)

TIDAL AMPLITUDE

Gill (1973, 1975), in seminal papers on tidal amplitude, has pointed out their ecologic significance in the intertidal region (Figure 154). He refers to the physical and biotic consequences of microtidal and macrotidal regimes. In brief, the macrotidal regime brings with it a far higher level of environmental heterogeneity (well-developed salt marshes, extensive mudflats, tidal sluiceways, etc.) and a lower expenditure of energy at any one locale, whereas the microtidal regime is characterized by the expenditure of the bulk of available surface water energy in one narrowly restricted area, with a consequent development of a more cliff-like shoreline, potholes, and other features dependent on the energy regime. The overall rarity of rocky bottoms in the geologic record prevents us from identifying many of the environments discussed by Gill. However, the presence or absence of the salt marsh (swamps of the tropical–subtropical areas occur in all tidal regimes) in the level bottom intertidal sequence gives one criterion for recognition of microtidal as opposed to macrotidal regimes in the past. The potentially greater taxic diversity of the macrotidal regime with its higher level of environmental heterogeneity, may help to explain otherwise anomalous diversity changes in assumed intertidal regions in various places. The importance of both the salt marsh and the swamp in generating a rich food supply for the nearshore-subtidal organisms may help to explain differing levels of shelly biomass in different regions. The importance of tidal amplitudes to our understanding of past environments has been essentially unstudied by the geologist and remains a rich field for potential research.

SEASONALITY

A great deal has been said about the role of environmental fluctuations as a variable in determining which taxa actually appear in an environment (Figure 155). It is clear from both experimental and observational data that the regularity and magnitude of environmental fluctuations play an important role in the establishment of presence or absence of taxa. Among the most important of the regular, large fluctuations are the seasons. For the northerly and southerly parts of the earth, the annual fluctuations in both light and temperature (Figure 156) are critical; because of its oceanic geography, the south temperate region, shows less fluctuation than does the north temperate region. On land, the presence of seasonal wet and dry seasons in the lower latitudes is also critical. In the marine environment, the low-latitude regions subject to monsoonal climates form another type of seasonality alien to many students. The monsoonal climate involves a major, annual shift in wind directions coordinated with a wet and dry season. This monsoonal phenomenon has important environmental consequences in terms of freshwater influx regulated by seasons for nearshore and intertidal organisms, and of rough versus quiet water for shallower depths. Deuser and Ross (1980) suggest that food supply seasonality in the abyssal oceanic realm is correlated with surface productivity seasonality.

Figure 157 (from van Steenis, in van Meeuwen *et al.,* 1960) illustrates the common misconception of the tropical belt as a region of everwet rainforest as a

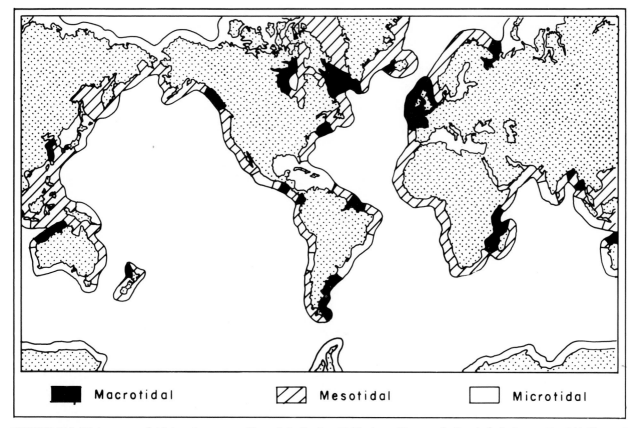

FIGURE 154. Major types of tidal environment. (From J. L. Davies, 1964, *Ann. Geomorph. Sonderheft, 8,* pp. 127–142, Figure 5. Copyright © 1964 by Gebruder Borntraeger.)

PEANUTS •   By Charles M. Schulz

FIGURE 155. Seasonality! (With thanks to C. M. Schultz, 1961, *Peanuts Parade 7.* Copyright © 1961 by Holt, Rinehart, and Winston.)

serious error. Also see Figure 158. Monsoonal wet–dry tropical seasonality is an evolutionary and ecologic force that must not be underestimated.

The role of seasonality in controlling ecologic diversification has been clearly described by Cain (1969):

> It occurred to me while studying birds in British Guiana that in the course of the year the common thrush in England passes through ecological niches appropriate to many different families of South American birds. When it is pulling up worms it is an ibis, when it is eating snails it is an Everglades kite, when it is boring into large fruit it is, I suppose, an icterid, when it is gulping down small fruit it is a cotingid or manikin (piprid), when it is taking caterpillars off leaves it is probably some form of tyrannid or formincariid. Now all of these activities can be staple modes of life throughout the year in the nonseasonal wet tropics. In England a bird that tried to live by any one of them alone would be extinct in a year. Specialization of niche therefore can go on to an extent which is almost impossible in the temperate zone, except, of course, to those migrants who can specialize in England in the summer (when England makes its nearest approach to tropical luxuriance) and then go to the tropics to maintain the same mode of life. I would like to suggest that the same stability gives an opportunity for plants also to specialize. . . seasonality is difficult to estimate. It must be done species by species [p. 234].

Cain's comments are phrased in terms of birds, but they are also applicable to the marine environment.

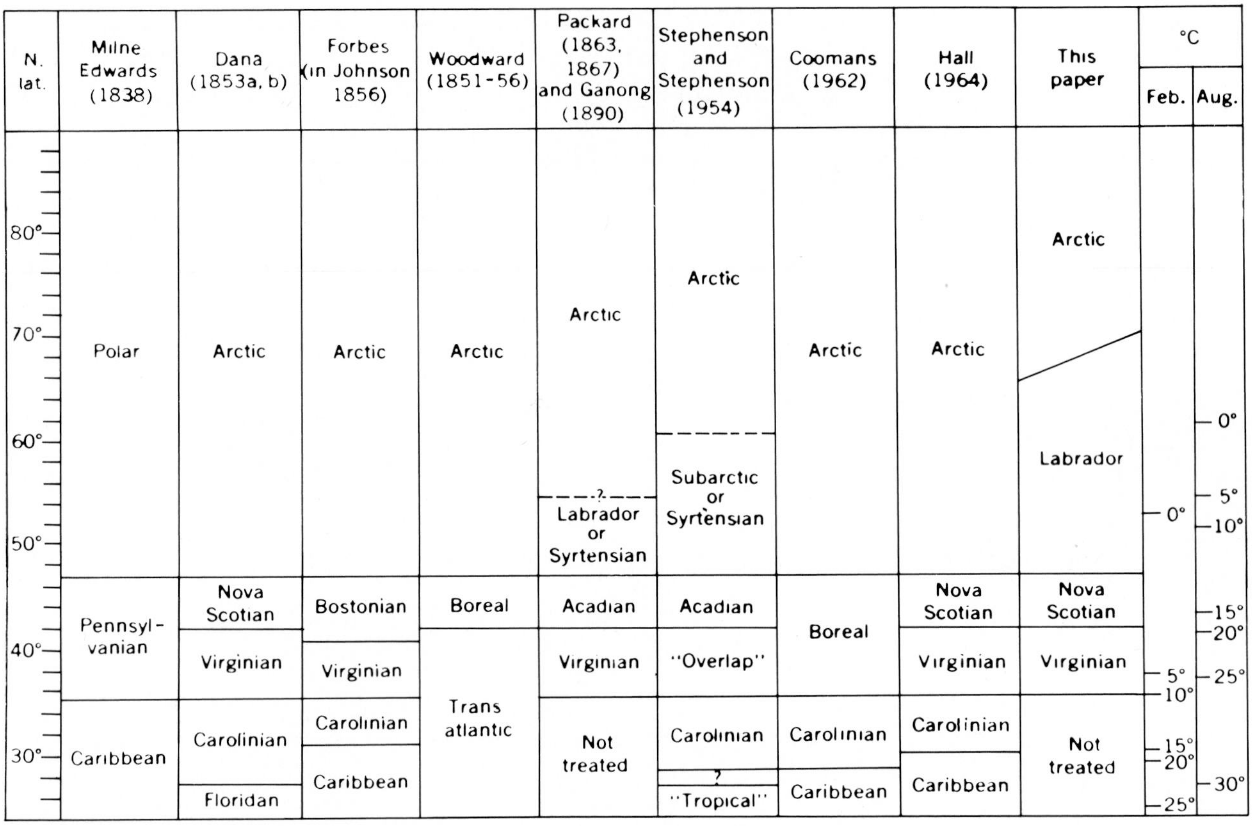

N. lat.	Milne Edwards (1838)	Dana (1853a, b)	Forbes (in Johnson 1856)	Woodward (1851–56)	Packard (1863, 1867) and Ganong (1890)	Stephenson and Stephenson (1954)	Coomans (1962)	Hall (1964)	This paper	°C	
										Feb.	Aug.
80°	Polar	Arctic	Arctic	Arctic	Arctic	Arctic	Arctic	Arctic	Arctic		
70°											
60°						Subarctic or Syrtensian			Labrador	0°	0°
50°					––.?–– Labrador or Syrtensian						5°
											10°
40°	Pennsyl-vanian	Nova Scotian	Bostonian	Boreal	Acadian	Acadian	Boreal	Nova Scotian	Nova Scotian		15°
											20°
		Virginian	Virginian		Virginian	"Overlap"		Virginian	Virginian	5°	25°
30°	Caribbean	Carolinian	Carolinian	Trans atlantic	Not treated	Carolinian	Carolinian	Carolinian	Not treated	10°	
										15°	
		Floridan	Caribbean			"Tropical"	Caribbean	Caribbean		20°	30°
										25°	

FIGURE 156. Note the marked decrease in seasonal temperature range correlated with decrease in latitude. (From J. E. Hazel, 1970, *U.S. Geological Survey Professional Paper 529-E, P. E1-E21,* Figure 4.)

Regions of high seasonality in the marine realm also tend to be populated by lower total numbers of predatory species and feeding types (i.e., fewer specialists, see Alverson *et al.,* 1964) than is the case in the more tropical regions. Inspection of the figures (see Figure 259, pp. 298–304) on larval settling times from the highly seasonal to the tropical regions provides additional data on the question of seasonality in terms of the food made available to croppers in different times of the year.

Richards (1952) points out that plant diversity in the tropics is lower in monsoonal regions than in the everwet regions.

In the monsoonal regions, the effects of seasonality on the marine environment should not be overlooked. The influx of fresh water has important effects on the shallow marine environment; these effects may be of significance in the larval (Paul, 1942) as well as in the adult stages of various organisms (also see McLusky *et al.,* 1975). The importance of changing wind directions must be considered as well. For example, the reef communities adjusted for a very turbulent, wind-induced environment in the Trades would not do well in a monsoonal regime.

There is also a certain seasonality in normal marine environments that may be reflected in the differing abundances and even differing taxa present in the sedi-

ment if sedimentation is high enough to provide an alternation of faunas reflecting seasonal changes.

Seshappa (1953) points out how seasonal effects (Figure 159) in a monsoonal environment decrease with increasing depth (at low latitude, of course, in his Indian examples). Similar changes with depth in regions where the seasonal effects correlate chiefly with light and temperature would also be predicted.

Hutchins (1947) makes the important point that the yearly temperature maximum and minimum act as controls on a regional basis (see Figures 99 and 100) (i.e., seasonality for both survival of the adults and reproductive activity). He reminds us that the maxima and minima for these two activities are seldom identical.

Frankenberg and Leiper (1977) and Croker (1977) provide recent examples of the characteristics of seasonal changes in benthic marine communities. Their data should be considered carefully by those trying to interpret data from the fossil record, where seasonality may be intricately involved.

Warner (1976) reviews the habits of the blue crab (*Callinectes*) in the Chesapeake Bay region. In this region, the crabs bury themselves in the mud during the winter. Under certain circumstances such a habit might be preserved in the fossil record.

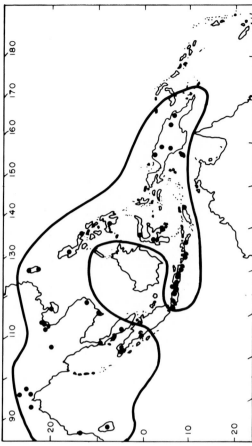

Map 2. *Smithia sensitiva* as an example of a species with preference for a feeble dry season as it occurs in N. Sumatra, N. Malaya, and W. Java, and scattered localities in New Guinea.

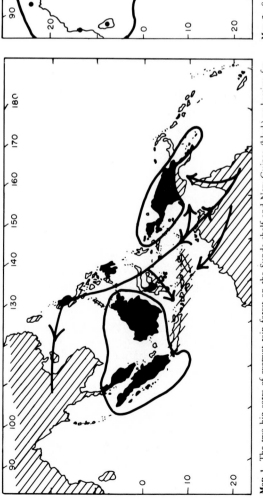

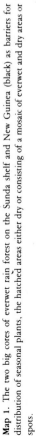

Map 1. The two big cores of everwet rain forest on the Sunda shelf and New Guinea (black) as barriers for distribution of seasonal plants, the hatched areas either dry or consisting of a mosaic of everwet and dry areas or spots.

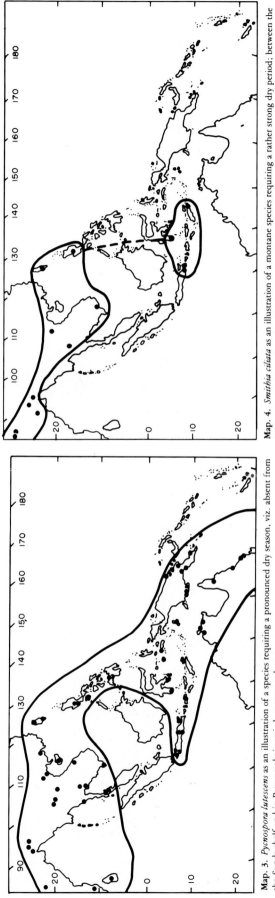

Map 4. *Smithia ciliata* as an illustration of a montane species requiring a rather strong dry period; between the Philippines and Java there is still a locality in Celebes. The area is semi-disjunct.

Map. 3. *Pycnospora lutescens* as an illustration of a species requiring a pronounced dry season, viz. absent from the Sunda shelf and in Papua only in strictly monsoonal spots.

FIGURE 157. Distribution of everwet rain forest and hatched areas showing either dry or mosaic of everwet and dry areas. (From C. G. G. J. van Steenis, 1960, *In* M. S. van Meeuwen, H. P. Nooteboom, and C. G. G. J. van Steenis, *Reinwardtia*, vol. 5, pp. 419–456, Maps 1–4, Indonesian Institute of Science.) Maps 2–4 show distribution patterns of taxa correlating with the climatic patterns.

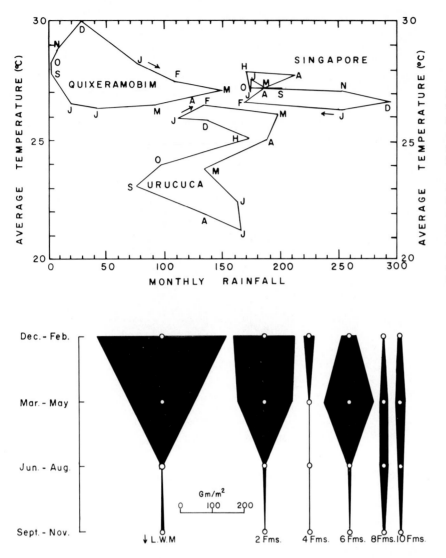

FIGURE 158. Hythergraphs showing variation of rainfall and temperature in a thorn–deciduous forest climate (Quixeramobim, 4°30′S, Brazil), a rain forest climate with seasonal differences in temperature (Urucuca, 14°30′S, Brazil) and a nonseasonal climate (Singapore, 1°18′N). Note the seasonal distinctions between the everwet and seasonal tropics. Letters refer to months of the year in order (J, F, M . . . D) (From P. T. Alvim, 1964, *In* M. Zimmermann, Ed., *Formation of Wood in Forest Trees,* Figure 1. Copyright © 1964 by Academic Press.)

FIGURE 159. Histogram showing the seasonal averages of the wet weights of animals (in gm/m²) on the sea bottom at different elevations during the year. (From G. Seshappa, 1953, *Proc. Nat. Inst. Sci. India,* 19, 257–279, Figure 3.)

ICE TRANSPORT AND DESTRUCTION

Ice is not normally thought of as a potent agent insofar as marine benthos are concerned. To date, no paleontologist has recognized its activity within the geologic record, but the following account should alert present and future paleontologists to its significance to analysis of potential ice-associated situations in the fossil record. In both the Arctic (Zenkevitch, 1963) and Antarctic (Hedgpeth, 1969; Shabica, 1972) floating ice in the intertidal and shallow subtidal zones is a potent agent for the destruction and removal of both infauna and attached epifaunal creatures from noncrevice environments (Dell, 1972). Lamb and Zimmermann (1957) describe the almost total lack of noncrevice benthic algae in the first 5 m of water in the Palmer Peninsula region of Antarctica, as well as the seasonal changes in the somewhat deeper-water benthic algal flora investigated down to a depth of about 40 m. They

also describe how the crevice algae in the first 5 m of water thrive during the summer, with little present except for a few corallines outside the crevices.

Ice also causes large seasonal variations in salinity (Shabica, 1972). In addition (Carey *et al.,* 1974; Reimnitz *et al.,* 1972; Shabica, 1972) grounded berg-ice has the capability of seriously ploughing up the shallow shelf region so as to seriously modify the benthic environment in a variety of ways, including the destruction of many organisms.

In nonpolar regions, floating winter ice is also a potent agent in modifying and destroying elements of the fauna through actual rasping and scraping action (see Colman and Segrove, 1955).

In addition to the activity of floating ice in the nonpolar regions, the action of freezing water in unusually cold spells (the "ice-winters" of northwestern Europe, see Kristensen, 1957) modifies the intertidal biota. Much of the biogeographic see–sawing north–south today in the Northern Hemisphere is due, in the intertidal and shallow-subtidal region, to unusually cold or warm seasons modifying the ranges of taxa and locally wiping out cold-loving or warmth-loving taxa (see Chapter 3, pp. 111–122).

Besides being aware of its role as a destructive agent in benthic environments, one must keep in mind that there are significant differences occurring in the benthos beneath snow-covered and snow-free ice areas (Hedgpeth, 1971). No matter how inconsequential a variable may appear to anthropocentric minds, the creatures of any given locale may employ the variable for purposes of diversification.

In the polar regions, the effects of anchor ice (Dayton *et al.*, 1969; Shabica, 1972) supplement the potential effect of grounded bergs and of pan-ice in destroying much of the shallow-water habitat.

However, an animal frozen into a block of anchor ice is not always dead. The work of Hargens and Shabica (1973) shows that a species of limpet has evolved a protective device (secretion of a coat of mucus) that permits the infrozen limpet to survive the experience.

Finally, Shabica (personal communication, 1976) relates that getting in and out of the water on an Antarctic shore in a wet suit while the water is filled with small brash-ice and bergy bits being moved about by an incoming swell is no small accomplishment; the bits are very hard; and being repeatedly struck by the blocks is distinctly unpleasant.

Thus, any shallow intertidal region of the past that is thought to have been ice-ridden should have low taxic diversity as one of its prime characteristics.

Tabulation of Bosence's (1976) Wave Tank Measurements of *Lithothamnium* Oncolite Competent Velocity[a,b,]

Specimen no.	Volume (ml)	Weight (gm)	Density (gm/ml)	Velocity for transport Mean	Velocity for transport Standard deviation
1	1.16	1.14	.98	10.2	
2	1.00	.37	.37	12.4	3.6
3	.46	.08	.17	11.4	1.3
4	.84	.61	.73	9.8	
5	.88	.26	.30	11.5	1.22
6	.75	.13	.17	12.2	.23
7	1.09	.87	.80	12.1	
8	1.53	.38	.25	11.5	1.02
9	1.30	.14	.11	12.0	2.13
10	.71	.63	.88	12.6	
11	.77	.50	.67	12.3	
12	.76	.26	.34	18.9	
13	2.41	1.75	.73	23.9	
14	1.41	.59	.42	16.2	3.1
15	1.32	.21	.16	11.0	
16	1.15	.60	.52	11.0	
17	1.65	.83	.51	26.4	
18	.49	.22	.45	17.5	
19	2.07	1.72	.83	9.1	.29
20	1.56	.50	.32	8.3	.73
21	2.30	.54	.23	8.3	.73
22	2.45	2.37	.97	8.2	
23	1.54	.58	.38	7.9	.69
24	2.31	.34	.12	10.1	2.75
25	2.73	1.85	.68	9.4	1.3
26	2.30	.47	.20	10.0	1.42
27	2.21	.36	.16	11,1	.60

[a]From D. W. J. Bosence, 1976, *Palaeontology, 19,* pp. 365–396, Table 1. Copyright © 1976 by the Palaeontological Association.
[b]Numbers 1–9 spheroidal forms; 10–18, discoidal forms; 19–27, ellipsoidal forms.

4

Communities and Their Characteristics

WHY COMMUNITIES?

We glibly speak about the flora and fauna characteristic of a particular time interval, but we seldom discuss whether or not this biota is distributed in a random or nonrandom manner. The geologist is aware that there are biofacies present (i.e., that fossils are not distributed in a random manner). We are well aware today that the biota is not distributed in a random, completely homogeneous manner. We are also well aware that each biotic patch is not entirely dif-ferent from every other patch in terms of both numbers of species present and their relative and absolute abundance. We are well aware, too, if we notice the distribution of organisms around us, that there are regularly recurring associations of plants and animals characteristic of every region and every environment. These regularly recurring associations are what the ecologist refers to as communities. The botanist even discusses phytosociology.

COMMUNITIES: THEIR DEFINITION AND RECOGNITION

The paleoecologist concerned with defining and recognizing communities within the fossil record must be cognizant of the attitudes adopted by the various factions that deal with living organisms (Figures 160–164). Ecologists, unfortunately, are not of one mind concerning the existence, definition, and recognition of communities. The major subdivisions of current conceptualizations of community definition may be categorized as follows:

1. Those who maintain that communities consist of organisms dependent and interdependent upon one another. Some of these factionalists view the resulting community as a superorganism that has a virtual life of its own—a living and breathing community.
2. Those who hold that communities are no more than chance aggregations of organisms conducting their affairs quite independently of one another—ships that pass in the night; apartment dwellers who have never been introduced to their neighbors.
3. Proponents of the middle ground who think that all communities show varying amounts of independence, dependence, and interdependence among their taxic constituents.
4. Definers of the community, regardless of their attitude toward dependence, independence, or interdependence, who reason that the dominant taxa (with considerable disagreement about the definition of dominance) should be used to define a community.
5. Those who define community by taxic composition, regardless of abundance.

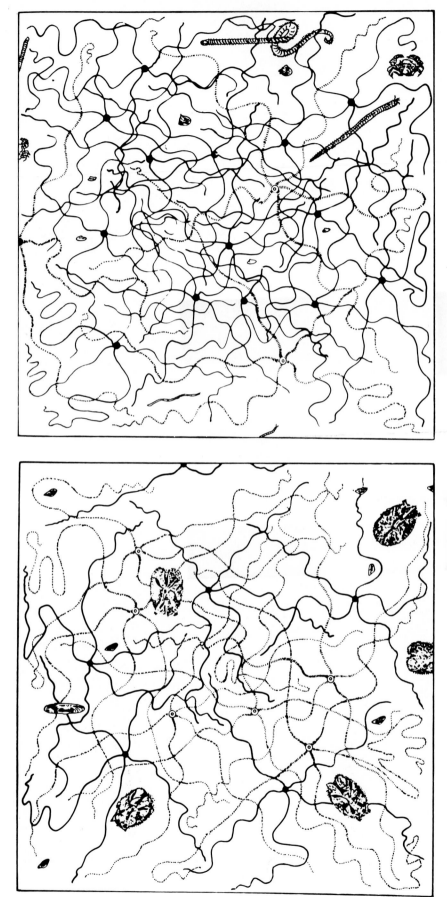

FIGURE 160. Diagram showing .25 m² of *Amphioplus-Ophionephthys* community. (From J. K. McNulty, R. C. Work, and H. B. Moore, 1962, *Bulletin of Marine Science of the Gulf and Caribbean*, *12*, pp. 204–233, Figure 8.)

FIGURE 161. Diagram showing .25 m² of *Amphioplus-Ophionephthys* community. (From J. K. McNulty, R. C. Work, and H. B. Moore, 1962, *Bulletin of Marine Science of the Gulf and Caribbean*, *12*, pp. 204–233, Figure 9.)

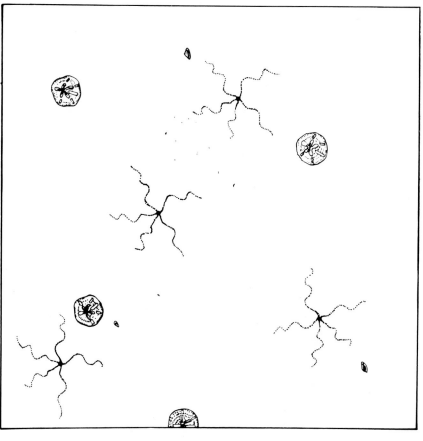

FIGURE 162. Diagram showing 2 m² of *Mellita–Tellina* community. (From J. K. McNulty, R. C. Work, and H. B. Moore, 1962 *Bulletin of Marine Science of the Gulf and Caribbean, 12,* pp. 204–233, Figure 11.)

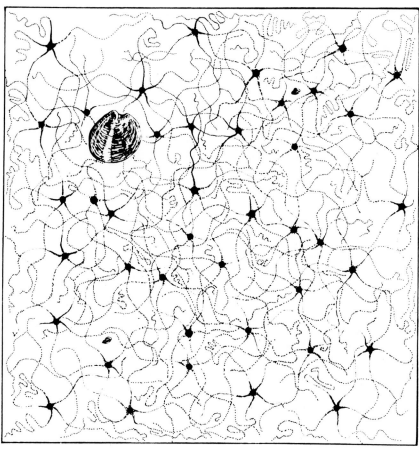

FIGURE 163. Diagram showing .10 m² of *Amphioplus–Dosinia* community. (From J. K. McNulty, R. C. Work, and H. B. Moore, 1962, *Bulletin of Marine Science of the Gulf and Caribbean, 12,* pp. 204–233, Figure 12.)

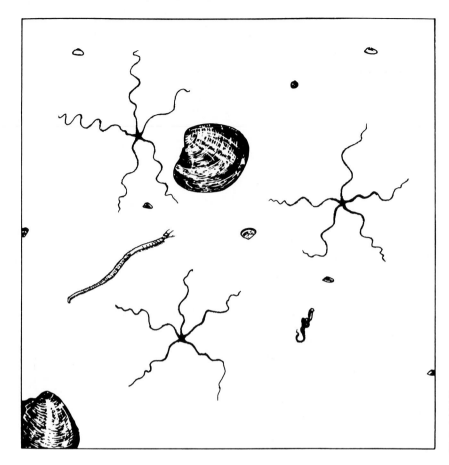

FIGURE 164. Diagram showing .25 m² of *Venus* community. (From J. K. McNulty, R. C. Work, and H. B. Moore, 1962, *Bulletin of Marine Science of the Gulf and Caribbean, 12,* pp. 204–233 Figure 14.)

6 . Those who define community by both taxic composition *and* relative abundance (opinions vary regarding absolute abundance).

7 . Those (Parker, 1974, 1976) who view communities of organisms as so intimately related to the physical environment that a subdivision of the environment without recourse to the organisms is sufficient to define community.

8 . Kauffman and Scott (1976) stress the "holistic" definition by which all taxa, large and small, at every trophic level, infaunal, epifaunal, nektic, pelagic, and planktonic should be included in the community. Scott (1976) and Parker (1976) discuss additional views of how best to define communities from the paleontologist's position.

9 . Finally, those individuals who, after years of experience in the field, "know" what is a community and what is not. They are the farmers in this field, that is, those who "know" when it is time to plough, sow, and reap, and who also know when it will rain. It is easy to dismiss their unquantitative, unscientific approach, which is based on mere experience, but they commonly turn out to know what they are talking about, just as most farmers do.

In light of these divergent views, is it understandable that the paleoecologist is uncertain as to the definition of the term *community*. The late Robert MacArthur (1971), summarized the community definition question well by an appropriate quote from Lewis Carroll: "Humpty Dumpty told Alice, 'When I use a word, it means just what I choose it to mean—neither more nor less' [pp. 189–190]."

Strictly speaking, the paleoecologist has difficulty in recognizing the spectrum of dependence, interdependence, symbiosis, coevolutionary, parasitic, and other relations that are readily apparent to the ecologist and that serve to support the various definitions of community. The paleoecologist is reduced to examining statistical data on relative abundance and presence or absence of taxa in an attempt to infer ecological interaction. At the same time, one must remain aware of the breadth of biological knowledge concerning community structure, which cannot always be inferred from paleontologic material, but which must be allowed for when drawing paleoecological conclusions from a sample. The paleoecologist has no difficulty in seeing hard-part taxic diversity, relative abundance, and even absolute abundance (if a sufficient number of graduate students, or children are available).

In the last decade, paleontologists have begun to study communities in a routine manner. Almost every issue of a paleontological journal contains at least one paper which describes a fossil community (see Boucot, 1975, 1978 for typical examples). Possibly the earliest reference to marine benthic communities in the fossil record is provided by Weller (1899).

In order to work out the relationships existing between the various local assemblages or local societies of organisms which were living during Kinderhook time in the present Mississippi Valley, it is our purpose to make a careful study of as many separate ones of these

fossil societies as can be secured. In each of these studies the fossils from a single horizon at a single locality will be discussed, that is, those organisms which we know actually lived together and formed a social community [p. 9].

Weller is here employing the phytosociological nomenclature developed near the end of the last century in northern Europe. Earlier paleontologists had commented on such things as oyster banks and coral reef associations, but never, as far as I can determine, with the community concept of the ecologist in mind.

THE PROBLEMS OF RECOGNIZING COMMUNITIES WHEN THERE IS TAXIC OVERLAP

Once it is understood that each taxon exists within a hyperspace region of environmental variables that correlate with changing abundances (from maximum to failure) that may, in principle, be contoured, it is obvious to all who have made observations in the real world that defining and recognizing communities in-

volves the question of overlapping taxic ranges involved with both the presence or absence of taxa and of their relative (as well as absolute) abundances (Figures 165 and 166). It is clear that there is an infinite number of communities that are potentially definable. Therefore, one defines communities for paleoecological purposes

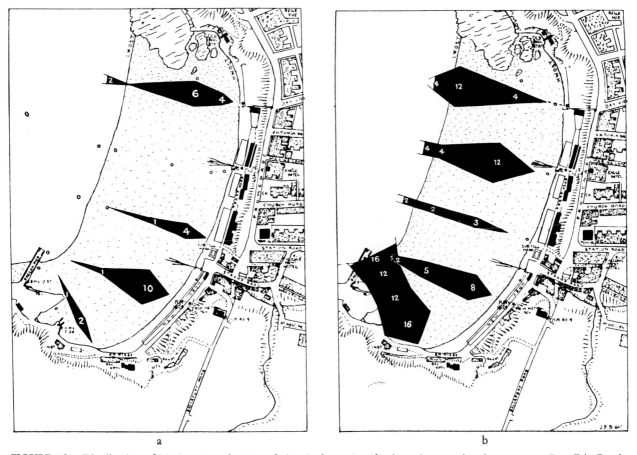

a   b

FIGURE 165. Distribution of *Nerine cirratulus* (*a*) and *Arenicola marina* (*b*) along the same beach transects at Port Erin Beach, Scotland. (From M. E. Pirrie, J. R. Bruce, and H. B. Moore, 1933, *Jour. Mar. Biol. Assoc. U.K., 18*, pp. 279–296, Figures 5–6.)

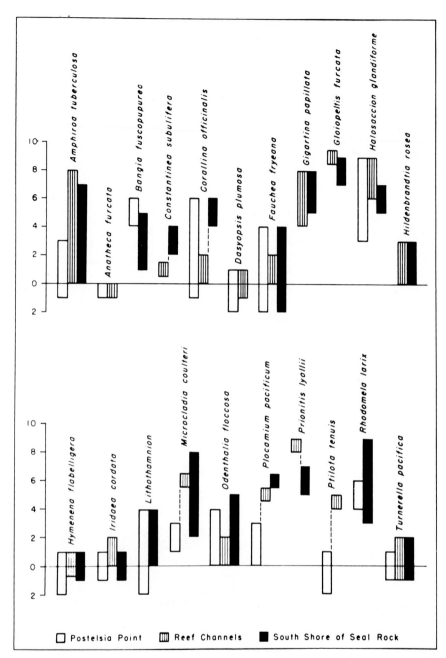

FIGURE 166. Diagram showing vertical distribution (in feet above or below the zero tide datum) of the principal red algae at three localities in the vicinity of Neah Bay, Washington. (From G. B. Rigg and R. C. Miller, 1949, *Proc. Calif. Acad. Sci.*, *4th ser.*, *26*, pp. 323–351, Figure 6.)

as recurring associations and relative taxic abundances that proved to be useful to the researcher. The basic questions concerning degrees of dependence, interdependence, and independence of concerned taxa must be dealt with separately by the paleoecologist.

Since we have defined community as a regularly recurring taxic association with certain taxic abundance levels, and since we have made an implicit assumption that physical factors in the community environment have remained approximately constant, we must contend with the fact that physical factors never remain precisely fixed for any length of time. Must there be a

different community for every hour, day, or week? Common sense and experience here tell us that hard winters kill most temperature-sensitive members of an intertidal community (see Kristensen, 1957, for a typical account); should we then appeal to a successional concept in which a different set of communities occurs before and after each killing-cold event? Or should we ignore the killing event if a minimum number of taxa are involved so that the community is not notably altered? Clearly there is no answer that will fit all situations.

Rigg and Miller (1949) present data for intertidal

algae (Figure 166) occurring under a variety of physical conditions; a good indication that communities cannot be defined too strictly as the same taxa reoccur in a variety of taxic associations under varying physical conditions. Had Rigg and Miller also included abundance data, their results would have further emphasized this point.

The literature is replete with accounts of short-term biotic changes at any one point, usually interpreted as due to a number of changing factors (see Ursin, 1952, for a typical example). There is real doubt that such short term fluctuations must be defined as the development and extinction of individual communities. Rigid statistical definitions lead to a state of "not seeing the forest for the trees." Mere data reduction, even with sophisticated techniques, is not the most intelligent path to an understanding of communities or to the behavior of natural phenomena. Mere data reduction is no substitute for cerebration.

Ekdale (1974a,b) discovered that dredge samples of both living and dead species commonly yield a larger number of dead than of living species. Although the dead species were commonly represented somewhere in the area they did outnumber the living species in specific samples. He interpreted that finding as the effect of a constantly changing environment in which the dead shells summed a number of ecologically different, but transitory, patches in order to yield a higher total number of species. If his finding, obtained from a subtropical area of biogenic sedimentation off the coast of Yucatan, is typical, the possibility that most fossil assemblages will consist of a summed shelly residue obtained from more than a single community must be considered. This question requires more careful investigation with modern samples to see if Ekdale's results are the rule in most marine environments rather than the exception. Chapter 4, pp. 335–337, gives an example of a similar type in the fossil record. Whether or not this phenomenon of summing a number of "patches," rapidly changing environments, and communities, into what amounts to one bed of shells ultimately sampled by the paleontologist, is the rule or the exception must be considered very carefully.

_ PERCENTAGE OF ORGANISMS WITH PRESERVABLE HARD PARTS _

Critical to the definition of fossil communities is whether or not the relation between shelly organisms capable of being preserved in the fossil record is close enough to that of the soft-bodied organisms incapable of being commonly preserved in the fossil record to yield consistent results. Data provided by Lawrence (1968), Stanton (1974, 1975), and Schopf (1978) suggest that although the actual ratio of shelly organisms to soft-bodied organisms may differ from community to community, there is no reason to suspect that absence of soft-bodied organisms from the fossil record invalidates the use of shelly organisms for purposes of community definition.

Olson (1976) summarized information concerning the percentage of organisms with preservable hard parts (including division of those with easily-preservable hard parts such as clams and snails, from the less easily preservable starfish and ophiuroids) plus those capable of leaving preservable trace fossils. Table 4.1 is a summary of his conclusions. His data were derived from a transect across the Oregon continental shelf and slope. Note that the percentages of preservable hard parts are in the 25–50% range. His data are in agreement with those of Craig and Jones (1966), Johnson (1965) and Warme (1969). It is obvious that benthic shelly preservable percentages in the 25–50% region are large enough to permit significant conclusions about the nature of communities, as well as the recognition of communities, although the missing soft-bodied percentage is high enough to necessitate caution in all instances.

Simpson's (1960) data for fishes (Table 4.2) shows the discrepancy among differing environmental factors involved in preservation of one group.

Craig and Jones' (1966) observation that the percentage of preservable organisms in at least one locality is significantly different for infaunal than for epifaunal organisms is a further complication.

Stanton (1976) and MacDonald (1976) present data evidencing the disparity between soft-bodied and shelly organisms present in modern environments. Their data, emphasizes the fact that attempts to reconstruct trophic relations on a quantitative basis have little hope of success. However, their data also clarify the definition of communities as a realistic problem in terms of shelly material when dealing with the fossil record.

For benthic ecologists interested in macrofauna, the almost total absence of annelids (except for the jaws of nereids [scolecodonts]) from the fossil record must be very disconcerting. The soft bodies of most annelids are, of course, responsible. But papers such as Hayward's (1977) and Pickerill and Forbes's (1978) should provide the benthic ecologist with confidence that annelids have been with us in the past as they are today.

DECOMPOSERS

Decomposition of the type important to the benthic marine realm may be divided into a consideration of soft parts, chitinous materials, calcareous materials

TABLE 4.1
Percentages of Benthic Species Capable of Leaving Body-Fossil and Trace-Fossil Records Behind—from the Oregon Continental Shelf [a,b]

Body fossils			Trace fossils		
Preservable	Maybe	Not preservable	Preservable	Maybe	Not preservable

Continental Shelf (0–200 m) / **Continental Shelf (0–200m)**

Body fossils			Trace fossils		
Echin. 6 ⎫	Aster. 22	Hydra. 20	Nemer. 1 ⎫ 5%	Scaph. 4	Hydra. 20
Brach. 2	Ophi. 6	Antho. 8	Polyc. 8 ⎭	Pele. 16	Antho. 8
Cirr. 2	Crabs 12	Nemer. 1		Holo. 2	Cirr. 2
Amph. 4 ⎬ 35%		Ceph. 2		Echin. 1	Amph. 4
Scaph. 4		Bryo. 7			Ceph. 2
Pele. 16		Holo. 3			Gastr. 26
Gastr. 25 ⎭		Shrimp 21			Bryoz. 7
		Gastr. 1			Brach. 2
Preservable and Maybe combined = 58%		Tunic. 1	Preservable and Maybe combined = 19%		Aster. 22
		Polyc. 8			Ophi. 6
					Echin. 5
					Holo. 1
					Shrimp 21
					Crabs 12
					Tunic. 1

Continental Slope (200–2800m) / **Continental Shelf (200–2800m)**

Body fossils			Trace fossils		
Cirr. 6 ⎫	Aster. 54	Hydra. 8	Nemer. 2 ⎫	Scaph. 7	Hydra. 8
Amph. 3	Ophi. 19	Antho. 13	Echiur. 2 ⎬ 19%	Pele. 9	Antho. 13
Scaph. 7	Crabs 15	Nemer. 2	Pogon. 3	Holo. 6	Cirr. 6
Pele. 9 ⎬ 23%	Crinoid 4	Ceph. 3	Polyc. 44 ⎭	Echin. 4	Amph. 3
Gastr. 27		Pogon. 3			Ceph. 3
Brach. 4		Echin. 2			Gastr. 29
Echin. 6 ⎭		Holo. 21			Brach. 4
		Shrimp 15			Aster. 54
		Gastr. 2			Holo. 15
		Tunic. 1			Shrimp 15
Preservable and Maybe combined = 57%		Polyc. 44	Preservable and Maybe combined = 29%		Crabs 15
		Echiur. 2			Crinoid 4
					Tunic. 1

Abyssal Plain (2800 + m) / **Abyssal Plain (2800 + m)**

Body fossils			Trace fossils		
Cirr. 6 ⎫	Aster. 12	Antho. 4	0%	Scaph. 3	Antho. 4
Scaph. 3	Ophi. 5	Shrimp 2		Pele. 1	Cirr. 6
Pele. 1 ⎬ 27%		Holo. 17		Holo. 5	Gastr. 1
Gastr. 1				Echin. 2	Brach. 1
Brach. 1					Aster. 12
Echin. 3 ⎭					Ophi. 1
Preservable and Maybe combined = 58%			Preservable and Maybe combined = 20%		Echin. 1
					Holo. 12
					Shrimp 2

Depth zonation	Total number of species	Percent species preservable as body fossils	Percent species leaving preservable trace fossils
Continental shelf	170	35–58	5–19
Continental slope	270	23–57	19–29
Abyssal plain	55	27–58	0–20

[a] From G. A. Olson, Geology 426 term paper, Oregon State University, 1976.

[b] Maybe, preservable only under optimum conditions; Amph., Amphipods; Antho., Anthozoans; Aster., Asteroids; Brach., Brachiopods; Bryoz., Bryozoans; Ceph., Cephalopods; Cirr., Cirripedes; Crabs, Crabs; Crinoid, Crinoids; Echin., Echinoids; Echiu., Echiuroids; Gastr., Gastropods; Holo., Holothuroids; Nemer., Nemertines; Ophi., Ophiuroids; Pogon., Pogonophora; Polyc., Polychaetes; Pele., Pelecypods; Scaph., Scaphopods; Shrimp, Shrimps; Tunic., Tunicates.

TABLE 4.2
Percentage of Genera of Recent and Known Fossil Teleosts Found in Various Environments[a]

Environment	Percentage recent genera		Percentage known fossil genera
Freshwater			
Still	2		
Both still and running	14	16	22[b]
Running only	14		.2[c]
Total	30		22
Marine			
Shallow	57		70
Deep	13		8
Total	70		78

[a] Reprinted from G. G. Simpson, *Evolution after Darwin, I. The Evolution of Life,* S. Tax, Ed., by permission of The University of Chicago Press. Copyright © 1960, Table 2.

[b] Unpublished data, rearranged, from Bob Schaeffer. Note the differences in representation between the fossil and the Recent situations. Found in lake deposits, but this presumably includes a representative proportion of genera that occurred in both lakes and streams.

[c] Genera found in streams but not in lake deposits.

(aragonite, calcite), noncalcareous plant materials, phosphatic materials (hydroxy–apatite, fluorapatite), and siliceous materials (opal). Other minerals and amorphous compounds are known from a variety of marine organisms (Lowenstam, 1973), but their general rarity shows that they are of limited importance to the fossil record.

Soft parts of animals and of plants, both marine and nonmarine, are quickly decomposed by a variety of bacteria, fungi, scavengers, and various predators. The unusual anaerobic depositional environments, such as the Burgess Shale of the Cambrian and the Bundenbacher Schiefer of the Devonian, with their exquisite preservation of unusual structures, serve to underline the rarity of this type of environment in the fossil record. The preservation of soft parts, although very unusual, reassures us that communities of the past were as rich in soft-bodied organisms as are modern ones.

The presence in the sea of chitin-destroying bacteria and fungi (Sindermann, 1970, pp. 149 and 160) presents another problem. Modern crab carapaces may be softened and ultimately destroyed by such bacteria either during life or shortly after death. The rarity of most fossil crustaceans may be due to the activities of such decomposing bacteria. Cobb (personal communication, 1979) mentions that eating of the cast exoskeletons is a common trait of many crustaceans, which could help to account for their overall rarity in the fossil record. However, some chitinous materials, such as the exoskeleton of the graptolites and the chitinozoa, are unusually resistant to degradation in

the normal marine environment. The rarity of eurypterid and limuloid materials in the fossil record suggests that they too may have easily decomposed exoskeletons. Probably, there are many varied types of chitinous materials possessing differing resistance to decomposing agencies. Until bulk macerations of marine materials are performed on a routine basis, including treatment with HF to eliminate siliceous materials, we will have a limited understanding of the preservation potential of chitin.

Nicol (1977) presented another view of the soft-bodied to hard-part problem by counting up and discussing the number of higher taxa having little likelihood of fossilization due to absence of hard parts and the number of living species of groups with hard parts having a reasonable chance of preservation as fossils (Table 4.3). Note that the percentage of marine organisms capable of preservation is reasonably high; high enough to constitute a sample if used with caution.

Farrow and Clokie (1979) have skillfully shown how boring algae within the modern photic zone first make a series of dense shell perforations; next algal-scraping chitons and limpets scrape off algae and shell material: the debris is a few tens of microns deep. Voigt (1977) illustrates some of the trace fossils left behind by such scraping. Runnegar et al. (1979) indicate that the fossil record of the chitons probably extends far back into the Early Cambrian. In addition to the calcareous decomposing potential of this boring algal–algal-scraping relationship, it should be realized that the relationship

TABLE 4.3
Animal Phyla Whose Species Are Soft-Bodied or Soft-Bodied with Hard Parts[a]

Phyla	No. of species	Phyla	No. of species
Soft-bodied			
1. Nematoda	50,000	15. Pogonophora	100
2. Platyhelminthes	13,000	16. Kinorhyncha	100
3. Tunicata	1,600	17. Onychophora	75
4. Rotifera	1,500	18. Chaetognatha	75
5. Rhynchocoela	750	19. Pentastomida	70
6. Acanthocephala	600	20. Calyssozoa	60
7. Gastrotricha	350	21. Mesozoa	50
8. Tardigrada	350	22. Cephalochordata	30
9. Sipunculoidea	330	23. Pterobranchia	25
10. Gordiacea	230	24. Phoronida	25
11. Echiuroidea	150	25. Priapuloidea	8
12. Myzostomida	150	26. Monoblastozoa	1
13. Ctenophora	100	Total	69,829
14. Enteropneusta	100		

Phyla	Species unlikely to be fossilized	Species likely to be fossilized	Total
Soft- or hard-bodied			
1. Arthropoda	983,000	17,000	1,000,000
2. Mollusca	7,000	51,000	58,000
3. Vertebrata	30,200	11,500	41,700
4. Protozoa	26,000	9,000	35,000
5. Coelenterata	5,500	4,500	10,000
6. Annelida	9,000	300	9,300
7. Echinodermata	4,300	1,700	6,000
8. Porifera	3,500	1,500	5,000
9. Bryozoa	1,000	3,000	4,000
10. Brachiopoda	0	300	300
Total	1,069,500	99,800	1,169,300
Totals of all species	69,829		1,239,129

[a]From D. Nicol, 1977 *Florida Scientist 40, 2,* pp. 135–140, Tables 1 and 2.

also provides tremendous potential for the generation of large volumes of fine-grained calcium carbonate, micritic debris.

Calcareous materials are normally quite resistant to decomposition except by boring organisms, which include varied fungi, algae, and bacteria.

Noncalcareous plant materials are easily decomposed by a variety of fungal, bacterial, and algal organisms in the sea.

Phosphatic materials, such as fish bones, are susceptible to decomposition (Brongersma–Sanders, 1949), but certain other phosphatic materials, such as shells of inarticulate brachiopods, are highly resistant to decomposition. Antia (1979, p. 112) suggests that fish bone in aerobic, that is, nonlaminated, environments may be in large part destroyed by the scavenging activities of marine invertebrates, including regular echinoids.

Ayyakkannu and Chandramohan (1971) provide information about the occurrences of phosphate-destroying bacteria and phosphatase in marine sediments; their data provide another possible explanation for the relative rarity of fish bones in the aerobic portions of the marine environment.

The possibility that varied modern decomposers, especially among the bacteria, were similarly active during the past, helps to explain some paleontological anomalies such as the overall rarity of fish and crustacean remains. Rare instances are known in which scavenging organisms have been preserved while still at work. From the Brazilian Cretaceous (Bate, 1972) well-

Ostracods

FIGURE 167. Note the impressions of the numerous ostracods associated with the large eurypterid prosoma (*Baltoeurypterus tetragonophthalmus* or *Eurypterus fischeri*) at the base of the photograph. Henningsmoen, written communication, 1978, indicates that most of these belong to *Beyrichia,* different instars, and that a few nonbeyrichiid paleocopes are present as well. Størmer, written communication, 1978, concludes that "In my opinion they were either scavengers or they lived in empty exuviae." The specimen Paleontologisk Museum, Oslo PMO Cat. No. 70696 comes from Stage 9b (about Middle Wenlock) Levretoppen 20, Gjettum, Baerum in the middle part of the Oslo–Asker district of southern Norway. The photograph is courtesy of Professor Størmer. Professor Henningsmoen comments additionally, "The preservation of the ostracodes is not good, but it seems that most or all of them are single valves. Some are certainly single valves. Some are oriented up, some down, but I do not know the original orientation of the slab, which I found loose."

preserved ostracodes have been found in close association with a fish that they were probably scavenging when they were entombed. Certain groups of modern ostracodes demonstrate similar behavior (also see Figure 167 for a Silurian example).

Størmer (1963) described organisms thought to have been active in decomposing a Carboniferous scorpion—included are a species of nematode as well as possible bacteria and fungi. Decomposers are not only a modern phenomenon.

BREADTH OF THE PALEONTOLOGIST'S COMMUNITY VERSUS THAT OF THE BIOLOGIST

Throughout this book it has been emphasized that the paleontologist does not deal with pristine community data available in principle to the ecologist. Is the disparity great enough to invalidate most attempts at community analysis? If not, to what extent are conclusions based on fossils alone useful? The routine absence of organisms lacking soft parts is obvious. Less clear to the novice may be the common situation in which most, if not all, microfossils remain unstudied, and fossil groups whose taxonomy is poorly understood remain commonly untreated; it is also common for groups requiring more difficult preparation techniques (e.g., thin sectioning) to remain unstudied. The biologist seldom, if ever, studies all of the microfauna, meiofauna, and macrofauna when performing community analysis. We must admit that the "worms" and echinoderms—including ophiuroids, echinoids, asteroids, and holothurians—plus the soft-bodied algae that are commonly studied today, remain largely untouched by the paleontologist. Is this a serious problem when trying to compare and contrast results obtained from modern and ancient environments? It is entirely reasonable to suspect that the community of the paleontologist may be broader than that of the biologist, covering more than one community in the latter's terms, but it is unlikely that there will be a difference in order of magnitude. We are talking about a paleontologist's community equivalent to two, three or possibly even four units recognized by the biologist—chaos does not reign.

There is also the problem of "condensation" of communities which was treated recently by Fürsich (1978). He points out that successional stages may be mixed by subsequent bioturbation, which results in misleading conclusions. His deductions parallel the observations of Warme et al. (1976), who pointed out that the dead fauna characteristic of a limited region commonly had a much higher diversity than the many small patches found in a living condition, despite the fact that the total number of species in the small patches was as great as that present in the dead fauna. Warme et al. did not suggest that each patch be considered as a separate community, but that the day-to-day dynamics operating in a small area might result in more species variability than would be encountered if the contents of the patch were summed over a longer period of time, thus smoothing out the many lesser perturbations caused by disease, predation, parasitism, and physical disturbances of various sorts.

The previous paragraphs make it clear that the community of the paleontologist is a broader, more time-averaged one than that of the biologist. But it is also clear that the community of the paleontologist is not merely a homogeneous mixture of shells derived from many communities scattered over a large region. The resolution achievable with fossils is less than that attained with living organisms, but not by a significant order of magnitude.

RELATIVE AND ABSOLUTE ABUNDANCE THROUGH TIME AT ANY ONE LOCALITY

Critical to the definition of communities as regularly recurring associations of taxa characterized by relatively fixed abundances of involved taxa is the determination as to whether such abstractions as "community" have biological reality. In the level-bottom environment, many data indicate great fluctuation in numbers of individuals belonging to any taxon, as well as some fluctuation of presence or absence of a given taxon, at any one locality through time, both seasonally (see Chapter 4, pp. 242–243 and Chapter 3, pp. 170–174) and nonseasonally (Tables 4.4–4.7; Figures 168–176). Therefore, rigidity about the precise limits used in determining abundance and taxic composition of any community is impractical. The alternative to laxity in definition is the provision of a differing community in time for every slight fluctuation of abundance or taxic presence, which may result in a separate community for each set of observations through time. How would the concept of a distinct community at each locale for each sequential set of observations help in understanding biological processes? Muus (1973) reviews some of the irregularities in larval abundance and settling for an area in Denmark; many workers have noted seasonal changes in taxic abundance as well as presence or absence (think only of the annual kelps as an example); and many have provided data on the nonsynchronized abundance relations observed with samples taken through the same and different years (Figures 174–176). Some taxic abundances, as well as presence or absence, are synchronized with one another, but a significant number are not. Even Petersen's (1914) data clarifies these relations.

Any understanding of community definition must cope with fluctuations of population size through time, particularly the nonseasonal, nonsynchronized fluctuations in population size of many species (see Chapter 4, pp. 317–320). Coe's (1956) summary is appropriate:

> Fluctuations in populations of marine invertebrates indicate that annual changes of considerable magnitude are of general occurrence. Some of these are evidently due to concurrent changes in environmental conditions (including chemical and physical alterations of the water), to enemies, to disease, and to changes in substrate and in ocean currents [also see Chapter 4, pp. 335–337]. Others are due to migration or introduction to new localities.
>
> Records of local populations of dinoflagellates, starfish, oysters, clams and mussels over long periods show that maximum and minimum populations have sometimes followed each other immediately, but that

populations of medium size are more numerous than those with either extreme.

Maximum and minimum populations generally occur at irregular and unpredictable intervals with no satisfactory indication of long term periodic or rhythmical cycles; it is concluded that the supposed periodicities are the result of random variations in the complex environmental conditions, including diseases.

The records show that the reproductive capacity of many marine invertebrates is such that relatively few individuals under favorable conditions can produce a maximum population in a single generation. Overpopulation has been frequent and has often been followed by a rapid decline in numbers, since the competition for space or food was too severe for survival.

Resurgent populations have sometimes resulted from the arrival of vast numbers of pelagic larvae transported from localities several miles distant by ocean currents. Except in bays and partially confined bodies of water, there is little chance that a large proportion of the larvae will settle in the immediate vicinity of their place of origin.

Introduction of a species to a new locality has sometimes resulted temporarily in a high rate of multiplication and territorial expansion, usually followed by a rapid decline in numbers. More frequently the introduced species has found the new locality un-

suitable for reproduction and has been rapidly eliminated. Catastrophic changes in the populations have been observed occasionally or at irregular intervals. These have resulted from excessive heat or cold, heavy rains and floods, heavy surf, adverse ocean currents, loss of associated vegetation, industrial wastes, excessive abundance of dinoflagellates, or epidemics of parasitic diseases, with the destruction of many or nearly all individuals of susceptible species over large areas. [p. 230]

The thrust of Coe's conclusions is that levels of dependence, interdependence, and independence, are variable from species to species. Thus, "perfect" definitions of particular communities should be constructed with care where abundances are involved.

The greatly varying abundances of cooccurring marine larvae (Muus, 1973) stresses the variability of associated taxa at different life stages.

Figures 174–176 provide an idea of the nature of these nonseasonal fluctuations. Keep in mind how such fluctuations will affect the diversity of a community from year to year, and how they will make a rigid community definition more confusing than useful. This is an area where experience and common sense may mean more than pure statistics.

TABLE 4.4
Nonsynchronous Changes in Faunal Parameters with Time[a]

Faunal parameters	Station 1	Station 2	Station 3	Station 4
Number of species				
Mean (1963 = 1964)	53.6	32.1	58.6	41.3
95% confidence limits	5.4	2.5	5.0	6.5
Mean (1967, 1969)	49.7	34.0	55.3	38.5
Range (1967, 1969)	44–53	30–37	51–64	29–48
Numbers of specimens (N)				
Mean (1963–1964)	2555.7	661.7	920.4	1994.4
Mean (1967, 1969)	2032.7	724.0	955.7	1822.3
Range (1967, 1969)	1197–3223	389–1122	544–1714	870–342
N (August–September)/N (January–February)	1.85	.76	1.95	1.90
Diversity, H				
Mean (1963–1964)	2.5	3.1	4.4	3.3
95% confidence limits	.4	.7	.1	.5
Mean (1967, 1969)	2.2	3.3	4.4	3.1
Range (1967, 1969)	1.6–2.8	2.1–4.1	4.1–4.7	3.0
Percentage dominance of first 4 species				
(1963–1964)	71.4	70.9	47.7	60.4
(1967, 1969)	76.5	60.0	42.8	71.8
Percentage concordance among rankings				
All dates	49.6	50.4	38.5	41.2
(Jan.–Feb.)	54.7	64.8	37.8	66.1
(Aug.–Sept.)	84.8	50.0	70.3	61.2
Standing crop (g/m²)				
Aug. 1964	7.56	18.86	9.81	9.65
Aug. 1967	4.04	16.50	5.72	2.91
Jan. 1969	6.26	18.44	8.00	13.68

[a]From E. Lie and R. A. Evans, 1973, *Marine Biology*, Table 1. Copyright © 1973 by Springer-Verlag New York, Inc.

TABLE 4.5
Seasonal Changes of Faunal Abundances[a]

Species	Jan. VI	Mar. VII	Mar. I	April VIII	May II	June IX	July III	Sept. IV	Sept. X	Oct. XI	Nov. V	Dec. XII
									Absolute numbers of individuals/400 cm² quadrat			
Lasaca cistula	7600	5500	6400	6600	11,100	6800	3700	4700	7600	5600	7400	1100
Littorina scutulata	337	113	188	326	541	95	1053	1149	482	772	444	1126
Balanus glandula	138	320	372	94	150	35	116	519	1619	1232	38	1043
Musculus sp.	7	6	876	6	36	53	319	0	2382	34	7	15
Littorina planaxis	0	545	10	8	33	1	12	106	94	808	7	0
Mytilus californianus	55	220	521	38	495	44	0	4	46	32	55	78
Dynamenella glabra	262	10	129	41	122	95	194	1	290	12	229	117
Chthamalus spp.	45	77	706	5	14	13	8	32	163	18	11	248
Syllis armillaris	123	1	21	66	140	54	127	25	22	24	78	2
Diaulota densissima	7	93	21	20	44	31	32	45	89	62	31	37
Aganopsis sp.	76	5	71	63	14	13	13	1	160	49	7	11
Syllis spenceri	43	82	58	24	41	35	39	1	17	39	53	37
Acmaea scabra	11	28	59	18	27	10	45	16	10	16	7	43
Acmaea digitalis	11	39	18	32	43	2	36	11	8	31	6	5
Mesostigmatid mites	15	29	0	51	24	3	3	2	1	52	37	3
Nemertopis gracilis	18	8	48	3	55	2	42	1	8	2	3	8
Tipulidae (larvae)	13	15	3	31	1	20	1	14	40	7	25	9
Hyale sp.	18	46	1	0	0	1	2	31	15	15	0	44
Oligochaete	1	0	7	19	31	1	5	68	5	0	3	0
Syllis vittata	14	1	29	13	1	19	2	2	7	0	22	0
Acmaea pelta	3	6	21	15	22	1	21	2	0	3	3	11
Allorchestes ptilocerus	1	0	21	0	0	15	0	0	7	0	18	2
Emplectonema gracile	0	0	2	5	12	1	10	1	3	2	6	13
Cyanoplax dentiens	0	0	5	3	0	2	12	0	1	1	3	0
Perinereis monterea	4	0	6	2	3	1	3	0	1	0	3	0
Pachygrapsus crassipes (juv.)	8	0	2	3	2	2	1	0	0	0	0	0
Nereis grubei	2	0	3	0	0	2	2	1	5	0	2	0
Notoplana acticola	0	0	1	0	0	1	0	0	1	0	3	0

[a]From P. W. Glynn, 1965, *Beaufortia, 12*, pp. 1–198, Table XXIII. Note that the changes do not appear to be necessarily cyclic.

TABLE 4.6
Nonsynchronous Changes in Faunal Abundances with Time[a]

Fauna	February		Fauna	May		Fauna	August	
	f	x̄		f	x̄		f	x̄
Elizabeth River stations (E6, E7, E8)								
Mya arenaria	83	112	Mya arenaria	78	210	Spiochaetopterus oculatus	78	19
Nereis succinea	100	35	Streblospio benedicti	100	71	Nereis succinea	78	7
Mulinia lateralis	100	57	Heteromastus filiformis	100	38	Heteromastus filiformis	56	6
Heteromastus filiformis	100	43	Nereis succinea	100	28	Paraprionospio pinnata	56	5
Spiochaetopterus oculatus	100	34	Spiochaetopterus oculatus	67	4	Sabella microphthalma	33	4
Sabella microphthalma	100	50	Sabella microphthalma	33	28	Molgula manhattensis	22	5
Paraprionospio pinnata	67	30	Mulinia lateralis	67	8			
Mud stations (B2, B4)								
Paraprionospio pinnata	100	66	Mulinia lateralis	100	242	Paraprionospio pinnata	100	44
Retusa canaliculata	100	23	Paraprionospio pinnata	100	80	Retusa canaliculata	100	35
Mulinia lateralis	50	697	Spiochaetopterus oculatus	83	40	Spiochaetopterus oculatus	100	25
Spiochaetopterus oculatus	100	15	Retusa canaliculata	100	25	Phoronis architecta	83	19
Balanus improvisus	50	9	Balanus improvisus	50	5	Ogyrides limicola	83	9
Ogyrides limicola	50	2	Ensis directus	83	8	Ampelisca abdita	83	3
Muddy-sand stations (A2, D4, D5)								
Spiochaetopterus oculatus	100	42	Spiochaetopterus oculatus	89	38	Spiochaetopterus oculatus	100	45
Paraprionospio pinnata	100	29	Paraprionospio pinnata	100	43	Phoronis architecta	78	32
Phoronis architecta	67	34	Phoronis architecta	67	40	Ampelisca vadorum	100	21
Pseudeurythoe pauci-branchiata	89	30	Pseudeurythoe pauci-branchiata	89	21	Paraprionospio pinnata	89	10
Ogyrides limicola	78	6	Retusa canaliculata	78	14	Pectinaria gouldii	78	7
Nereis succinea	56	6	Unciola irrorata	89	98	Pseudeurythoe paucibranchiata	62	5
			Polydora ligni	67	22	Nereis succinea	62	10
Sand stations (D1, D2, D3, F1, F2, F3, F4, F5)								
Retusa canaliculata	96	84	Heteromastus filiformis	96	79	Spiophanes bombyx	100	191
Polydora ligni	79	54	Polydora ligni	87	29	Phoronis architecta	87	62
Unciola irrorata	63	50	Unciola irrorata	79	49	Ampelisca verrilli	79	35
Ampelisca vadorum	71	24	Ensis directus	92	31	Heteromastus filiformis	96	36
Ensis directus	50	9	Retusa canaliculata	87	22	Spiochaetopterus oculatus	83	41
Ampelisca vadorum	54	18	Ampelisca vadorum	75	26	Ampelisca vadorum	100	69
Nephtys magellanica	75	13	Nephtys magellanica	63	15	Glycera dibranchiata	100	28

[a] From D. F. Boesch, 1973, *Marine Biology, 21*, pp. 226–244, Table 5. Copyright © 1973 by Springer–Verlag, New York.

TABLE 4.7
Nonsynchronous Changes in Local Dominance of Principal Organisms on Monthly Collectors[a]

Month of exposure[b]	1943	1945
January–February		Encrusting bryozoans Barnacles Bugulae Tunicates
February	Barnacles	Barnacles Bugulae
March	Barnacles Tunicates Bugulae	Barnacles Tunicates Encrusting bryozoans
April	Barnacles Tunicates	
June–July	Tunicates Barnacles	Barnacles Tunicates
July–August	Barnacles Tunicates	Barnacles Tunicates
August–September	Barnacles Tunicates	
September		Barnacles Tunicates
October		Barnacles
November		Barnacles Bugulae Encrusting bryozoans
November–December	Barnacles Tunicates Bugulae	
December	Barnacles Tunicates Encrusting bryozoans	Barnacles Bugulae
December–January	Barnacles Tunicates Encrusting bryozoans	

[a]From C. M. Weiss, 1948, *Ecology 29,* pp. 153–172, Table III.
[b]Notation of 1 month, e.g. January indicates exposure of January 1–February 1. Notation of 2 months, e.g. January–February, indicates exposure of January 15–February 15.

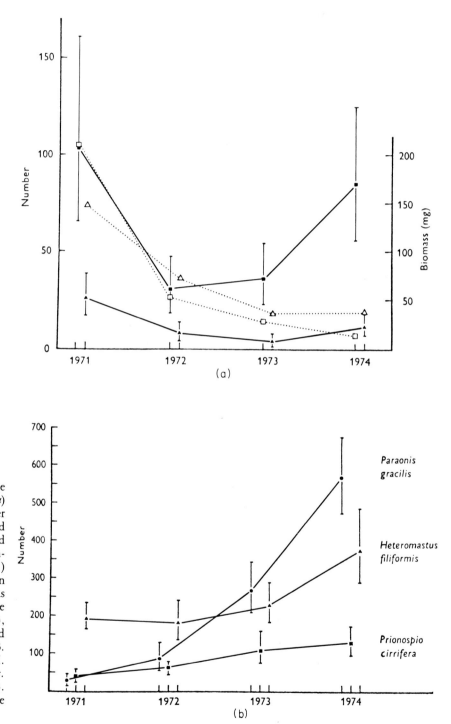

FIGURE 168. Nonsynchronous change in faunal abundances with time. (*a*) Variation in numbers per square meter (solid symbols), expressed as derived means and 95% confidence limits, and biomass per square meter (open symbols) for *Ammotrypane aulogaster* (□) and *Abra nitida* (△). (*b*) Variation in numbers per square meter expressed as derived means and 95% confidence limits for *Paraonis gracilis* (●), *Heteromastus filiformis* (▲) and *Prionospio cirrifera* (■). (From J. B. Buchanan, P. F. Kingston, and M. Sheader, 1974, *Jour. Mar. Biol. Assoc. U. K., 54,* pp. 785–795, Figures 2–3. Copyright © 1974 by Cambridge University Press.)

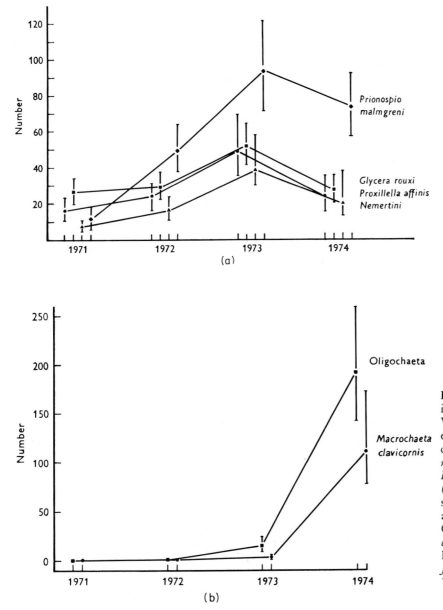

FIGURE 169. Nonsynchronous changes in faunal abundances with time. (*a*) Variation in numbers per square meter expressed as derived means and 95% confidence limits for *Prionospio malmgreni* (▲), *Glycera rouxi* (■), *Praxillella affinis* (●), and *Nemertini* (▲). (*b*) Variation in numbers per square meter expressed as derived means and 95% confidence limits of Oligochaeta (■) and *Macrochaeta clavicornis* (●). (From J. B. Buchanan, P. F. Kingston, and M. Sheader, 1974, *Jour. Mar. Biol. Assoc. U. K., 54,* pp. 785–795, Figures 4–5. Copyright © 1974 by Cambridge University Press.)

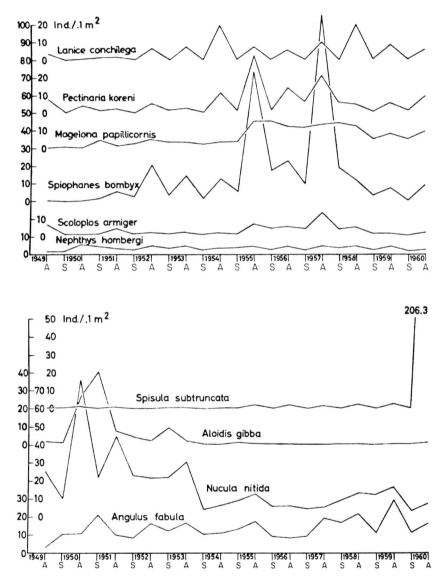

FIGURE 170. Nonsynchronous yearly changes in taxic abundance. S-Spring; A-Autumn. (From E. Ziegelmeier, 1963, *Veroffentlichungen Inst. Meerforsch. Bremerhaven, Meersbiol. Symp., 23,* pp. 101–114, Abb. 6–7.)

196

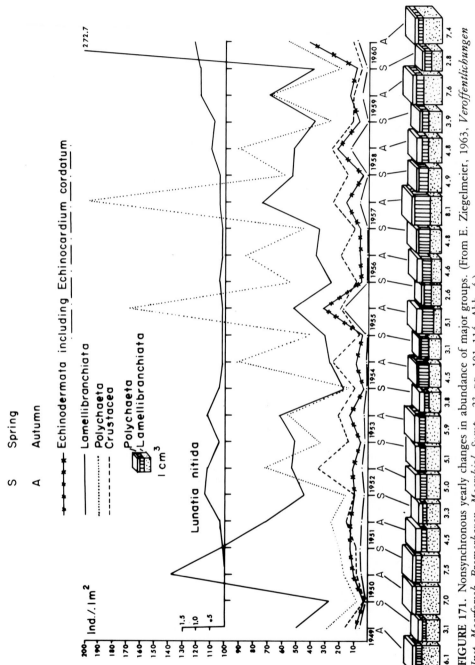

FIGURE 171. Nonsynchronous yearly changes in abundance of major groups. (From E. Ziegelmeier, 1963, *Veroffentlichungen Inst. Meerforsch. Bremerhaven, Meeresbiol. Symp., 23*, pp. 101–114, Abb. 4.)

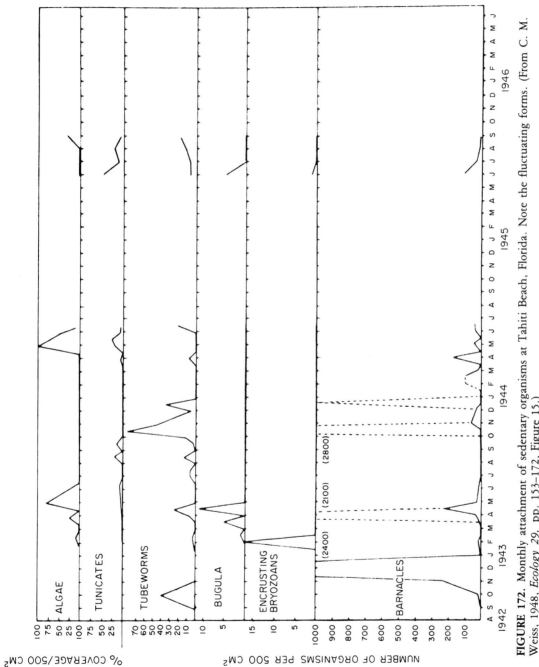

FIGURE 172. Monthly attachment of sedentary organisms at Tahiti Beach, Florida. Note the fluctuating forms. (From C. M. Weiss, 1948, *Ecology 29*, pp. 153–172, Figure 15.)

197

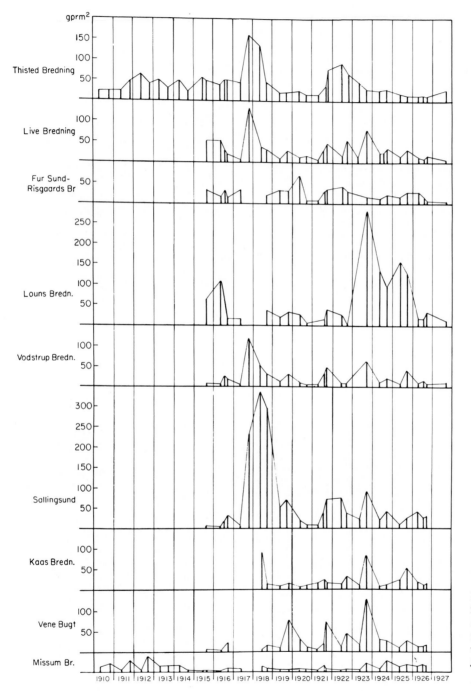

FIGURE 173. Fluctuations in first class plaice-food from a number of localities in the Limfjord, Denmark. (From H. Blegvad, 1928, *Rept. Danish Biol. Station, 34,* pp. 35–52, Figure 1.)

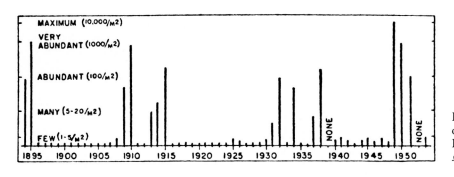

FIGURE 174. Population fluctuations of *Donax gouldi* at La Jolla. (From W. R. Coe, 1956, *Journal of Marine Research, 15*, pp. 212–232, Figure 7.)

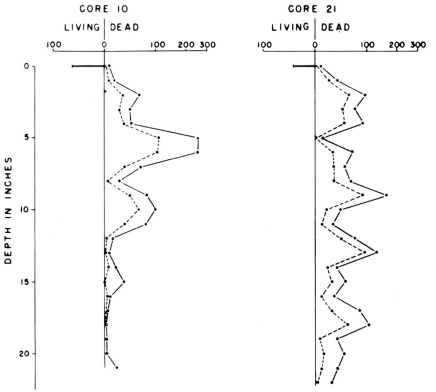

FIGURE 175. Number and vertical distribution of dead and living *Gemma* in two cores and the area immediately above the core site (dotted lines are for articulated valves; solid lines for total number of "paired" valves). (From W. H. Bradley and P. Cooke, 1959, *U.S. Fish & Wildlife Service, Fisheries Bull., 58*, pp. 305–334, Figure 17.)

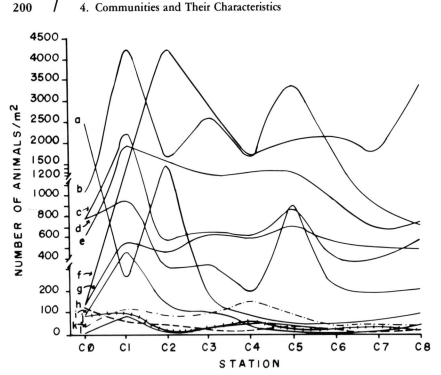

FIGURE 176. Local abundance differences for benthic animals at five stations near Halifax, Nova Scotia (*a*) Oligochaete species B; (*b*) *Hydrobia* sp.; (*c*) *Macoma baltica;* (*d*) *Mya arenaria;* (*e*) oligochaete species A; (*f*) *Pygospio elegans;* (*g*) *Chiridotea caeca;* (*h*) *Scolecolepides viridis;* (*i*) oligochaete species C; (*j*) *Heteromastus filiformis;* (*k*) *Eteone longa;* (1) *Polydora ligni.* (From E. L. Mills, 1969, *Journal of the Fisheries Research Board of Canada, 26,* pp. 1415–1428, Figure 1. Copyright © 1969 by Dept. of Fisheries and the Environment, Scientific Information and Publications Branch.)

COMMUNITY DEFINITION AND SPECIMEN SIZE

Marine ecologists are aware that organisms constituting the bulk of the benthic biomass are represented by relatively few individuals (see Reish, 1959, for data illustrating this point), and those represented by many individuals generally represent a small fraction of the biomass. These comments refer to standing-crop biomass; not to productivity. The experience of the ecologist is duplicated by the paleontologist (see Fürsich's data in Hallam, 1975b) who finds that large shells dominate any fossil collection but are few compared with small ones (the shelly biomass relations are similar to the standing-crop biomass relations of the ecologist). Therefore, when defining communities from the fossil record, concentrate on the less conspicuous smaller shells rather than on the obvious but few large species. This, of course, is a practical consideration since as many of the available fossils as possible should be examined. Thus, brachiopods may tend to be overemphasized in contrast to the small ostracodes. Amsden (1975) has reviewed a situation of this sort and drawn the appropriate conclusions. One cannot carry a microscope into the field; if a species is

reasonably abundant and large, there is no reason to reject it when defining communities because minute taxa are more plentiful as individuals in the shelly biomass. If carried to absurdity this principle would require us to name many communities of planktonic organisms for the palynomorphs available for study following laborious extraction rather than after the readily apparent graptolites, agnostids, or other pelagic organisms.

For the ecologist, it is enough to define communities in terms of living organisms. The paleoecologist must consider the great variety of body fossils, the shells, bones, pollen grains, bits of wood and leaves, how the trace fossils can best be utilized, the tracks, trails, burrows, and other evidences of activity, and must even consider evidence afforded by coprolites, if one is fortunate enough to find and recognize them (see Bronnimann, 1972, for a summary of the usefulness of anomuran crustacean coprolites; for a compilation of evidence regarding coprolites of the past see Hantzschel *et al.,* 1968).

CROPPING, PREDATION, AND COMMUNITY STRUCTURE

Of the various feeding methods (reviewed by Walker and Bambach, 1974) in the marine benthos, predation and cropping are of particular interest because they are important variables in the control of biotic diversity

and abundance in any given area. Kuris (1974) points out that parasitic castrators in the sea, as well as parasitoids on land, have a similar effect in helping to control abundance and community composition (In

Chapter 4, p. 282, comments are made, with references, about how parasitic castrators that affect gastropods might be recognized in the fossil record). For the purpose of discussion, we define predation as carnivorous feeding involving search and prey selection. Cropping is a less selective carnivorous feeding habit in which relatively numerous prey organisms are browsed in a similar manner to that of an herbivore feeding on varied plants. It is entirely reasonable when considering communities of the past to deduce that cropping and predation undoubtedly played a significant role in determining the recorded taxic abundances and distributions.

In the Recent, the effect of cropping and predation are conspicuous in both controlled experiments and field observations.

Much experimental evidence indicates that many taxa do not commonly exist in certain basically favorable environments because of predator and cropper activity. Warren's (1936) elegant experiment in which *Mytilus edulis* was shown to thrive in shallow subtidal waters if protected from starfish predation by wire netting is an excellent example. Sammarco, Levington, and Ogden's (1974) discussion of the Randall Halo, an area adjacent to a coral reef where seagrass is absent because of sea urchin grazing and fishes coming out a short distance from the protection of the reef is another example of such control. Wells's (1961) discussion of brackish-water oysters emphasizes that they thrive because their stenohaline predators are excluded from the environment.

Virnstein (1977) has performed caging experiments showing that the population density of prey species may be diminished by predators, such as crabs and fishes, whereas some cooccurring low trophic-level species (some of the infauna in particular) are not susceptible to predation effects. Predation could result in widely differing relative abundances which might well mislead the paleontologist; again suggesting that excessively narrow community definitions may create

confusion despite their initially attractive statistical "precision."

Bakus (1969) provides an overview of the shallow-marine literature. He points out that the presence of fish with either demersal flatfish morphology or with the appropriate pavement-type crushing teeth indicates that predation on bottom shellfish has existed since at least the Devonian (Figures 188–189). Cloud (1959) offers an excellent summary of the coral "nibbling" proclivities of the tropical parrot fishes, as well as of their importance in generating sediment (Figure 190). The overall ecologic impact in the tropical–subtropical regions of the parrot fishes and similarly behaving types should not be underestimated. The rarity and low abundance of such factors as intertidal barnacles and even various noncalcareous algae in these warm regions may be due to the activity of these fishes (see Newman, 1960; Stephenson and Searles, 1960). Bakus (1964) summarizes the evidence favoring low densities of exposed, rocky-bottom barnacles and of soft algae in the reef environment, with the data favoring cropping by reef fish as the cause. He also summarizes the evidence suggesting that in nonreef tropical environments there may be a higher density of both rocky-bottom type barnacles and of soft algae due to the absence of the right types of cropping fish (see also Stephenson and Searles, 1960).

Figures 177–196 and Tables 4.8–4.15 provide a variety of evidence from the present and from the fossil record illustrating various possible types of predation affecting marine invertebrates.

Predation, by definition, has the capability of seriously disturbing community structure. This is particularly true in communities of relatively sedentary, low trophic-level deposit-feeding marine invertebrates and of low trophic-level suspension- and filter-feeders. A sudden incursion of a demersal fish or active invertebrate carnivore may locally remove a significant fraction of certain taxa, seriously disrupting community structure both in terms of abundance, and in number

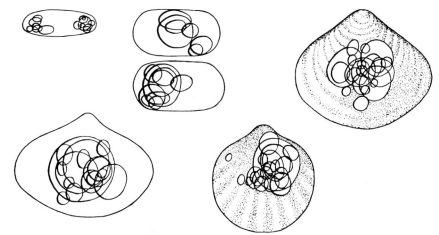

FIGURE 177. Diagram showing locations of most frequent drilling on some Niger Delta ostracodes and bivalves. (From R. A. Reyment, 1971b, *Introduction to Quantitative Paleoecology,* Figure 28. Copyright © 1971 by Elsevier Scientific Publishing Company.)

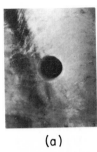 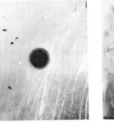  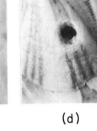

(a)   (b)   (c)   (d)   (e)

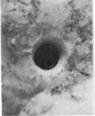   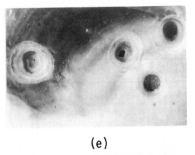

(f)   (g)   (h)   (i)   (j)

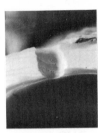 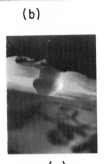 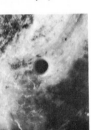  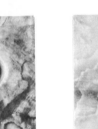

(k)   (l)   (m)   (n)   (o)

 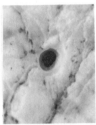

(p)   (q)   (r)   (s)   (t)

 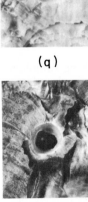  

(u)   (v)   (w)   (x)   (y)

FIGURE 181. (a) Boring by *Eupleura caudata* (Say, from lower Chesapeake Bay, Va. Boring in *Crassostrea virginica* (Gmelin) from Bogue Sound, N. C.; cylindrical borehole with negligible bevel. OD (outside diameter) 1.5 mm, ID (inside diameter) 1.1 mm. USNM 673535. (b-k) Borings by *Eupleura caudata etterae* Baker, from Chincoteague Bay, Va. (b) Boring in *Crassostrea virginica* (Gmelin) from Bogue Sound, N.C.; cylindrical borehole with slight bevel. OD 1.4 mm, ID .9 mm. USNM 673536. (c) Boring in *Crassostrea virginica* (Gmelin) from Bogue Sound, N.C.; parabolic borehole with sharp edge. OD 1.5 mm, ID .7 mm. USNM 673537. (d) Boring in *Balanus venustus niveus* Darwin from Bogue Sound, N.C.; parabolic borehole with countersunk edge. Note irregularity of the surfaces of the hole imposed by the sculpture of the barnacle valve. OD 1.4 mm, ID .9 mm. USNM 673538. (e) Borings in *Crassostrea virginica* (Gmelin) from Bogue Sound, N.C.; from inside the valve in the area of the adductor muscle scar outward toward a live oyster attached to the outside of the oyster valve being bored. Snails were crowded in observational aquarium; only one of the four holes is complete. Range of OD 1.6–2.3 mm. USNM 673539. (f) Longitudinal section of boring in *Crassostrea virginica* (Gmelin) from Bogue Sound, N.C.; hole is slightly inflated in parts of the shell consisting of chalky calcite and constricted at the outer and inner openings where the shell is of a hard translucent nature. OD 1.3 mm, ID .7 mm, MD (middle diameter) 1.6 mm, depth 1.8 mm. USNM 673540. (g) Longitudinal section of boring in *Crassostrea virginica* (Gmelin) from Bogue Sound, N.C.; a boring similar to that in (f) except that the inner and outer strata of translucent calcite are thicker. The inner opening is narrowly constricted by a wide shelf. OD 1.3 mm, ID .8 mm, MD 1.6 mm, depth 1.5 mm. USNM 673541. (h) Longitudinal section of boring in *Crassostrea virginica* (Gmelin) from Bogue Sound, N.C.; the two strata of soft chalky calcite at the inner end of the hole have been excavated slightly more than the remaining translucent shell. MD 1.6 mm, ID .9 mm, depth 2.4 mm USNM 673542. (i) Longitudinal section of boring in *Crassostrea virginica* (Gmelin) from Bogue Sound, N.C.; in homogeneously translucent calcite, a uniform nearly cylindrical borehole with a slight inflation at the inner end. OD 1.1 mm, ID .9 mm, MD 1.2 mm, depth 1.5 mm. USNM 673543. (j) Longitudinal section of boring in *Crassostrea virginica* (Gmelin) from Bogue Sound, N.C.; hole is an oblique paraboloid. Drilling was first carried out at right angles to the shell surface and then veered to one side to effect entrance into the mantle cavity of the prey. OD 1.3 mm, ID .6 mm, depth 1.6 mm. USNM 673544. (k) Longitudinal section of boring in *Crassostrea virginica* (Gmelin) from Bogue Sound, N.C.; markedly stratified structure of the shell contributes to the irregularity of the surface of the hole. As in (j), the hole was first bored perpendicular to outer shell surface, possibly until preliminary penetration was effected and then in the direction of the oyster mantle cavity. OD 1.2 mm, ID .5 mm, MD 1.6 mm, depth 1.6 mm. USNM 673541. (l) Boring by *Ocenebra erinacea* (Linné) from Plymouth, England. Boring in *Crassostrea virginica* (Gmelin) from Bogue Sound, N.C.; incomplete hole bored on the inside of the oyster valve. OD 1.4 mm. USNM 673545. (m) Boring by *Ocenebra japonica* (Duncker) from Puget Sound, Wash. Boring in *Crassostrea virginica* (Gmelin) from Bogue Sound, N.C.; typical slightly parabolic muricid hole, appearing slightly naticid in shape because of the thin shell of the prey. OD 1.1 mm, ID .6 mm. USNM 673546. (n-u), (x) Boring by *Urosalpinx cinerea follyensis* Baker from Chincoteague Bay, Va. (n) Boring in *Crassostrea virginica* (Gmelin) from Bogue Sound, N.C.; hole is characterized by wide inner opening and slightly irregular sides. OD 1.9 mm, ID 1.2 mm. USNM 673547. (o) Boring in *Crassostrea virginica* (Gmelin) from Bogue Sound, N.C.; edge of outer opening conspicuously countersunk, slight shelf in inner opening. OD 2.0 mm, ID 1.0 mm. USNM 673548. (p) Boring in *Crassostrea virginica* (Gmelin) from Bogue Sound, N.C.; cylindrical hole with countersunk outer edge, shelf at inner opening, opening to one side. OD 2.1 mm, ID .9 mm. USNM 673549. (q) Boring in *Crassostrea virginica* (Gmelin) from Bogue Sound, N.C.; hole with conspicuous crescent-shaped shelf. OD 2.0 mm, ID 1.6 mm. USNM 673550. (r) Boring in *Crassostrea virginica* (Gmelin) from Bogue Sound, N.C.; parabolic hole with irregularity on one side, resembling a naticid borehole. OD 2.3 mm, ID 1.4 mm. USNM 673551. (s) Boring in *Crassostrea virginica* (Gmelin) from Bogue Sound, N.C.; hole irregularly elliptical parabolic OD 1.4 × 2.0 mm, ID .6 × .9 mm. USNM 673552. (t) Boring in *Crassostrea virginica* (Gmelin) from Bogue Sound, N.C.; oblique, spherical, parabolic hole with small inner opening to one side. OD 1.3 mm, ID .4 mm. USNM 673552. (u) Boring in *Crassostrea virginica* (Gmelin) from Bogue Sound, N.C.; highly irregular form of parabolic hole probably resulting from heterogeneity of structure and form of shell. OD 1.3 mm, ID .9 mm. USNM 673553. (x) Longitudinal section of boring in aragonitic shell of snail *Murex fulvescens* Sowerby from Beaufort Inlet, N.C.; the homogeneity of the shell material permitted drilling of the smooth-surfaced nearly cylindrical hole. OD 1.6 mm, ID 1.2 mm, depth 2.5 mm. USNM 673554. (v), (w) Boring by *Urosalpinx cinerea follyensis* Baker from Wachapreague Bay, Va. (v) Boring in small thin-shelled rapidly growing *Crassostrea virginica* (Gmelin) from Cape Cod, Mass.; oyster shell at site of perforation is composed of hard translucent calcite, hole is elliptical parabolic. OD 1.7 × 2.0 mm, ID 1.0 × 1.2 mm. USNM 673555. (w) Longitudinal section of boring illustrated in (w). Depth of hole .6 mm. USNM 673555. (y) Boring by *Urosalpinx cinerea* (Say) from Burnham-on-Crouch, England. Boring in *Crassostrea virginica* (Gmelin) from Bogue Sound, N.C.; spherical parabolic hole, predominantly so shaped because of location in thin-shelled oyster valve. OD 1.2 mm, ID .6 mm. USNM 673556. (From M. R. Carriker and E. L. Yochelson, 1968, Recent gastropod boreholes and Ordovician cylindrical borings, *U.S. Geol. Surv., Prof. Paper 593-B,* Plate 1.)

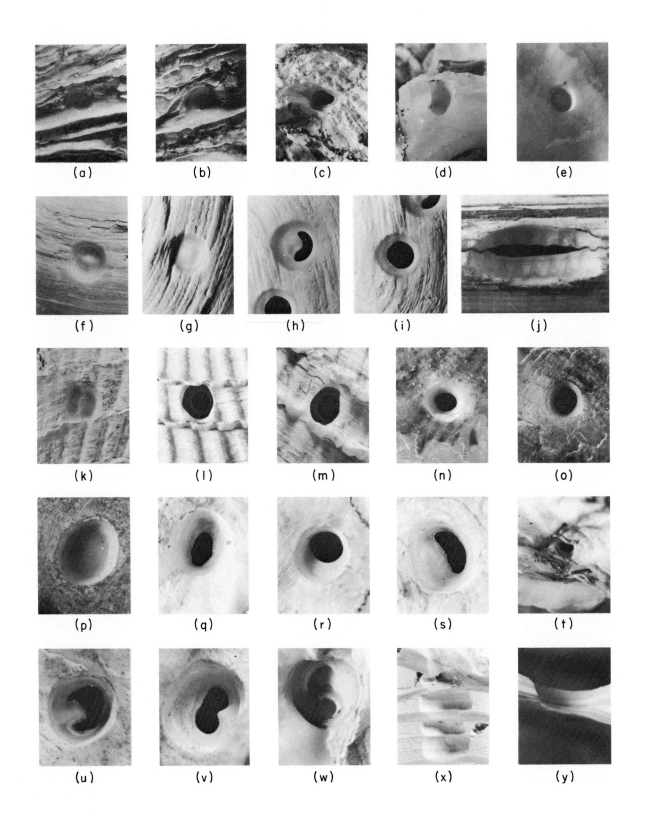

(a)   (b)   (c)   (d)   (e)

(f)   (g)   (h)   (i)   (j)

(k)   (l)   (m)   (n)   (o)

(p)   (q)   (r)   (s)   (t)

(u)   (v)   (w)   (x)   (y)

FIGURE 182. (*a–d*) Borings by *Thais haemostoma floridana* Conrad, from Bogue, Sound, N.C. (*a*) Boring in *Crassostrea virginica* (Gmelin) from Bogue Sound, N.C.; hole drilled typically at the juncture of the valve edges. Drilling frequently proceeds into the opposing valve, leaving a slight concavity before the snail extends its proboscis into the cavity of the bivalve to feed. MD 1.6 mm. USNM 673557. (*b*) Boring in *Crassostrea virginica* (Gmelin) from Bogue Sound, N.C.; hole similar to that in (*a*) but larger. MD 2.0 mm. USNM 673557. (*c*) Boring in *Crassostrea virginica* (Gmelin) from Bogue Sound, N.C.; hole is highly irregular as a result of the structure and configuration of the oyster shell in the vicinity of the borehole. OD 1.6 × 2.0 mm, ID .8 × 1.2 mm. USNM 673558. (*d*) Boring in *Crassostrea virginica* (Gmelin) from Bogue Sound, N.C.; hole was drilled at the edge and juncture of the two valves of the oyster, but "misdirected" into the lower valve for some distance (shown here) before the snail turned into the cavity of the oyster to feed. MD 1.6 mm. USNM 673559. (*e*) Boring by *Thais haemastoma* Linné from Bimini, Bahamas. Boring in *Crassostrea virginica* (Gmelin) from Bogue Sound, N.C.; incomplete hole bored obliquely from inside the oyster valve. A small live oyster growing on the outside of the bored valve probably stimulated selection of the site by the snail. OD 1.4 × 1.6 mm, ID 1.1 mm. USNM 673560. (*f–i*) Boring by *Polinices duplicatus* (Say) from Cape Cod, Mass. (*f*) Boring in *Mya arenaria* Linné from Cape Cod, Mass.; incomplete spherical parabolic hole with central boss in the bottom. OD 2.1 mm. USNM 673561. (*g*) Boring in *Mya arenaria* Linné from Cape Cod, Mass.; irregularity of shell surface causes irregularity on the edge of the hole. OD 2.3 mm. USNM 673562. (*h*) Boring in *Mya arenaria* Linné from Cape Cod, Mass.; the borehole is almost through the shell and emphasis of rasping in the interradial region has left the central boss. OD 2.7 mm. USNM 673563. (*i*) Same specimen as in (*h*) but showing the spherical parabolic borehole complete. OD 2.4 mm, ID 1.5 mm. USNM 673563. (*j*) Boring by *Murex fulvescens* Sowerby from Beaufort Inlet, N.C. Boring in *Mercenaria mercenaria* (Linné) from Bogue Sound, N.C.; hole is drilled typically at juncture of two valves. Note persistence of shell pattern at the bottom of the elliptical parabolic hole, probably indicating differences in hardness. OD 3.6 × 7.2 mm. USNM 673564. (*k–n*) Borings by *Murex florifer arenarius* Clench and Perez Farfante from Alligator Harbor, Fla. (*k*) Boring in *Chione cancellata* (Linné) from Bogue Sound, N.C.; incomplete hole reflecting the costa over which it was drilled. OD 2.3 mm. USNM 673565. (*l*) Boring in *Chione cancellata* (Linné) from Bogue Sound, N.C.; illustrates effect of the cancellate sculpture of the substratum on the form of the hole. OD 1.9 mm, ID 1.0 mm. USNM 673566. (*m*) Boring in *Chione cancellata* (Linné) from Bogue Sound, N.C.; this clam is larger than that in (*l*), but the effect of the sculpture of the shell on the hole is still evident. Hole is deeply inflated. OD 1.6 × 2.3 mm, ID 1.2 mm. USNM 673567. (*n*) Boring in *Crassostrea virginica* (Gmelin) from Bogue Sound, N.C. Note that a hole bored by this snail in smooth shell resembles the spherical parabolic hole of naticids. OD 2.0 mm, ID 1.2 mm. USNM 673568. (*o*) Boring by *Murex pomum* Gmelin from Alligator Harbor, Fla. Boring in *Crassostrea virginica* (Gmelin) from Bogue Sound, N.C.; hole is spherically parabolic, and the steepness of the curve of the sides is intermediate between the typically naticid boreholes. OD 2.0 × 2.0 mm, ID 1.2 × 1.5 mm. USNM 673569. (*p*) Boring by *Bedeva haleyi* (Angas) from Port Jackson, Australia. Boring in *Crassostrea virginica* (Gmelin) from Bogue Sound, N.C.; the irregular sculpture of the surface of the oyster shell produced a somewhat disfigured borehole. OD 1.0 mm, ID 0.8 mm. USNM 673570. (*q–y*) Borings by *Murex brevifrons* Lamarck from Puerto Rico. (*q*) Boring in *Crassostrea virginica* (Gmelin) from Bogue Sound, N.C.; incomplete elliptical parabolic hole with concave bottom. OD 3.4 × 3.9 mm, ID 2.3 × 2.7 mm. USNM 673571. (*r*) Boring in *Crassostrea virginica* (Gmelin) from Bogue Sound, N.C.; elliptical parabolic hole with pronounced shelf. OD 2.8 × 4.1 mm, ID 1.2 × 2.0 mm. USNM 673572. (*s*) Boring in *Crassostrea virginica* (Gmelin) from Bogue Sound, N.C.; oblique nearly spherical parabolic hole. OD 2.9 × 3.1 mm, ID 1.6 × 1.8 mm. USNM 673573. (*t*) Boring in *Crassostrea virginica* (Gmelin) from Bogue Sound, N.C.; oblique elliptical parabolic hole with pronounced shelf. OD 3.9 × 4.3 mm, ID 1.0 × 2.6 mm. USNM 673574. (*u*) Boring in *Crassostrea virginica* (Gmelin) from Bogue Sound, N.C.; nearly spherical parabolic hole with conspicuous crescentic shelf. OD 3.7 × 3.9 mm, ID 1.7 × 2.3 mm. USNM 673575. (*v*) Boring in *Crassostrea virginica* (Gmelin) from Bogue Sound, N.C.; elliptical parabolic hole with slightly crescentic shelf. Hole bored between two high ridges, which are out of focus in photograph, on the surface of the shell of the oyster. OD 3.9 × 4.8 mm, ID 1.4 × 2.5 mm. USNM 673576. (*w*) Boring in *Crassostrea virginica* (Gmelin) from Bogue Sound, N.C.; hole drilled at juncture of upper valve, to the right, and lower valve, to the left. The snail "mistook" the inner valve for continuation of the upper valve and bored into it a short distance, shown by the small circular hole at bottom, before discovering oyster flesh between the valves to the right (between large intermediate level oval boring and small circular boring beneath). OD 3.5 × 3.6 mm. USNM 673577. (*x*) Longitudinal section of boring in *Crassostrea virginica* (Gmelin) from Bogue Sound, N.C.; the section was ground slightly off center to emphasize the alternation of chalky and translucent strata of the oyster shell and the deeper excavations by the snail into the chalky strata. MD 3.0 mm, depth 6.3 mm. USNM 673578. (*y*) Longitudinal section of boring in *Crassostrea virginica* (Gmelin) from Bogue Sound, N.C. This hole is typical for the species; it illustrates the slightly oblique spherical parabolic shape with conspicuous shelf, inner opening to one side of center, and gentle countersinking of the outer edge. The perforation was made almost entirely in translucent calcite; to the right of the hole is seen stratification of chalky and translucent shell. OD 4.7 mm, ID 1.8 mm, depth 2.2 mm. USNM 673579. (From M. R. Carriker and E. L. Yochelson, 1968, *U.S. Geol. Surv., Prof. Paper 693-B*, Plate 2.)

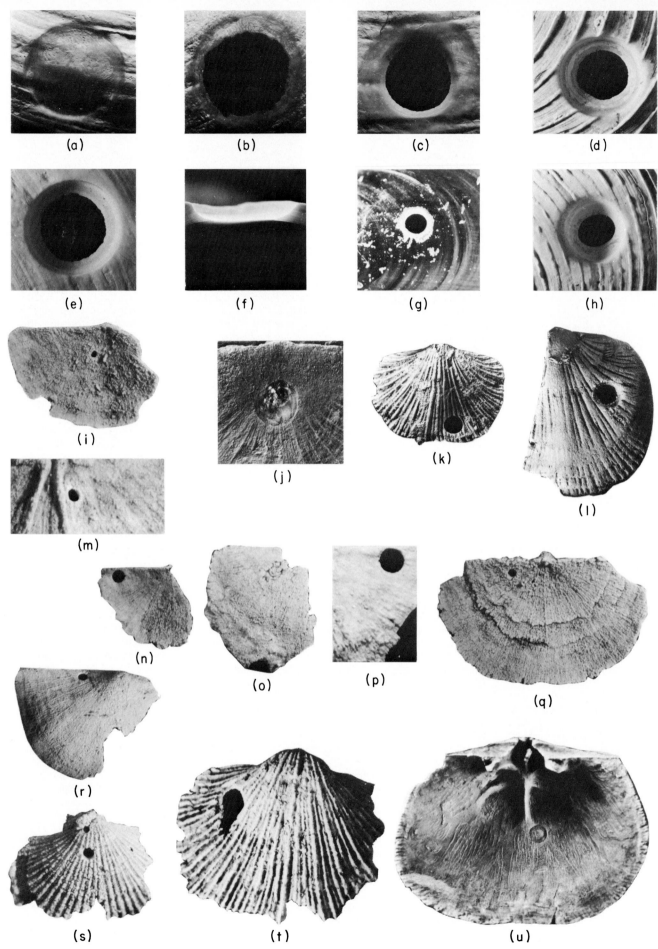

FIGURE 183. (*a–f*) Borings by *Lunatia heros* (Say) from Cape Cod, Mass. (*a*) Boring in *Mya arenaria* Linné from Cape Cod, Mass.; note persistence of valve sculpture and central boss at the bottom of the hole. OD 5.0 mm. USNM 673580. (*b*) Boring in thin-shelled *Mya arenaria* Linné from Cape Cod, Mass.; note irregularity of inner opening, due perhaps to thinness of the shell. OD 6.4 mm, ID 4.7 mm. USNM 673581. (*c*) Boring in moderately thick shell of *Mya arenaria* Linné from Cape Cod, Mass. OD 6.6 mm, ID 3.4 mm. USNM 673582. (*d*) Boring in *Mercenaria mercenaria* (Linné) from Cape Cod, Mass.; illustrates the classical form of the spherical parabolic naticid borehole in a prominently ridged valve. The beveling of the hole includes the shell ridges OD 4.4 mm, ID 2.8 mm. USNM 673583. (*e*) Boring in thick-shelled *Mya arenaria* Linné from Cape Cod, Mass.; a typical spherical parabolic naticid borehole in smooth shell. Note the gentle beveling of the outer edge of the hole. OD 5.0 mm, ID 3.6 mm. USNM 673584. (*f*) Longitudinal section of boring in *Mya arenaria* Linné from Cape Cod, Mass.; illustrates the graceful parabolic curve of the typical naticid borehole. OD 5.9 mm, ID 3.4 mm, depth 1.2 mm. USNM 673585. (*g*) Boring by *Lunatia triseriata* (Say) from Cape Cod, Mass. Boring in *Mytilus edulis* Linné from Cape Cod, Mass.; shows the beveling of the edge of the outer opening as a result of rasping of the periostracum. OD 1.6 mm, ID 1.1 mm. USNM 673586. (*h*) Boring by *Natica severa* (Gould) from Korea. Boring in *Mercenaria mercenaria* (Linné) from Bogue Sound, N.C.; interior surface of hole is very smooth; outer edge is moderately beveled; outer opening is circular and inner opening is elliptical. In this hole the parabolic form changes from spherical to elliptical inward. OD 3.3 × 3.3. mm, ID 1.6 × 1.9 mm. USNM 673587. (*i*) Boring in *Rafinesquina* cf. *R. alternata* Emmons from Waynesville shale, from cuts 14 and 15 on Big Four Railroad, east of Weisburg, Ind. Boring in brachial valve; this is the original drawing of fig. 1E of Fenton and Fenton (1931). Carnegie Museum of Pittsburgh no. 9833/7094. (*j*), (*k*) Boring in *Dalmanella* from Waynesville shale, near junction of Indiana State Roads 1 and 46, Cedar Grove quadrangle, Indiana. (*j*) Boring in brachial valve. USNM 155010. (*k*) Boring in brachial valve. USNM 155011. (*l–r*) Borings in *Sowerbyella* from the Shipshaw Formation, 1 1/4 miles below Chute aux Galets, Quebec. (*l*) Small boring in brachial valve. USNM 155012. (*m*) The same specimen from the interior, enlarged. (*n*) Intermediate size boring in brachial valve. USNM 155013. (*o*) Broken boring on edge of brachial valve. USNM 155014. (*p*) The same specimen as (*n*) from the interior, enlarged. (*q*) Small boring in brachial valve. USNM 155015. (*r*) Small boring in pedicle valve; distinctly oval in outline. USNM 155016. (*s*) Boring in *Hebertella* from Ottawa Formation, 1.2 miles west of Odessa interchange on Highway 401, Ontario. Small boring in pedicle valve; the smaller hole above may be artificial. USNM 155017. (*t*) Boring in *Taphrorthis peculiaris* Cooper from the lower one-third of Pratt Ferry Formation of Cooper (1956),.2 mile south of Pratt's Ferry, Blocton quadrangle, Ala. Boring in pedicle valve; the lower half of the hole is broken. This specimen is figured by Cooper (1956) as pl. 38, fig. 19. USNM 117984b. (*u*) Boring in *Mimella globosa* (Willard) from the Benbolt Formation of Cooper and Prouty (1943), from the roadside .25 mi southwest of New Bethel Church, Hilton quadrangle, Virginia. Incomplete boring in interior of brachial valve; this specimen is figured by Cooper (1956) as plate 89, Figure 10. USNM 117038b. (From M. R. Carriker and E. L. Yochelson, 1968, *U.S. Geol. Surv., Prof. Paper 693-B,* Plate 3.)

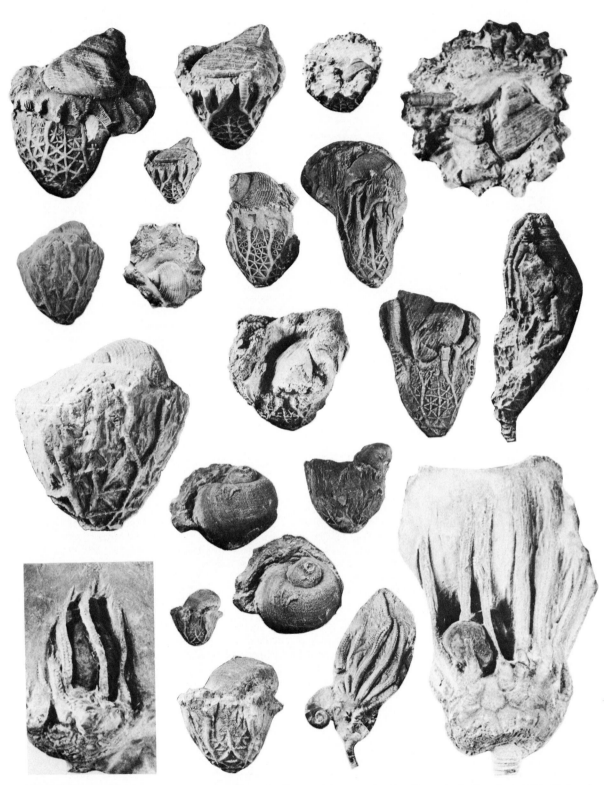

FIGURE 184. Coprophagous platyceratid gastropods in feeding position on crinoids. (From A. L. Bowsher, 1955, University of Kansas Paleontological Contribution, *Mollusca*, Article 5, pp. 1–11, Plate 1.)

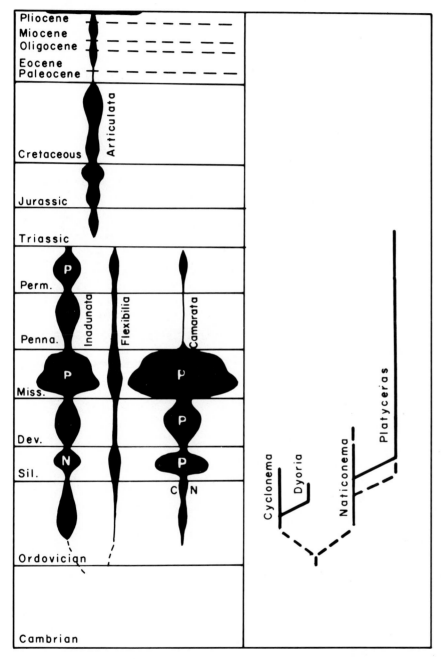

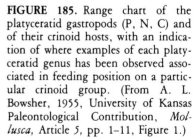

FIGURE 185. Range chart of the platyceratid gastropods (P, N, C) and of their crinoid hosts, with an indication of where examples of each platyceratid genus has been observed associated in feeding position on a particular crinoid group. (From A. L. Bowsher, 1955, University of Kansas Paleontological Contribution, *Mollusca,* Article 5, pp. 1–11, Figure 1.)

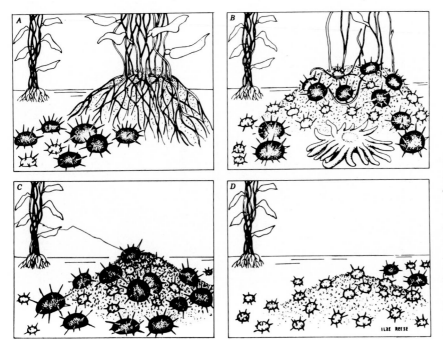

FIGURE 186. A "herd" of sea urchins grazing on the giant kelp *Macrocystis pyrifera;* the light-colored species *Strongylocentrotus purpuratus* appears to follow the dark-colored *S. franciscanus* (the asteroid *Pycnopodia helianthoides* may be an echinoid predator). The sequence *a–d* indicates arrival at feeding site, active feeding, and subsequent departure. (From R. A. Boolootian, 1966, *In* R. A. Boolootian, Ed., *The Physiology of Echinodermata,* Figure 8.4. Copyright © 1966 by Wiley–Interscience.)

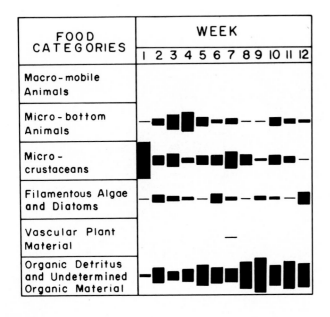

FIGURE 187. Food of pinfish taken from *Zostera* beds near Beaufort, North Carolina, at weekly intervals throughout the summer of 1961. (From R. M. Darnell, 1970, *American Zoologist, 10,* pp. 9–15, Figure 1. Copyright © 1970 by The American Society of Zoologists.)

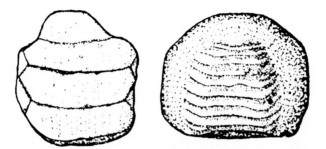

FIGURE 188. Typical demersal, bottom-feeding type crushing teeth of rays. (From A. S. Romer, 1945, *Vertebrate Paleontology*, Figure 53. Copyright ©1945 by University of Chicago Press.)

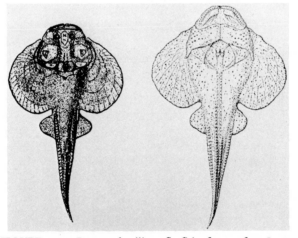

FIGURE 189. Bottom-dwelling flatfish form of a Lower Devonian type (*Gemuendina*). (From A. S. Romer, 1945, *Vertebrate Paleontology*, Figure 40. Copyright © 1945 by University of Chicago Press.)

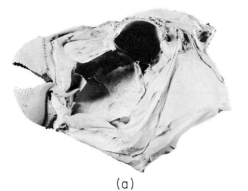

(a)

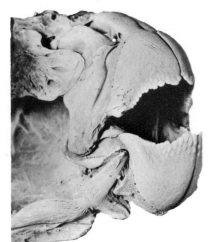

(b)

(c)

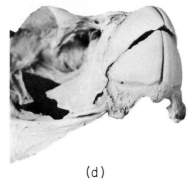

(d)

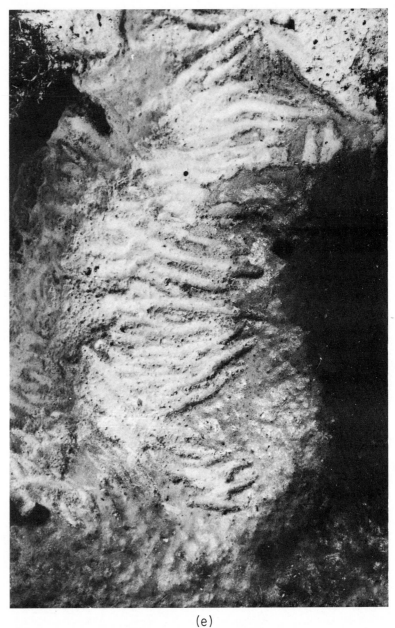

(e)

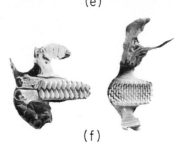

(f)

FIGURE 190. (*a*), (*b*), (*f*) Skull (× 1), beak (× 2), and elements of pharyngeal dentition (× 1) of a parrot fish (*Scarus* sp.) from Onotoa Atoll, Gilbert Islands. USNM fish coll.164322. (*c*), (*d*) Skull (× 1) and beak (× 2) of the puffer fish *Arothron hispidus* (Lacepede) (identified by J. C. Randall) from Onotoa Atoll, Gilbert Islands. USNM fish coll. 164320. (*e*) Beak markings (× 3) of parrot fish on living surface of hydrocoralline *Millepora* sp. from seaward reef front at Carysfort Light, east coast of Florida. (From P. E. Cloud, Jr., 1959b, *U.S. Geol. Surv., Prof. Paper 280k,* Plate 130.)

FIGURE 191. Evidence of predation by fish (flaking and chipping of bivalve margins) from the Permian of Wyoming. (Left): Valve fragment of *Scaphellina* (× 1) with distinctive prelithification damage along ventral margin—possibly due to fish predation. (Right): Valve of modern *Placuna placenta* (× 1 1/2) damaged by fish predation. (From D. W. Boyd and N. D. Newell, 1972, *Journal of Paleontology, 46*, pp. 1–14, Plate 2, Figures 3 and 4. Copyright © 1972 by Society of Economic Paleontologists and Mineralogists.)

FIGURE 192. Evidence of predation by crabs on gastropod shells, and a crab caught in the act. (From G. A. Bishop, 1975, *In* R. W. Frey, Ed., *The Study of Trace Fossils,* Figure 13-3. Copyright © 1975 by Springer–Verlag New York, Inc.)

FIGURE 193. Shell fracturing presumably due to Crustacea. The shells are from dredge samples from the middle central bay (bottom clay, depth about 25 m). 1, 2, *Chlamys opercularis* (L.); 3–14, *Venus striatula* (da C.); 15–23, *Nassarius semistriatus* (Brocchi); 24–28, *Turritella communis* Risso; 3, one valve crushed, the other still attached to it by the ligament; 23, shell repaired after fracturing as in 15 and 16; 24, shell three times repaired, fourth time successfully broken and consumed by predator. (From G. C. Cadée, 1968, *Zool. Verhandl., 95*, pp. 1–121, Figure 43. Copyright © 1968 by Rijksmuseum van Natuurlijke Historie, Leiden.)

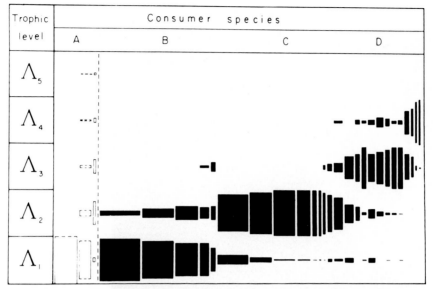

FIGURE 194. Model of the ecosystem showing sources of energy for the various consumer species in relation to trophic levels. Category (*a*) includes various microorganisms which, for convenience, have been combined into three groups (those which derive energy entirely from breakdown of vegetation, those which receive their energy by attacking certain chemical bonds common to both plants and animals, and those which specialize on certain compounds present mostly in animal matter). Category (*a*) should probably equal the widths of all remaining categories combined. Category (*b*) includes mostly zooplankton (and a few benthic) species which harvest the phytoplankton, but which also take in a certain quantity of other microorganisms. Category (*c*) includes detritus feeders (which are presumed to derive more nourishment from the microbes than from the substrate) as well as certain browsers on the zooplankton. Category (*d*) includes the higher predators representing both the broad multivores and the specialists, but for all of which the energy sources must be allocated to the proper trophic levels. (From G. J. Bakus, 1969, *Int. Rev. General and Exper. Zool., 4*, pp. 275–369, Figure 4, after Darnell, 1968. Copyright © 1969 by Academic Press.)

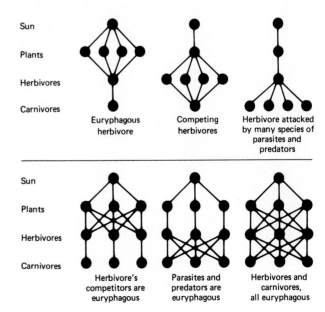

FIGURE 195. Various patterns of three-level trophic web organization. Each dot represents a species, and the lines between dots indicate that the species in the lower level consumes the species on the level above. (From G. J. Bakus, 1969, *Int. Rev. General and Exper. Zool., 4*, pp. 275–369, Figure 3, after Watts, 1968. Copyright © 1969 by Academic Press.)

217

TABLE 4.11
Relation between Bored and Unbored Bivalves Near the Eddystone[a]

Bivalve	Size (mm)	Total bivalves bored by *Natica*	Total bivalves unbored	Percentage bored	
Glycymeris glycymeris	>15.0	0	5	0.0	
	5.0–15.0	25	156	13.8	
	2.5– 5.0	171	545	23.9	21.5
	1.5– 2.5	33	130	20.3	
Astarte triangularis	1.5– 2.5	248	267	48.2	48.2
Gafrarium minimum	5.0–15.0	30	120	20.0	
	2.5– 5.0	81	529	13.3	14.9
	1.5– 2.5	28	143	16.4	
Chione ovata	5.0–15.0	21	72	22.6	
	2.5– 5.0	122	392	23.7	20.9
	1.5– 2.5	34	205	14.2	
Chione fasciata	>15.0	0	2	0.0	
	5.0–15.0	63	62	50.4	
	2.5– 5.0	143	301	32.2	32.8
	1.5– 2.5	21	99	17.5	
Cardium ovale	5.0–15.0	2	2	50.0	
	2.5– 5.0	20	250	7.4	7.2
	1.5– 2.5	3	69	4.2	
Cardium scabrum	5.0–15.0	7	29	19.4	
	2.5– 5.0	20	73	21.5	20.9
	1.5– 2.5	5	19	20.8	
Pseudamussium similis	5.0–15.0	24	37	39.3	
	2.5– 5.0	88	286	23.5	25.6
	1.5– 2.5	0	3	0.0	
Nucula spp.	5.0–15.0	5	22	18.5	
	2.5– 5.0	15	70	17.6	16.8
	1.5– 2.5	0	7	0.0	

[a]From J. E. Smith, 1933, *Jour. Mar. Biol. Assoc. U.K.*, **18**, pp. 243–278, Table 7.

TABLE 4.12
Percentage by Volume of the Major Groups of Food Organisms from the Digestive Tracts of 80 Hogfish[a]

Animal group	Percentage
Pelecypoda	42.6
Gastropoda	39.7
Scaphopoda	.6
Crustacea	
Brachyura (crabs)	6.1
Anomura (hermit crabs)	4.9
Amphipoda	1.0
Cirripedia	.5
Echinoidea	4.6

[a]From J. E. Randall and G. L. Warmke, 1967, *Caribbean Journal of Science*, **7**, pp. 141–144, Table 1.

"crabs," and starfish of the past to which we should attend, rather than the more spectacular higher trophic-level organisms.

Contrary to Wigley's conclusion (personal communication, 1976), which was based largely on outer- and middle-shelf work, is that of researchers on the inner-shelf and even intertidal areas where large vertebrates, skates and rays in schools, can devastate clam flats by removing the clams, then crushing the shells (see Bigelow and Schroeder, 1953b; Herald, 1967; Orth, 1975; and Smith, personal communication, 1976). It is obvious in light of this information, that bottom-feeding fishes in some instances will thoroughly disrupt the community—even to the extent of digging to some depth in their search for shellfish—whereas, in other instances, predation is minimal. Another situation where each case must be studied and resolved separately is one where there are large amounts of shelly debris that might have passed through the gut of a shellfish-crushing fish. It might alert one to the local importance of this type of predation. An additional nearshore–estuarine example of piscine predation is provided by the Drumfish (*Pogonias*), an important predator, with a varied assortment of invertebrate predators of the brackish-water oysters along the eastern coast of the U.S. (see Smith,

TABLE 4.13
Mollusks Identified from the Digestive Tract of the Hogfish[a]

Species	Frequency[b]	Species	Frequency[b]
Pelecypoda		Gastropoda *(Continued)*	
Aequipecten gibbus	2	*Bulla striata*	2
Aequipecten mucosus	1	*Bursa thomae*	5
Americardia media	5	*Calliostoma jujubinum*	1
Anadara notabilis	2	*Cerithium* sp.	13
Antigonia listeri	1	*Cerithium algicola*	1
Arca sp.	1	*Cerithium litteratum*	6
Arca imbricata	3	*Cerithium muscarum*	1
Arca zebra	1	*Columbella mercatoria*	11
Barbatia cancellaria	1	*Conus* sp.	3
Barbatia domingenis	1	*Conus jaspideus*	2
Chama sp.	2	*Conus juliae*	1
Chama macerophylla	2	*Conus mus*	1
Chama sarda	3	*Coralliophila* sp.	1
Chione sp.	1	*Coralliophila abbreviata*	3
Chione paphia	4	*Coralliophila caribaea*	5
Chlamys imbricata	1	*Crassispira* sp.	1
Codakia costata	1	*Cyphoma gibbosum*	1
Diplodonta sp.	1	*Drillia albinodata*	1
Echinochama arcinella	1	*Fissurella* sp.	1
Glycymeris sp.	4	*Leptadrillia splendida*	1
Glycymeris pectinata	6	*Mitra nodulosa*	1
Glycymeris undata	1	*Modulus modulus*	1
Isognomon alatus	1	*Muricopsis oxytatus*	1
Laevicardium sp.	4	*Nassarius albus*	9
Laevicardium laevigatum	5	*Natica* sp.	2
Lucina pennsylvanica	2	*Oliva* sp.	1
Macrocallista maculata	1	*Oliva reticularis*	4
Ostrea sp.	7	*Olivella* sp.	1
Papyridea soleniformis	2	*Polinices* sp.	1
Pecten sp.	2	*Polinices lacteus*	4
Pecten ziczac	5	*Polystira* sp.	1
Phacoides radians	1	*Pusia* sp.	1
Pinctada radiata	3	*Risomurex roseus*	4
Pinna carnea	4	*Rissoina* sp.	1
Pitar fulminata	1	*Strombus* sp.	6
Pteria colymbus	1	*Strombus costatus* (juvenile)	4
Trachycardium sp..	3	*Strombus gigas* (juvenile)	10
Trachycardium isocardia	3	*Tegula* sp.	2
Trachycardium magnum	5	*Tegula fasciata*	1
Trachycardium muricatum	4	*Terebra* sp.	1
Gastropoda		*Thais* sp.	1
Anntillophos sp.	1	*Tonna maculosa*	1
Antillophos candei	1	*Trigonostoma rugosa*	1
Astraea sp.	5	*Trivia* sp.	2
Astraea caelata	1	*Turbo* sp.	1
Astraea phoebia	1	*Vasum* sp.	1
Astraea tuber	2	*Xenophora conchyliophora*	3

[a]From J. E. Randall and G. L. Warmke, 1967, *Caribbean Journal of Science*, 7, pp. 141–144, Table 2.
[b]Frequency indicates the number of fish containing mollusk species listed.

TABLE 4.14
Percentage of Fragments[a] in the Total Number of Entire Shells Plus Fragments[b,c]

Species	Oceanic zone			Outer bay		
	Number of samples	Total of shells and fragments	Percentage of fragments	Number of samples	Total of shells and fragments	Percentage of fragments
Spisula sp.	3	1053	28.9			
Venus ovata	4	883	24.5			
Venus fasciata	2	358	13.9			
Tellina donacina	4	1433	38.6			
Venus striatula	1	470	14.5	4	845	34.8
Chlamys opercularis				2	493	68.3
Myrtea spinifera				3	517	32.3
Nassarius pygmaeus				3	566	31.3
Nassarius semistriatus				3	526	46.4
Turritella communis				1	980	83.5

[a] Hinge fragments in bivalves, top whorls in gastropods.
[b] From G. C. Cadee, 1968, *Zool. Verhandl. 95*, pp. 1–121, Table 3. Copyright © 1968 by Rijksmuseum van Natuurlikjke Historie, Leiden.
[c] This gives an indication of the fraction of shells destroyed by predators in this case.

TABLE 4.15
Data on Bored *Dicoelosia* from the Wenlock of Baillie–Hamilton Island[a]

Collection number	C 22184	C 12272
Total Dicoelosia	3274	1971
Dicoelosia with boreholes	27	41
Bored brachial valves	3	2
Bored pedicle valves	3	8
Bored articulated specimens	21	31
Percentage of total specimens bored	.64	2.1
Mean diameter of boreholes (mm)	.33	.31
Standard deviation (mm)	.11	.075

[a] From D. M. Rohr, 1976 *Journal of Paleontology 50*, pp. 1175–1179, Table 1. Copyright © 1976 by Society of Economic Paleontologists and Mineralogists.

1904, for a typical account of drumfish predation, in which the shells are reported to be ground up).

As we tend to think of large predators as the most important in the community, we also tend to think that the most conspicuous prey is important in the total community structure. DiSalvo (1973) points out the importance of bacteria as food for many marine organisms. This suggests that predation or cropper-mediated species changes at even the microscopic level might have great impact upon the entire community structure.

Several experiments (Dayton, 1971, 1972, 1973, 1975a,b; Dayton and Hessler, 1972; Dayton *et al.*, 1974; Lubchenco and Menge, 1978; Menge, 1972; Neudecker, 1979; Paine, 1963b, 1966, 1969; Vance,

1979) have shown clearly that in some situations the activities of predators act to maintain high taxic diversity and low dominance amongst the lower trophic-level organisms, whereas in other cases activities of predators and croppers have the opposite effect (Stephenson and Searles, 1960). *Acanthaster planci* (Endean, 1973; Frankel, 1977) is notorious with regard to the effect of explosive, catastrophic predation.

The effect on taxic diversity by either predation or cropping will depend on the prey organisms' methods of partitioning resources. If in the absence of a predator a particular species dominates some limited resource, selective predation against the monopolizer should lead to higher diversity.

The effect of random cropping depends upon

whether locally perturbed areas of the environment are repopulated successionally. If a cropper clears areas by random cropping of all species, and the bare locale is repopulated in an orderly sequence of species, a community containing cropped patches of various ages will have a high diversity due to the successional stages that are summed in sampling. For a discussion of successional mosaic diversity see Paine and Levins (in preparation).

The question of food specificity by cropper and predator (the [un]willingness and ability of each to switch from one food material to another as the local situation demands) is important. There is a complete spectrum, from omnivorous types (Tables 4.16 and 4.17; see Figure 195, p. 217) to very prey-specific types: Kohn's (1971) and Kohn and Nybakken's (1975) specific species of *Conus* are good examples, as is Landenberger's (1968) *Pisaster* species. The paleoecologist must exercise care in defining communities because interactions with croppers and predators may affect the relative abundances of the lower trophic-level organisms. Some prey species are cropped by a great variety of predatory species (Edwards, 1969b, cites the

fact that an intertidal gastropod is taken by octopods, crabs, other gastropods, starfish, and possibly even fishes, as well as being eaten by birds). These observations further complicate the task of defining community relations.

Branch (1976), in a discussion of sympatry amongst limpets, emphasizes the important point that competition between specialists and generalists involves different problems. The specialists retreat further and further into their more restricted niches, where they rule supreme. The generalists behave in an opportunistic manner by taking advantage of varied environmental conditions involving widespread distribution of their numbers, both locally and geographically. The basic differences between the two types of competition at every stage in life and in every stage of the reproductive cycle should be kept in mind when analyzing community and biogeographic data.

Prey specificity and priority is an active area of research and speculation in behavioral ecology that is not fully open to the paleontologist. In terms of preferred prey size (Dodson, 1970) predators generally favor prey in or near their own size range. Few lions

TABLE 4.16
Feeding Data on Muricid Gastropods[a]

Muricid predator	Prey species	Method(s) of feeding
Chicoreus ramosus (as *Murex fortispina*)	*Ostrea cristagalli* and *Arca* sp.	Using ceratus as lever
C. brevifrons	Fed unselectively on unfamiliar species of mollusks	Boring through shell or "smothering"
Murex tribulus	Carrion-unselective	Meat-cutting
Haustellum haustellum	Carrion-unselective	Meat-cutting
Muricanthus radix	Various bivalves and barnacles, apparently unselective	Chipping bivalve lip-margins, using outer apertural lip
Muricanthus fulvescens	Various bivalves, favoring *Crassostrea virginica*	Chipping valve margins, using outer lip; or prying, using powerful foot and shell; or boring through shell of prey
Phyllonotus pomum	Large specimens of *C. virginica*	Boring through shell of prey
P. erythrostomus	Various bivalves, apparently unselective	Prying, using powerful foot and outer apertural lip
Ocenebra erinaceus	Various bivalves and gastropods	Boring through shell of prey
Urosalpinx cinerea	Various bivalves and barnacles, with preference for young *C. virginica*	Boring through shell of prey
U. perrugata	Various bivalves, little apparent preference	Boring through shell of prey
U. tampaensis	Bivalves, particularly *Brachidontes exustus*	Boring through shell of prey
Eupleura caudata	Bivalves, preference for *C. virginica*	Boring through shell of prey
Favartia cellulosa	*Brachidontes exustus*	Boring through shell of prey
Calotrophon ostrearum	Barnacles, especially *Balanus amphitrite niveus*	Boring through opercular plates

[a]From G. E. Radwin and A. D'Attilio, 1976, *Murex Shells of the World*, Table 2. Copyright © 1976 by Stanford University Press.

TABLE 4.17
Observed Diets of Eight Predatory Gastropods[a]

Prey	Predator							
	Pleuro-ploca	F. tulipa	F. hunteria	B. con-trarium	B. spira-tum	Murex	Sinum	Polinices
Gastropods								
Pleuroploca gigantea	—	—	—	—	—	—	—	—
Fasciolaria tulipa	13	1	—	—	—	—	—	—.
F. hunteria	6	12	—	—	—	—	—	—
Busycon contrarium	17	2	—	—	1	—	—	—
B. spiratum	4	6	—	—	—	—	—	—
Murex florifer	12	5	1	—	—	—	—	—
Sinum perspectirum	—	1	—	—	—	—	—	—
Polinices duplicatus	1	2	—	—	1	—	—	—
Conus floridana	—	1	—	—	—	—	—	—
Turbo castaneus	—	1	—	—	—	—	—	—
Urosalpinx sp.	—	2	1	—	—	—	—	—
Nassarius vibex	—	2	8	—	—	—	—	—
Pelecypods								
Mercenaria campechiensis	—	—	—	4	—	—	—	3
Chione cancellata	—	8(2)[b]	13(7)[b]	115	1	81	—	4
Macrocallista nimbosa	—	—	—	29	—	—	—	4
Cardita floridana	—	—	2(2)[b]	12	—	5	—	—
Aequipecten irradions	—	2	—	—	1	—	—	—
Trachycardium muricatum	—	1	—	—	—	—	—	—
Laevicardium mortoni	—	—	—	1	1	—	—	—
Mactra fragilis	—	1	—	1	3	—	1	1
Anomia simplex	—	—	2	—	—	1	—	—
Ostrea equestris	—	—	2	—	—	1	—	—
Modiolus americanus	—	—	4	5	—	—	—	—
Noetia ponderosa	—	1	1	4	—	—	—	—
Lucina floridana	—	—	—	1	2	—	1	1
Ensis minor	—	—	—	—	—	—	—	—
Atrina rigida	10	—	—	—	—	—	—	—
A. serrata	3	—	—	—	—	—	—	—
Polychaetes								
Diopatra cuprea	—	—	16	—	—	—	—	—
Onuphis magna	—	—	15	—	—	—	—	1
Owenia fusiformis	—	—	—	—	—	—	—	—
Barnacle								
Chthamalus sp.	—	—	1	—	—	—	—	—
Other								
Carrion	—	4	10	—	5	1	—	—
Total individuals	66	52	76	172	15	89	2	14
Total species represented	8	17	13	9	8	5	2	6

[a]From R. T. Paine, 1963b, *Ecology 44*, pp. 63–73, Table II. Copyright © 1963 by the Ecological Society of America.
[b]Number in parenthesis indicates number of bivalves that initially had been drilled by *Murex*.

search after mice, and few weasels tackle moose. In the oceans, this rule of thumb is complicated by filter feeding and suspension feeders, which extract very small "prey" from large volumes of water. However, these carnivores are seldom considered to be predators. Preserved animals morphologically adapted to seek, to manipulate, and consume prey may be cautiously assumed to prey upon organisms of the same general size (see Reyment, 1967; Walne and Dean, 1972).

Prey density is also a critical feature in determining the specificity and priority of the predator. Holling (1966) has reviewed some of the many aspects of prey–predator relations in terms of prey density. In terms of increased taxic diversity, predation upon the most abundant fauna should increase diversity by decreasing dominance.

In terms of prey specificity remember that certain prey is attractive to almost all predators (Loesch, 1953, details how birds, crabs, snails, and others move in on masses of available *Donax* in the intertidal, to "have a good feed"). As pointed out by McGowan (1971), omnivorous predators diminish population-size fluctuations of prey, affecting community definition in terms of abundance of prey species.

Bishop (1975) has carefully reviewed the predation problem from the paleontological view and suggested that the evidence may be usefully arranged in terms of search, capture, penetration, ingestion, digestion, and defecation. He provides excellent illustrative materials covering much of this field. However, the paleontologist has been restricted to a qualitative field. The paleontologist is denied a routine examination of the stomach contents necessary for a firm understanding of predator–prey relations. However, predation has been a factor since the Cambrian and no trophic history can ignore its varied potential.

Bite marks on bivalves, snails and other invertebrates are common evidence of predation (see Figures 191–193; Table 4.18).

Shell borings have great potential for providing insight into past predator–prey relations. Carriker and Van Zandt (1972) published an excellent account of the shell-boring behavior of a muricid gastropod—including the range in size and depth of the boreholes (minute holes in minute snails and large holes in larger ones, with a great range in variability of diameter for individuals, as well as a maximum depth range). Some predators generate boreholes at very specific sites on their preferred prey whereas the reverse is true for others (see Figure 4 in Carriker and Van Zandt, 1972, for an almost random scatter of muricid borehole sites over an oyster species, and compare this with Berg and Nishenko's 1975, highly regular site locations, as well as Sohl, 1969; see also Figures 178 and 179, p. 202). Rohr (1976) summarized much of the data, indicating how carnivorous gastropod bore-holes of the post-Paleozoic may be recognized, something of the frequency with which such bore-holes are recognized as a cause of death in some bivalves, and some of the problems of interpreting predator-caused bore-holes other than the typical post-Paleozoic gastropod types. The clustering of bore-holes on a particular portion of the shell of a species is a good reason to suspect predation as the cause rather than the random holes made by boring organisms looking for a suitable hard substrate in which to live (Figures 177–183 and 196; Tables 4.8–4.11 and Tables 4.15–4.16). However, shell borings are not infallible indicators of predation because as Orr (1962) described, a bivalve symbiont capulid-gastropod actually drills a hole through the host shell in order to share the host bivalve's incoming food supply. Finally, the studies of Richards and Shabica (1969) indicate that caution is needed in assigning a predatory origin to all holes in fossil shells; there is strong evidence that some holes are due to sediment-boring organisms and have no relation to predation.

Unusual preservation situations occasionally provide insight into predator–prey relations. Zangerl and Richardson (1963) relate the activities of various fishes to themselves and to associated shelly invertebrates. Evidence for bites, and so on, are provided (see also the mosasaur–ammonite bite story in Kauffman and Kesling, 1960).

An excellent example of schematizing trophic relations is Bowsher's (1955) treatment of platyceratid gastropods of the Middle Ordovician through Permian (Figures 184 and 185). Through a combination of the known stratigraphic ranges of the coprophagous gastropods and the overlapping time ranges of their crinoid hosts in the context of the known specimens in which the snails have been found in contact with anal regions of crinoids, the validity of the relation was established. Well's (1947) comment about the common association of pelmatozoan limestones in the Paleozoic with abundant platyceratid gastropods lends further support to Bowsher's conclusions. It is of interest to know that the post-Paleozoic Capulidae have a gross form (see Caullery, 1952) similar to that of the platyceratids, but they parasitize varied echinoderms and other gastropods, externally. They do not appear to have been merely coprophagous.

The study of coprolites is one "concrete" method of learning something about both predation and cropping. The memoir by Hantzschel, El-Baz, and Amstutz (1968) summarizes much of what has been accomplished. The bulk of the coprolite data deals with macroscopic contents such as bones and plant fragments; very little of it deals with microscopic remains, and there is remarkably little from the marine or freshwater realm—chiefly the remains of terrestrial tetrapods. (However, see Bronniman, 1972, for data on anomuran coprolites with a record extending back to the Triassic.)

A paleontologically important factor, although one that is very difficult to evaluate, is the size-selective habits of predators. Today, for example, cod, flatfishes, skates, rays and so on, dine on benthic shellfish among other prey. But they do not eat every size, as their mouths are not large. Some of the bottom-feeding fishes crush the shells of their prey (Chao, 1973), whereas others pass them through the gut entire (Bigelow and Schroeder, 1953a). It is clear that bottom feeders that are size selective and crush their prey may badly prejudice the fossil record by removing a significant percentage of the small specimens.

However, there is an accumulation of fisheries data suggesting that predation affecting the early growth stages of many commercial shellfish is more intense than that affecting later growth stages. For example, Galtsoff and Loosanoff (1939) report that small starfish successfully deal with oyster spat, but that larger oysters are relatively safe; Nelson (1931) reports that oyster drills preferentially attack oyster spat; Loosanoff and Engle (1940) report a high correlation between abundance of starfish spat and death of oyster spat; Needler (1933) reports on selective destruction of small oysters by starfish; Lunz (1947) reports that the blue crab (*Callinectes*) selectively destroys small oysters; Mobius (1877) reports that crabs and drills selectively attack oyster spat; Moore and Pope (1910) report that the oyster drill selectively attacks the spat rather than the adult oyster. If oysters are typical, it is clear that size-

TABLE 4.18
Examples of Shelly Benthos Treatment by Predatory Fish

Excrete shells entire	Crush and break shells	Dissolve shells
(Arntz, written communication, 1977) cod, dab, bivalves	(Arntz, written communication, 1977) flounder, plaice, bivalves	
	Bray and Ebeling, 1975 white seaperch, bivalves	Bray and Ebeling, 1975 white seaperch, bivalves
	Bray, written communication, 1977 white seaperch 5 of 55 had bits of shell in hindgut (avg. percentage volume = 71.6), i.e. some material did not dissolve.	
	Cadee, 1968 unspecified taxa	
Chao, 1973 cunner, bivalves		
Collette, written communication, 1976 Batrachoididae (toadfishes), snails		
	Hiatt and Strasburg, 1960 *Echidna zebra* (moray eel); crushing teeth, regular echinoids	
	Gynocranium griseus (snapper); crushing teeth, clams, urchins	
	Monotaxis grandoculis; crushing teeth, clams, urchins	
	Mulloidicthys samoensis; snails, bivalves (small ones uncrushed), echinoids many wrasses (Labridae); heavy pharyngial teeth, snails, bivalves, chitons, echinoids, ophiuroids	
	Haemulon flavolineatum (French grunt), chitons, bivalves, ophiuroids, echinoids	
	Haemulon macrostomum (Spanish grunt), echinoids	
	Haemulon parra (Sailors choice), gastropods, bivalves, ophiuroids	
	Haemulon plumieri (White grunt), echinoids, gastropods, ophiuroids, bivalves	
	Haemulon sciurus (Bluestriped grunt), bivalves, echinoids, ophiuroids, gastropods	
	Archosargus rhomboidalis (Sea bream), gastropods, bivalves	
	Calamus bajonado (Jolthead porgy), echinoids, bivalves, gastropods	
	Calamus calamus (Saucereye porgy), ophiuroids, bivalves, echinoids, gastropods, chitons	
	Calamus penna (Sheepshead porgy), gastropods	
	Calamus pennatula (Pluma), ophiuroids, bivalves, gastropods	
	Diplodus caudimacula (Roundspot porgy), gastropods, chitons	
	Gerres cinereus (Yellowfin mojarra), bivalves, gastropods, ophiuroids	
	Equetus punctatus (Spotted drum), gastropods	
	Mulloidichthys martinicus (Yellow goatfish), bivalves, ophiuroids, chitons	

Excrete shells entire	Crush and break shells	Dissolve shells
	Pseudupeneus maculatus (Spotted goatfish), bivalves	
	Malacanthus plumieri (Sand tilefish), ophiuroids, chitons, echinoids	
	Trachinotus falcatus (Permit), gastropods, echinoids, bivalves	
	Trachinotus goodei (Palometa), gastropods, bivalves	
	Bodianus rufus (Spanish hogfish), ophiuroids, echinoids, gastropods, bivalves	
	Halichoeres bivittatus (Slippery Dick), echinoids, gastropods, ophiuroids, bivalves, chitons	
	Halichoeres garnoti (Yellowhead wrasse), ophiuroids, gastropods, bivalves, echinoids, chitons	
	Halichoeres maculipinna (Clown wrasse), gastropods, bivalves, chitons	
	Halichoeres poeyi (Black-ear wrasse), gastropods, ophiuroids, echinoids, chitons, bivalves	
	Halichoeres radiatus (Puddingwife), bivalves, gastropods, echinoids, ophiuroids	
	Hemipteronotus novacula (Pearly razorfish), gastropods, bivalves	
	Hemipteronotus splendens (Green razorfish), gastropods, bivalves	
	Lachnolaimus maximus (Hogfish), bivalves, gastropods, echinoids	
	Thalassoma bifasciatum (Bluehead), gastropods, ophiuroids	
	Labrisomus guppyi (Shadow blenny), chitons, gastropods	
	Labrisomus nuchipinnis (Hairy blenny), gastropods, ophiuroids, echinoids	
	Xererpes fucorum, small gastropods	
	Micrometrus minimus, females, small gastropods	
	Quast, 1968	
	Pile Perch (*Rhacochilus vacca*), crushed snails, bivalves, ophiuroids, irregular echinoid	
	Randall, 1967	
	Dasyatis americana (Southern stingray), crushed bivalves	
	Aetobatis narinari (Spotted eagle ray), crushed gastropods and bivalves, may eject most shelly debris orally	
	Holocentrus rufus (Longspine squirrelfish), gastropods, ophiuroids	
	Holocentrus vexillarius (Dusky squirrelfish), gastropods, chitons	
	Anisotremus surinamensis (Black margate), echinoids, gastropods, ophiuroids, bivalves	

(Continued)

TABLE 4.18 *(Continued)*

Excrete shells entire	Crush and break shells	Dissolve shells

<table>
<tr><td></td><td>

Anisotremus virginicus (Porkfish), ophiuroids, bivalves, gastropods
Haemulon album (Margate), echinoids, bivalves, ophiuroids, gastropods
Haemulon carbonarium (Caesar grunt), gastropods, echinoids, chitons
Haemulon chrysargyreum (Smallmouth grunt), bivalves, gastropods
Trigger fish; heavy teeth, snails, echinoids, clams
Trunk fish, clams, snails
Puffers; platelike teeth, snails, bivalves
Hobson, 1974
  Monotaxis; molariform jaw teeth, snails, echinoids
  Wrasses (Labridae); strong pharyngial teeth, clams, snails, echinoids
  Trigger fishes; strong teeth, echinoids, snails
  Balloonfishes; beak, heavy plates, echinoids, snails, bivalves
  Spiny puffers; beak, heavy plates, snails, echinoids

</td><td></td></tr>
</table>

McEachran, written communication, 1977 skates, bivalves (periostracum remains)

Orth, 1975
  rays, bivalves
Shumway and Stickney, 1975
  Cunner, barnacles, bivalves, snails
J. W. Smith, personal communication, 1976 cownose ray, bivalves
  Sciaenops ocellatus, bivalves
  Pogonias cromis, bivalves

David Stein, personal communication, 1977, *Psychrolutus phrictus* (cottid), snails, clams

Wendy Gabriel, personal communication, 1977, *Microstomas pacificus* (Dover sole), 1 cm snails, (*Mitrella gouldi*) in last 5 cm of intestine; variable occurrence; also thin bivalve fragments up to 5 mm
Schafer, 1972, p. 98
  Raja clavata, Acipenser sturio, Conger conger, Gadus callarias, Melanogrammus aeglefinus, Molva molva, Mullus surmuletus, Anarhichas lupus, A. minor, Scophthalmus maximus, Hippoglossus hippoglossoides, Pleuronectes platessa, Limanda limanda, faeces packed with brittle star skeletal parts.
Merton C. Ingham, 1963, term paper done for W. G. Pearcy, *Platichthys stellatus,* pharyngial teeth, small bivalve fragments (generically identifiable) in stomach and entire length of gut (average fragment size decreased through gut).

Excretes shells entire	Crush and break shells	Dissolve shells
	Bigelow and Schroeder, 1953a, p. 504–505, wolffish (*Anarhichus*) crush or swallow entire snails, clams, crabs, mussels, starfish, echinoids, viselike molars Johnston, 1954 　*Gobiesox maeandricus*, snails Mitchell, 1953 　*Gibbonsia elegans*, small gastropods 　*Dactylopterus volitans* (Flying gurnard), bivalves 　*Balistes vetula* (Queen triggerfish), echinoids, bivalves, ophiuroids, gastropods 　*Canthidermis sufflamen* (Ocean Triggerfish), echinoids, 　*Lactophrys bicaudalis* (Spotted Trunkfish), ophiuroids, echinoids, asteroids 　*Lactophrys trigonus* (Trunkfish), bivalves, echinoids, asteroids, gastropods 　*Sphaeroides spengleri* (Bandtail puffer), bivalves, gastropods, echinoids, ophiuroids 　*Canthigaster rostrata* (Sharpnose puffer), gastropods, bivalves, echinoids, asteroids 　*Chilomycterus antennatus* (Bridled burrfish), gastropods 　*Diodon holacanthus* (Spiny puffer), gastropods, bivalves, echinoids 　*Diodon hystrix* (Porcupinefish), echinoids, gastropods, bivalves 　*Ogcocephalus nasutus* (Shortnose batfish), gastropods, bivalves	
Earl Krygier, personal communication, 1977, *Salmo clarki clarki* 1 to few mm *Littorina* type gastropods; observed in intestines; in coastal Oregon streams, upstream from coastal waters where snails were ingested (would have led to transport of a marine snail shell into the nonmarine environment)		
		Waldo Wakefield (written communication, 1979) *Platichthys stellatus* (starry flounder) 393 mm. 12 cm from anus (c. 35 cm tract) 1 22 cm *Dendraster eccentricus* (sand dollar)-plates beginning to separate. *Parophrys vetulus* (English sole) 328 mm. 6–9 cm from anus (c. 25 cm tract) 5 sand dollars from 11–14 mm diameter with spines removed, plates intact. 2 7–8 mm *Macoma* spp. some solution apparent. 11–15 cm. from anus 4 sand dollars 8–13 mm diameter with aboral spines intact (contrast to more posterior specimens), plates intact. 1 complete small gastropod.

(Continued)

TABLE 4.18 (*Continued*)

Excrete shells entire	Crush and break shells	Dissolve shells
		Anarrichthys ocellatus (wolfeel) 1397 mm; 16–20 cm. from anus (c. 31 cm tract), gastropod shell fragments—no solution apparent; shells used by hermit crabs present in stomach. *Isopsetta isolepis* (buttersole) 323 mm from anus; 2 bivalve shell fragments, largest dimension 23 mm, no solution apparent. 316 mm from anus (22 cm tract) one shell fragment with longest dimension 26 mm. 12 cm from anus; one bivalve shell fragment; longest dimension 17 mm.
	Bell *et al.*, 1978 Three species of monacanthid; fragment bryozoans, but the pieces remain undissolved in rectum.	

selective predation, crowding, and competition for space by other species, is largely responsible for the bulk of juvenile mortality. If this is so, a sigmoid mortality curve should result.

Consistent with the aforementioned possibility is Dayton's (1971) report that the barnacle *Balanus*, as well as the mussel *Mytilus*, in some intertidal situations reach a "size refuge" from which certain predators are unable to effectively take them. Dayton (1975a, p. 150) reports additional molluscan examples of this type with a starfish as predator.

Ropes (1969) and Walne and Dean (1972) provide data on the size-selective habits of a crab, which involves its predation on a variety of bivalves. Above a certain size, the bivalves are safe from crab predation. Hibbert (1977) relates something of the complex predator relations characteristic of *Mercenaria*, the hardshell clam, and the fact that many of its predators do not attack the large individuals. Feare (1970) reports on a snail whose young attack small mussels.

Hornell (1922) reports that several teleosts favor small pearl oysters as opposed to the larger individuals, although he also reports that rays in this area are effective predators of the larger bivalves.

Edwards (1975) describes the correlation between predator size and prey size in *Polinices*, the moon snail, and its bivalve prey *Mya* (his Figure 1) (see Chapter 4, p. 270 for additional examples).

Edwards and Huebner (1978) further document the greater mortality by predation among the young of *Mya* and the correlation between prey size and predator size, emphasizing that above a certain threshold size limit, many shells are almost immune from further predation.

Moore (1913) describes how the drill *Urosalpinx cinerea*, in addition to being cannibalistic, switches from thick-shelled prey to a thinner-shelled specimen if such is made readily available—again suggesting that smaller prey are preferred and that mortality rates will be affected accordingly.

Cameron (1907) comments on how the drumfish (*Pogonias*) on the Louisiana coast prefers small oysters (under 5 cm long), generally ignoring larger shells.

There has been little systematic effort to study predator–prey relations of the past. However, there are a wealth of natural history observations concerning predator–prey relations such as those discussed by Carter (1968). Some of this raw data, I hope, will be systematically organized to see if it is of value to us in our understanding of evolution. Part of the difficulty in defining predator–prey relations of the past is caused by the rarity of predators as compared with the low trophic-level creatures on which many of them prey. Add to this the poor chance of preservation afforded echinoderm predators because their plates disintegrate so easily after death, the difficulty with which carnivorous gastropods of the past are recognized as such (unless they belong to the same genera or families as do thoroughly modern carnivorous types), and the difficulty of being certain that all cephalopods of the distant past, such as Ordovician nautiloids, were carnivorous. Biologists have applied much time and effort to the study of both predation and cropping and have developed concepts about various groups that may be of value when considering the fossil record.

Most of the discussion applies to predation. Very little has really been said about cropping. Bromley (1975) describes the characteristic traces made during the crop-

TABLE 4.19
Data Showing the Shift from the Predominantly
Planktotrophic Mode at Low Latitudes to the Nonplankto-
trophic Mode at Higher Latitudes (Prosobranch Gastropods)[a]

Region	Species with pelagic development (%)
South India	91
Bermuda Isles	85
Persian Gulf	75
Canary Islands	68
New Caledonia	57[b]
South West coast of Africa	31[c]
Portugal, West Spain, Bay of Biscay to Arcachon	63
Finistaire and Brittany to Cherbourg and English Channel coast	67
South coast of Ireland	65.2
West coast of Ireland	66.3
South West England and Wales plus East Ireland	68.1
Irish Sea	68.2
North coast of Ireland	68.4
North West coast of Scotland	63.4
Southern part of North Sea	57.1
East and North East Scotland to Cape Wrath including Orkney and Shetland Islands	61.2
Danish South frontier, North Sea north of Dogger Bank, Skagerrak and Inner Danish waters plus West Norway to Stavanger	60.4
West coast of Norway north of Stavanger to Bergen and Trondheim areas	58.8
Lofoten Isles area, Norway	45.5
West Finmark, Norway	34.4
East Finmark to Murmansk, West Murman	16.7
West and South Iceland	27.3
Labrador	8.3
North and East Iceland	7.5
White Sea	7.2
South Eastern Canadian Arctic	6.9
Baffin Island, Canadian Arctic	less than 5.0
East Greenland	0

[a]From S. A. Mileikovsky, 1971, *Marine Biology 10*, pp. 193–213, Table 1. Copyright © 1971 by Springer-Verlag New York, Inc.
[b]Insufficient data.
[c]The influence of the cold Guinean Current.

ping activities of both fossil and recent regular echinoids, and Voigt (1977) describes the similar evidence of chiton and gastropod cropping from the fossil record and from the present. Little effort, to date, has been put into searching for such evidence. It is reasonable to expect that more intensive investigation would not go unrewarded. The frequency and abundance of such activity might provide us with another

tool with which to estimate trophic relations during the past.

The problems of cropping and predation immediately bring to mind the companion concepts of standing crop and productivity. Productivity has to do with the rate of production of biomass (or shelly material insofar as the paleontologist is concerned), whereas standing crop refers to the shelly biomass (for the paleontologist) present at any instant in time. The standing crop may have been produced at a low or high production rate. One should not fall into the trap of regarding a mass of shells as absolute evidence for a high production rate. Estimation of productivity and production rates is difficult for the geologist; it is the other side of the coin of estimating biomass levels and nutrient levels of the past—problems that the geologist has still not solved satisfactorily.

The complexity introduced into any community history drawn by a paleoecologist is increased when interanimal interaction such as predation is considered. Not only are the prey organisms directly affected by physical parameters, but they are indirectly affected by the physical parameters that affect their predators. The depth-correlated distribution of demersal fish off West Africa (Longhurst, 1958), and off western North America (Alverson et al., 1964) are examples. The fish feed extensively on the benthic fauna; thus, the invertebrate benthic community is in part controlled by the depth preferences of the fish. Since most marine organisms begin life small and grow larger, there may be considerable change in predator–prey interaction during the organisms' life time. Taylor and Chen (1969) detail an extreme prey-predator change; the complete reversal of roles. Large octopuses eat small scorpion fish, whereas large scorpion fish eat small octopuses. Birkeland (1974) reviews other examples of potential prey–predator feedback involving changing life stages (see also Figures 194–195).

An estimate of predator selectivity may be had by recourse to studies such as Sheehan and Lesperance (1978), which show's that in some shell beds only one or a few species are bored in a predatory manner. Adegoke and Tevesz (1974) and Taylor (1970) present some semiquantitative data on the frequency of muricid and naticid boring in an Eocene fauna. The data indicates various levels of prey selectivity on the part of the different predators involved. (It should be noted, however, that these authors did not distinguish the predators above the family level). Connell (1970) has shown how some intertidal organisms find a refuge from essentially subtidal predators. Feare (1970) presents additional evidence concerning a whelk vis-à-vis a predatory, subtidal set of crabs.

STRUCTURALLY SELECTIVE PREDATION

In shelly organisms characterized by more than one skeletal element, the rule is to find departures from the 1:1 ratio in number of elements encountered. Every paleontologist dealing with such organisms can think of many examples where one valve or one element is unusually rare or abundant as compared with others. With Paleozoic brachiopods, for example, many of the stropheodontids are characterized by a relative rarity of brachial valves as compared with pedicle valves—a ratio of 1:20 is not unusual; with Silurian pentamerine brachiopods, a pedicle valve to brachial valve ratio of 20:1 or even greater is common. Such relations are probably best explained by selective predation—with the predator destroying one element preferentially over another. If differential postmortem transport were the explanation we would expect to find unusually high concentrations of the commonly low abundance element, but such is rarely the situation. It is hard to appeal here to differential solution or scavenging activity.

PALEOECOLOGICALLY SIGNIFICANT PREDATORS

Of great concern to benthic marine paleoecology is the overall effect of predation in altering the aspect and shelly biomass residue represented by the fossil record. Any predatory activity that preferentially removes one taxic component or size grade relative to another should be considered. Predation that results in crushing or dissolution beyond taxonomic recognition, or in the scattering of shelly debris beyond the original area of occurrence, must be considered. Experiments have shown that highly prey-specific predators will tend to increase the diversity of benthic communities if they feed on a dominant species, but that omnivorous predators will reduce diversity in situations of varying prey dominance (see Porter, 1972, for a summary statement).

Predator–prey relations are very difficult to determine from the fossil record; that is, in the absence of stomach contents or any opportunity to observe the actual processes. Therefore, the paleoecologist is reduced to trying to extrapolate into the past predator—prey relations observed now. In extrapolating such relations attention is, of course, paid to actual paleontologic evidence of predation such as shell breakage, boreholes, and to relative abundance relations between potential predator and prey taxa. It must continually be kept in mind, however, that trophic relations postulated for the fossil record are seldom "proved." In the following section, a series of comments will be presented regarding the potential of various predatory groups, with evidence from the present and the past.

Marine biologists commonly agree that the major shellfish predators of the present include the fishes, starfish, various malacostracans, and gastropods. Many other groups of marine invertebrates also are predacious at some trophic level, but present evidence suggests that those we have mentioned are the major predators worthy of serious consideration.

There is little evidence that any of the marine invertebrate taxa are consistently taking their prey far enough away from the capture site to materially affect community structure as observed in the fossil record. This is not, however, necessarily the case with the fishes. The starfish and gastropods normally do not materially affect their prey by biting, chipping, or otherwise crushing or seriously mutilating the shells (there are a few gastropod exceptions), but the malacostracans may badly break the shells, as do some shell-crushing fishes.

Dayton et al. (1974) provide a detailed view of the trophic complexities encountered in a benthic community complex in terms of the interactions of various predators with each other and with prey taxa; their story is a warning against overly simplistic paleoecologic conclusions.

Gastropods

Carnivorous gastropods operate in several ways. Some types gain entry to their prey prior to feeding by means of boreholes (see Rohr, 1976, for an account and an introduction to the extensive literature on the subject). Other gastropods such as the conids have darts capable of injecting poison into their prey (see Kohn, 1959, 1971; Kohn and Nybakken, 1975, for an introduction to the literature). Still others (Figure 192) rasp with their shell margins at the edge of the prey shell until the flesh is exposed to radular activity, or they force the prey valves apart with projections ("ceratus") of their own shell. There are some predators that pry with their feet (see Table 4.16). The feeding preferences of carnivorous gastropods are varied. Some ingest both herbivorous and carnivorous gastropods, others prefer bivalves, and a few feed on small fish (some of the conids, for example). The bulk of the Jurassic and younger shell-boring gastropods belong to the families Naticidae and Muricidae (Sohl, 1969), which generate a characteristically countersunk borehole in the shells of their prey (Figure 181). Such countersunk boreholes are not recognized in the pre-Jurassic. However, Rohr (1976) reviews the possibilities for ascribing some of the parallel-sided boreholes, located statistically in favored portions of the prey species (Figures 177–179), to the activities of older groups of long-extinct gastropods. Rohr's evidence is, of course, circumstantial, as his potential pre-Jurassic predaceous, boring gastropods have never been seen in the act of boring. Should his conclusions be in error, his data would, nevertheless, indicate the existence of some pre-Jurassic group(s) of predators capable of generating parallel-sided boreholes in favored portions of the prey shells. Ausich and Gurrola (1979) present additional evidence and discussion about the possibilities of pre-Jurassic predaceous gastropods.

Kornicker et al. (1963) found that in one prey species snails alternately bore opposing valves. When

Amphipods

Immediately foll
may be particularly
that by forms not
Templeman (1954)
newly molted lobst
nonmolting lobster

Fishes

Fishes were prob
to the Late Silurian
However, after the
should be consider
organisms. Demers
fish form (see Figu
if equipped with
pharyngial crushing
may have eaten sh
appears as early as
question for the
predators crushed
whole, dissolved th
or were capable o
ments. It is also i
piscine predators ra
and digesting to
fragments or entir
munity in localiti
source community.
voiding-rate for she
on this critical info
time various shellf
munities, and how
unit of time. Groo
pleuronectiformes
such times would
benthic shell-fish o
tensive. The pres
documented as far
1952).

Fossil fish them
deposits. The rar
something about
scavengers and dec
an original rarity.
an excellent accoun
remains in both t
ment. Any predat
should keep her d

An additional c
nontropical fishes
spend a juvenile
from that where tl
1971). The paleoe
various fishes may
with both growth
Schwartz (perso

combined with studies of site specificity in borehole location, this observation will reveal more information about predation as contrasted with random boring-organism activity.

Paine (1963b) in a study of eight cooccurring gastropods, makes a number of interesting points. He suggests that the lower trophic-level predators are more prey-specialized than others. He emphasizes the correlation between prey size and predator size, and also provides figures representing omnivorous behavior shown by the taxa studied (Table 4.17). Note (Figure 197) the food webs indicated by his data. An application of Paine's conclusions to the fossil record suggests that care should be taken in assigning most predation to any one carnivore within a specific environment, unless the evidence is compelling. Smith's data (1933), when added to the Kohn data, supplies more information on the degree of prey selectivity shown by some snails. Vermeij (1977b) provides additional historical perspective on snail predation. Sohl (1967, p. 12) provides examples from the fossil record (Cretaceous) of correlations between prey and predator, as well as between parasite and host; these data present a strong case for interaction, and are the best available without actually viewing the organisms. The report by Taylor and Taylor (1977) that gastropod prey specificity decreases with increase in latitude is of interest to the paleontologist. Dudley and Vermeij (1978) and Thomas (1976) provide quantitative data on the frequency of gastropod predation of a bivalve and of a gastropod genus; their quantitative specificity should be emulated to develop understanding of past patterns of predation in comparison to those of the present.

Cephalopods

There is little evidence implicating most cephalopods in major predatation on the modern shelly marine benthos. The activities of the octopods, which are reported to enjoy many of the malacostracans (in common with fishes having similar tastes) may be an exception. Some of the octopods are reported to litter the areas in front of their lairs with shelly remains of their malacostracan prey. Little is known about the feeding preferences of the pre-Cenozoic cephalopods, but a predaceous life style favoring shelly benthos is reasonable for many groups of ammonoids and some of the nautiloids. There are no available reports emphasizing relations between fossil shelled cephalopods and piles of shelly benthic debris that could be interpreted as stomach contents, feces, or prey accumulations of any other sort. Some octopods (Pilson and Taylor, 1961) generate minute, tapered boreholes in the prey shells, and many octopods are reported to employ their beaks and poison glands for immobilizing their prey. Arnold and Arnold (1969), and Wodinsky (1969) further describe *Octopus* boreholes, as well as providing further information on their feeding habits. The discussion includes varied

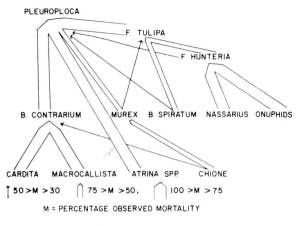

FIGURE 197. Food relations of six predatory gastropods (*Pleuroploca, Fasciolaria tulipa, Busycon contrarium, Murex, B. spiratum, F. hunteria*) and their principal prey. Compare with Table 4.17 data. (From R. T. Paine, 1963b *Ecology, 44*, pp. 63–73, Figure 3. Copyright © 1963 by the Ecological Society of America.)

gastropods, bivalves (Hartwick and Thorarisonsson, 1978), and even chitons. Arnold and Arnold (1969) explain that at least one octopod is selective regarding the position in which it will drill a particular snail. They provide experimental data indicating the morphologic cue that leads the octopod to a precise drilling location. Robba and Ostinelli (1975) provide evidence of Pliocene octopod molluscan-predation by means of borehole morphology. Stanton and Nelson (1980) cite similar evidence of octopod predation from the Texas Eocene. There is little data concerning the feeding habits of the extinct groups of cephalopods. However, Schwarzbach (1936) illustrates a Mesozoic ammonite specimen that suggests some might have fed on small ammonites; this was indicated by a mass of broken aptychal material in the living chamber of a specimen preserved with its aptychus in place.

Echinoderms

Among the modern echinoderms, starfish (Feder and Christensen, 1966) are implicated as important predators of the shelly benthos. Ophiuroids today include many small to minute elements of the bentho-pelagic fauna in their diet, but there is little data implicating them as major predators on the larger shelly benthos. There the bivalves, gastropods or brachiopods, to name only a few, are numerous. The starfish feed in a variety of ways, including the complete ingestion of their shelly prey; in most cases they leave the remains of their meals close to the capture site. The regular echinoids, although including carnivores, appear to be chiefly herbivorous, whereas the irregulars are chiefly deposit feeders, although on occasion they take live animal food.

Starfish p
lished, but t
planci (see F
counts and v
periodicity t
the possibili
tinctions of
related to si
convincing
crediting suc

Dayton *ei*
starfish pred
vironments :
of smaller p
ing the relat
substrate (F
record for ec
Bishop (197
starfish and
as evidence
meister (19
shells in tl
Ctenophora
predatory b
Eocene or e

Spencer a
dovician Sc
gastropod p

Kier and
predation
munication
that I shov
cidaroid ecl
cookies."

Decapods

Decapod
or by peeli
become ava
193) or by i
soft parts o
tion is dou
directly by
dages. Tur
damage to
in the inte
Kubacska (
that leaves
stomatopoc
niques; the
than the c
decapods l
ble that at
gnathobase
predators.
in terms o
evidence f
gastropods

tion on shelled animals probably is more important to fishes on sand bottoms than to fishes on coral reefs. Smith (personal communication, 1978) also concludes that fish predation in the tropical reef environment is important.

That fish predation on the benthos living on tropical level-bottoms is significant is evident from reports on tropical pearl fisheries where destruction of pearl oysters and other shellfish by fish is important. (Herdman, 1905, Ceylon; Herdman, 1906, India; Southwell, 1913, Ceylon; Villadolid and Villaluz, 1938, Philippines). Some of these authors also record that invertebrate predators are active, although no assessment is provided of the relative importance of fish as compared to invertebrates.

Hornell (1922) reports that, on the Indian pearl banks, fish predation (by rays, trigger fish, and *Lethrinus*) is significant, as is gastropod and starfish predation. The trigger fish (*Balistes*) and *Lethrinus* favor the smaller pearl oysters; the rays prefer larger ones. No information is provided about the relative importance of invertebrate versus vertebrate predators, but Hornell implies that the fish are more important (a single sample is reported in which one of fourteen shells had been drilled by snails). It is clear that there is much predation by fish in this tropical level-bottom environment. Herdman (1906) emphasizes that drill predation is chiefly on the spat, and that ray predation is on the larger specimens; with the rays breaking the shells in a characteristic manner. Herdman also suggests that starfish predation may be important, and that both crabs and octopus take pearl oysters. Finally, Southwell (1910) provides some limited semiquantitative data from a Sinhalese hard-bottom ("calcrete") type level bottom that strongly suggests that fish predation on bivalves is greater than invertebrate predation. However, generalizing from such limited data is unwise.

Reyment (1967, Table 30) provides data on the frequency of snail predation-caused boreholes suggesting that, off the delta of the Niger, snail predation is the major cause of mortality. Reyment's data, in addition to consideration of what is known of pearl oysters, implies that on soft level-bottoms, invertebrate predation in the tropics dominates, but that on the hard bottoms ("calcrete" of Hornell, Southwell, and Herdman) fish predation dominates. But the sample is too small to permit any reliable conclusions regarding the relative importance of fish versus invertebrate predation in tropical areas.

Invertebrate predation on the early growth stages of the macrobenthos (i.e., when they are very small shells) is clearly an important factor, probably more so than predation from fishes. Possibly fish predation becomes most important when the macrobenthos reaches larger sizes. Young *et al.* (1976) remark that in Florida waters a particular seagrass community may be characterized by decapods preying heavily on the macrobenthos, which in turn are heavily preyed upon by fishes. The effect of fish predation on the fossil record obviously has been complex.

Finally, care must be taken when interpreting postmortem artifacts as due to predation. For example, Weiler (1929) provides evidence suggesting that the occurrence of isolated heads and trunk regions in some fossil fish is most likely the result of postmortem weakening of the skeleton followed by current sorting. One should be very careful not to suggest that a disparity in counts of hard parts (left and right valves, cephalons and pygidia, etc.) belonging to a particular species reflects a predilection of a hypothetical predator for the part in short supply unless there is really convincing evidence such as bite marks, characteristic traces of some sort, or other reliable indications of selective predatory behavior.

ECOTONES, ECOCLINES, AND THE DEFINITION OF COMMUNITIES

Once a community is defined as a regularly recurring taxic association, with the condition that relative abundances of taxa remain within reasonable limits, we, as samplers, are faced with two polar situations. The first occurs when two groups of regularly recurring taxic associations share a common boundary with a limited, narrow area of overlap in which mixing of taxa characteristic of the two community units occurs, or even in which additional taxa characteristic of the ecotonal region may occur. The second situation occurs when we have two end points with widely differing taxic associations, between which there is a complete, infinitely gradational ecocline. In other words, the first situation makes community recognition and definition easy since it is a simple recognition of unlike entities (except for the narrow ecotonal region that has its own unique character); in the second situation the recognition of communities may be accomplished only after arbitrary divisions of an ecocline have been made. The ecoclinal situation may be treated as either a single community with polar areas that are almost completely unlike, or as a number of intergradational units that have been distinguished purely for the purposes of descriptive convenience. There are, of course, gradations between the ecoclinal and the ecotonal extremes that also must be dealt with.

Examples of the ecotonal situation include the boundary between level-bottom and rocky-bottom communities, whereas ecoclinal situations include, for example, the gradational changes from one level-bottom community to another as the shelf region is crossed in an area of average conditions, with depth, temperature, and other parameters gradually changing. Parker (1975) provides an excellent ecoclinal example, as does Barnard (1963).

McIntosh (1963) presents a number of examples of ecotonal and ecoclinal data, chiefly from the nonmarine environment. They emphasize the commonly gradational and intergrading nature of taxic distributions that makes the definition of communities statistical in nature (his prize is an illustration of the taxic gradients present in three species of intestinal parasite abundances observed as one travels down the gut of a sheep).

Bloom, Simon, and Hunter (1972) describe a relationship between suspension- and filter-feeding benthos, which in turn correlates with sediment grain size; the relationship is an ecoclinal type divided into three communities. Two of the communities are characterized by a superdominance at either end of the spectrum; the third is taken from the middle range of the ecocline.

Harger (1968, 1969, 1970, 1971, 1972) has provided abundant experimental and observational data from an ecoclinal area where two species of *Mytilus* coexist. *M. californianus* is adapted to rough water in a variety of ways, whereas *M. edulis* is adapted to quiet water in which a moderate amount of sedimentation is operating. In the ecotonal area, excessively rough water inevitably strips *M. edulis* from its essential substrate; in quiet intervals *M. californianus* is either smothered by overgrowing *M. edulis* or by sediment from which it is unable to escape (Figure 198). This is, incidentally, a case study in which careful consideration of shell morphology would probably provide no clue to the habits of either species, although the high correlation of *M. edulis* with muddy sediment and *M. californianus* with nonmuddy sediment would provide the clue as to quiet and rough water.

Working across an intertidal–shallow subtidal level-bottom region, Johnson (1970) documents the thoroughly ecotonal nature of the taxic distributions, as have many observers working on both level-bottom and rocky substrates in strictly intertidal surveys. Johnson also documents, although without comment, the markedly higher taxic diversity of the subtidal as contrasted with the intertidal.

That ecoclines and ecotones are significant factors is emphasized by Newman (1979). He found that, on a regional scale off the coast of Oregon, certain endemic barnacles are found only in an ecoclinal (biogeographically speaking) area. These barnacles are relict species, formerly far more widespread. In other words, the ecoclinal region represents a region of environmental overlap, with taxa on either side pressed to some of their ecologic limits (as is sometimes evident from their phenotypic variants) in a manner that permits additional taxa (relicts in Newman's example) to take advantage of a situation of "weakness" in order to exist. Elsewhere the endemic taxa would evidently be unable to compete successfully; the ecocline serves as an ecologic refuge.

It is critical when employing the ecotonal or ecoclinal concept in understanding a taxic gradient that one be reasonably certain that the taxic gradient represents a climax situation rather than successional stages, which can also initiate a taxic gradient. Jones (1969) presents data, interpreted by him as ecotonal (see his Figures 21–54 for an example and his "Conclusions"). No evidence is available that differing successional community levels are not involved. It might be argued that communities are communities regardless of the successional level they have reached, so that an ecotone is an ecotone or an ecocline is an ecocline whether or not one is dealing with a climax situation or with one of varying successional levels. A region subject to a uniform environmental regime may have a patchy set of communities because the patches have been affected by

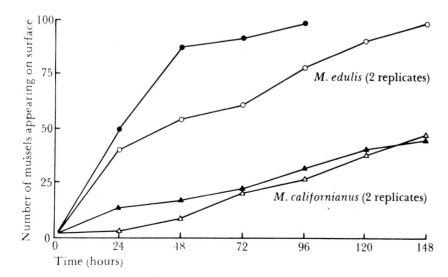

FIGURE 198. Comparison between numbers of *M. edulis* and *M. californianus* crawling clear of 5 cm of road gravel placed over 100 individuals of each species. Two replicate runs are recorded. *M. edulis* was able to maneuver clear of the gravel more readily than *M. californianus*. (From J. R. E. Harger, 1972, *American Scientist*, 60, pp. 600–607, Figure 2.)

catastrophes at different times. This situation must be distinguished from one in which there is a "permanent" patchy distribution of physical variables initiating durable ecotonal or ecoclinal distributions, provided the perturbing local catastrophes do not occur with consequent development of a uniform climax community. How is this dilemma solved? For the ecologist, the question entails experimental work to find successional communities in each area and by examination of enough transects located in a biogeographic unit to determine that the local possibilities of catastrophe have been distinguished from transectional changes in environment. For the paleoecologist, it means the examination of stratigraphic sections for repeats that may represent successional events, plus the study of enough transects to lower the probability of confusing successional stages with ecotonal–ecoclinal situations. In other words, it is a sampling problem in which common sense must be employed.

An example is provided by an area of muddy substrate on which a successional series of communities has arisen through time as the result of developing shelly substrates on the mud. Contrast this with an identical muddy substrate, parts of which are intermittently subject to storms that bring in more mud and

smother the shelly substrate dwellers on the mud, resulting in a patchy time distribution of muddy and shelly substrate organisms rather than a uniform vertical succession of muddy followed by shelly organisms.

The ecoclinal–ecotonal question is one of scale. On a scale of 1 to 1 million, an ecotone may cover many miles, whereas if the scale is 1 to 100, the ecotonal spread would cover many study areas and remain undistinguished.

Another approach to the recognition of ecotonal situations is provided by Schindel and Gould's (1977) recognition of character displacement in the fossil record. They demonstrated that two species of the same genus that normally occur in separate communities show good evidence for character displacement in ecotonal situations. There was a distinct narrowing of variation shown by two species of land snails and far less overlap in morphology than was present in nonecotonal situations.

Fenchel (1977, Figure 199) provides a fine example of character displacement among congeneric gastropods. A simple set of data like Fenchel's could be collected from readily available areas within the geologic record.

DISTANCE NECESSARY FOR SIGNIFICANT BENTHIC CHANGE

The coenologist *cum* geologist working in many regions is accustomed to a degree of stratigraphic–community continuity extending for many kilometers or even hundreds of kilometers on parts of certain platforms. Such continuity is more typical of platform than of continental margin-geosynclinal type environments. Coastal regions are commonly more environmentally patchy than are offshore regions. Miyadi (1941b) has discussed a Japanese situation in which the benthos differed significantly among five bays in close proximity. As one moves laterally, one must be prepared for surprises. Benthic community studues in the level-bottom environment provide a fair amount of resolution even with the equipment necessary for sampling purposes in the subtidal environment, whereas the patchiness and small-scale complexity of the reef environment have delayed detailed community studies even in the available intertidal region.

ENVIRONMENTAL HETEROGENEITY, COMMUNITY DEFINITION, ANIMAL SIZE, AND COMMUNITY OVERLAP

Many authors have dealt with the variation in range covered by a vagrant creature, whether a fish, an octopus, a cow, or a mouse, from the point of view of community definition (see Smith and Tyler, 1973, for an example). A problem in community definition is that certain larger taxa roam widely to satisfy various requirements that cannot be met by a small, environmentally uniform spot. Thus they occur sporadically within the territory of a number of other communities. Each of these smaller area communities survives in a

uniform environment, in contrast to the vagrant, which covers a great variety of environments. The wandering cow will cover a variety of terrain that a particular mouse would never enter. From the cow's point of view, patches of food, cover, shade, and so on are scattered about. Therefore, a good correlation should exist between the size of a vagrant animal and the degree of environmental complexity that it experiences as a desirable range. This conclusion has long been apparent for placental mammals of varied types, but it

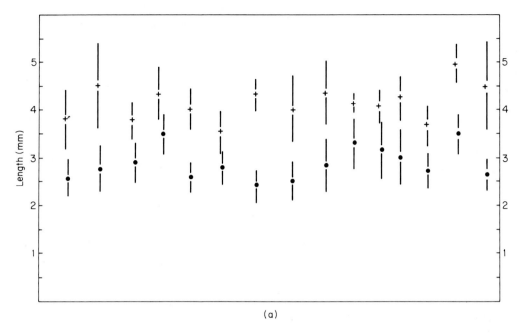

(a)

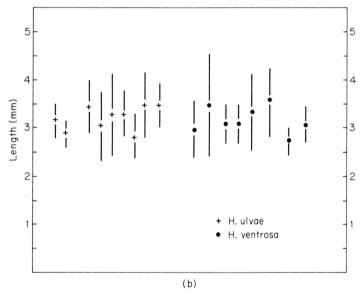

(b)

FIGURE 199. Average shell lengths of *Hydrobia ulvae* and *H. ventrosa* (b) from 17 localities where they occur allopatrically and (a) from where they coexist. The bars indicate one standard deviation. Note the evidence of character displacement. (T. Fenchel, 1977, *In* B. C. Coull, Ed., *Ecology of Marine Benthos,* Figure 6. The Belle Baruch Library in Marine Science, 6. Copyright © 1977 by University of South Carolina Press.)

also appears to be true for aquatic organisms, considering the piscine interests of such people as Smith and Tyler (1973). Therefore, we must be willing to define community extent in a variety of ways for organisms of differing sizes and habits in the same region; to define communities that have varying degrees of spatial overlap; and to recognize that the amount of environmental heterogeneity included within the area of a single community varies from fairly uniform to widely

varied (with migratory animals the problem will be increased manyfold, but is basically similar).

Intimately involved here are the habits and characteristics of the organisms belonging to various guilds. Ecologists working on the organisms belonging to a particular guild tend to define their communities in terms of that guild. This tendency results in overlapping definitions of communities by different groups of specialists.

COMMUNITIES DOMINANTLY CHARACTERIZED BY SPECIES BELONGING TO ONE MAJOR GROUP

In a later discussion, it will be made clear that communities are commonly defined in terms of the taxonomic group that interests the definer (i.e., students of brachiopods see and define brachiopod-based communities, students of birds define bird-based communities, etc., without saying too much about the other organisms that may be abundant).

However, Carney (personal communication, 1976) has pointed out the very obvious but basic, relation concerning the number of species and the abundances of major taxic groups in communities. The obvious fact is that in many communities, the dominant taxa belong to the same class or phylum. This is commonly the case whether it be clams, snails, seaweeds, echinoderms, brachiopods, annelids, and so on. It is commonly the case as well when dealing with fossils. The student of fossil brachiopod communities seldom interacts with the student of fossil molluscan communities or with the student of fossil coral and stromatoporoid communities. There are, of course, situations where this is not true, and there are myriad situations where the less common taxa are assigned to a variety of higher taxa. It is normal to find a wide variety of higher taxa represented in most high diversity communities among the rarer elements. Thus, when examining the fossil record, it is common to find communities dominated by brachiopods, trilobites, tetracorals, stromatoporoids, echinoderms, and so on, with the rarer species belonging to other major taxa. It is uncommon to find that three or four dominant species in a community belong to different phyla or classes. This is one of the reasons, in addition to preser-

vational peculiarities, why brachiopod collectors may be attracted to localities different than those of interest to the student of corals or crinoids.

A reasonable explanation of this relation is the probability that any particular set of physical parameters will define an environment best suited to a particular major group of organisms. In other words, each class or phylum, has certain general requirements that, taken together, are relatively unique to it and provide it with certain competitive advantages given the right set of circumstances. Therefore, it may be that there are environments more favorable to corals than to molluscs, to clams than to snails, to echinoderms than to crustaceans, and so on in terms of dominance within a community.

Another aspect of this community definition–taxic group variable is that specialists may be looking only at their own groups. If this is the case, it may involve organisms operating at only one trophic level (Figure 195). Obviously, this will provide a biased understanding of the community. However, attending to only one trophic group (the term guild covers this usage fairly well) may be the only practical procedure if other trophic levels and groups are poorly and/or erratically represented because of rarity, poor preservation, lack of understanding of their taxonomy, difficulty of extraction, and so on. This is not meant to suggest that better understanding would not be achieved if all the potential parts of the community were studied, but rather it is an acknowledgement that the ideal (Kauffman and Scott's holistic goal) is seldom possible in the real world.

SEASONALITY, MIGRATORY ANIMALS, AND THE DEFINITION OF COMMUNITIES

The definition of a community assumes a certain level of taxic stability. In regions where seasonality is a factor, one must consider the seasonal changes in the community caused by migration. For the paleontologist, the problem of seasonal migration will not be apparent if mortalities of migrants do not occur; if they do occur they may not suggest that migration was involved. Rather, a distribution pattern in which a particular taxon is found in a variety of regions may suggest eurytopy, a gross misinterpretation of the migrants' essential stenotopy. The only remedy for such misinterpretation is to investigate the population age structure for indications of a predominance of one age group over another in the various regions.

That the concept of seasonal changes due to migration is a real one in the marine world is indicated by Wells (1961), who indicates that the blue crab,

Callinectes, withdraws from the brackish-water region into deeper, more normal salinity conditions during the winter. Some fishes are known to have similar habits.

Hesse (1979) describes seasonal migrations in the gastropod *Strombus gigas,* and seasonal aggregations occurring during the winter in the warm waters of the Turks and Cocos Islands. Because the typical biologist has shown little concern with the extent of migrations affecting marine benthos, we are inclined to believe that most marine benthos are nonmigratory. This impression may be incorrect, and the paleontologist should be aware that there is little evidence on the subject.

Nishimoto and Herrnkind (1978) have reviewed the literature dealing with the annual migrations of *Callinectes* (also see Bainbridge, 1961). Fischler and

Walburg (1962) report that the male blue crab does not appear to migrate, remaining chiefly in brackish water, but that the female migrates into fully marine conditions to spawn, and the young migrate back into brackish waters. This cyclic behavior would, in principle, provide a clue to migratory behavior if relative abundances of young, adult female, and adult male were available as fossils—it is also convenient that crabs are easy to sex because of their carapace morphology.

Harville and Verhoeven (1978) suggest that although most West Coast dungeness crabs stay relatively close to one spot during the year in a few areas tagging and recovery data suggests seasonal migration in an inshore–offshore manner. Venema and Creuteberg (1973) pro-vide information on the seasonal migrations of a Dutch crab.

Herrnkind and Kanciruk (1978); Kanciruk and Herrnkind, (1978) describe mass migrations of spiny lobsters, and Enright (1978) reviews the migration and homing capabilities of other marine invertebrates.

Whatley and Wall (1975) suggest that there may be seasonal migrations among some ostracoda.

Mattison *et al.* (1977) summarize and provide references to much of the literature regarding movements by echinoids. Some of these movements are in search of food; others are in response to stimuli such as winter cold (see Elmhurst, 1922, for a typical example).

MIGRATION AND GROWTH STAGE

One might think the fact that a form utilized near-shore, shallow-water nurseries and then migrated to offshore, deeper-water regions would be virtually impossible to establish in the fossil record. However, Christensen (1976) and Stevens (1965) have shown that for some Cretaceous belemnite species such is the case. They have shown that small specimens of some forms occur almost exclusively in nearshore facies, together with larger ones, but that the largest ones occur only in offshore facies. The possibility of obtaining this type of information for a variety of vagrant taxa, using similar evidence, is worth considering. Their conclusions find circumstantial support from the earlier paleother-mometry work done on belemnites (Urey *et al.*, 1951), which established that the earlier (i.e. young growth stage) portions of the belemnite test gave higher temperatures (i.e. possibly warmer, nearer shore water) than was true for the later (i.e. older growth stage portions) that might have been deposited in deeper, colder, more offshore water. Stevens (1965) has also shown that belemnite breeding probably occurred in a squid-like manner inshore followed by death.

SPACING AND COMMUNITY DEFINITION

In studying and defining communities the paleontologist, in common with the benthic marine ecologist, deals with discrete samples. It is critical that there be some knowledge of whether or not individual specimens comprising the sample are aggregated in a manner that might affect the nature of the sample. It is well known today that benthic organisms are not always distributed over their area of occurrence in a perfectly random manner. Mussels form blankets, oysters form blankets, scallops tend to be scattered about, and some terebratuloid brachiopods form clumps that employ older individuals as substrate for younger ones. Many species occur in a scattered, spaced-out pattern. In order to adequately sample these different distributional types, some attention must be paid to their modes of occurrence. Larval substrate selection is heavily implicated for the sessile benthos, and a variety of larval and postlarval mechanisms is implicated for the vagrant benthos. There is a general tendency, among the sessile benthos, for blankets dominated by one species to occur in the more turbulent environments, such as the mussels and oysters of the present, as well as the pentameroid brachiopods of the older Paleozoic.

Infaunal deposit feeders and many vagrant epifaunal forms tend to space themselves out. High diversity communities are, in general, characterized by a spacing out in a more or less random manner of the various species both sessile and vagrant. All these factors must be considered when defining and recognizing communities. It is also obvious that species tending to occur in patches during life will appear to be far less patchy in the fossil record under conditions of low sedimentation rate. The fossil record will tend to amalgamate patches, particularly if there has been a small amount of postmortem transport, so as to mask the originally patchy distribution present during life. Thus, community definition made on the basis of fossils may suggest a generally higher diversity than was actually present at many places during life. However, if the community is defined to include a number of patches the question becomes one of semantics rather than substance.

Gage and Coghill (1977) provide an excellent summary introduction to the problems of spacing encountered with marine benthos.

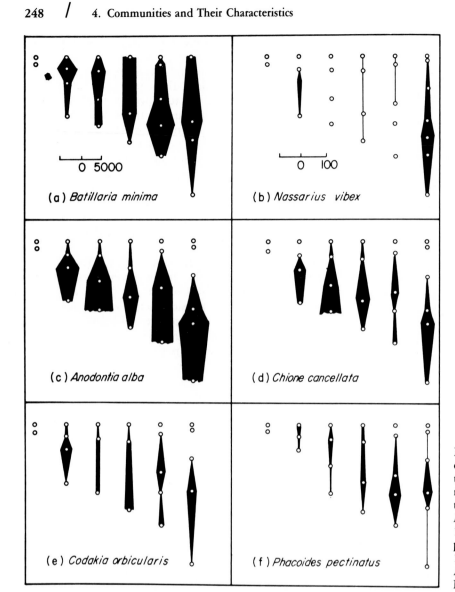

(a) *Batillaria minima*

0 5000

(b) *Nassarius vibex*

0 100

(c) *Anodontia alba*

(d) *Chione cancellata*

(e) *Codakia orbicularis*

(f) *Phacoides pectinatus*

FIGURE 205. Numerical distribution of the dominant molluscs along transects. The scale lines represent numbers per m²; the bivalve distributions are drawn to the same scale as *Nassarius*. (From H. B. Moore *et al.*, 1968, *Bulletin of Marine Science, 18*, pp. 261–279, Figure 7. Copyright © 1968 by Rosenstiel School of Marine & Atmospheric Sciences, University of Miami.)

shelf with the shelf margin as the outer limit, and 6 as possibly shelf-margin region or upper bathyal. It was then shown (Figure 206) how a previously unstudied community could be fitted into the Benthic Assemblage scheme by reference to cooccurring, previously studied communities on both sides, above, and below.

For example, the previously unassigned *Striispirifer* Community of the Rochester Shale (Figure 206) must be assigned to a Benthic Assemblage. The Rochester Shale containing the *Striispirifer* Community grades eastward into the Herkimer Sandstone. The western member of the Herkimer also possesses a *Striispirifer* Community fauna, but the eastern member is characterized by a fauna with attributes of Benthic Assemblages 1 and 2. Westward the Rochester Shale grades into the Fossil Hill Formation, which contains some beds with a Benthic Assemblage 3 Pentamerinid Community fauna. Vertically the Rochester Shale is underlain by the Irondequoit Limestone, which has a

Benthic Assemblage 4 Stricklandid Community fauna (*Costistricklandia*) in the lower, Rockway Member, and reeflike bodies that suggest Benthic Assemblage 3 (all of the Silurian reeflike bodies are considered to lie in this position) for the upper member. Overlying the Rochester Shale is the Lockport Group that lacks any "key" genera, indicating a specific Benthic Assemblage, but is itself overlain by the Salina Group containing Benthic Assemblage 1 fauna. Summing up these data one brackets the *Striispirifer* Community in the Benthic Assemblage 3 position despite the fact that it lacks any previously studied taxa indicating such an assignment. Absence of the widespread Benthic Assemblage 2 genus *Salopina* of the *Salopina* Community from the Rochester and western member of the Herkimer is another argument for placing the *Striispirifer* Community in Benthic Assemblage 3. The absence of *Dicoelosia*, common to taxonomically diverse deeper-water communities, is a good argument for excluding

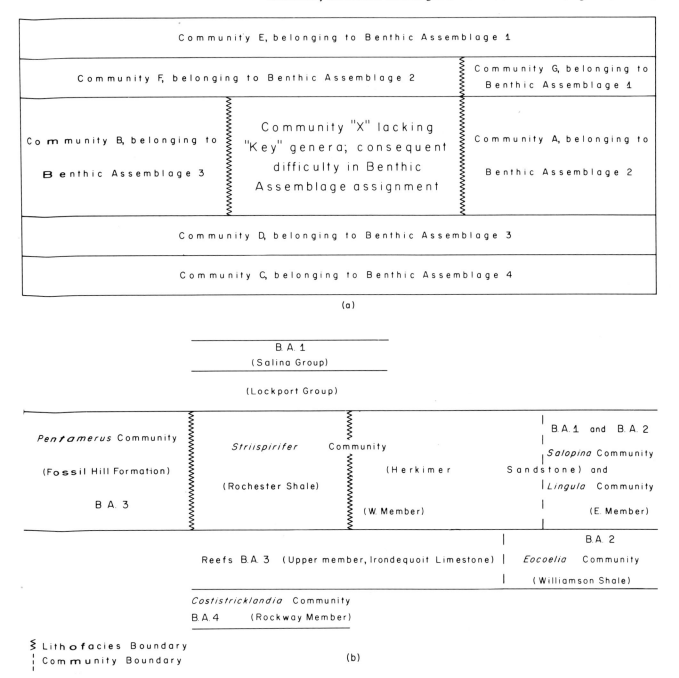

(a)

(b)

Lithofacies Boundary
Community Boundary

FIGURE 206. (*a*) The Diagram indicates how community "X", which lacks "key" taxa that permit it to be assigned to a Benthic Assemblage, may be bracketed into a Benthic Assemblage position by reference to adjacent communities whose Benthic Assemblage position has been previously determined. (*b*) The lower diagram shows relations of the *Striispirifer* Community which lead to the conclusion that a Benthic Assemblage 3 position is reasonable. Other interpretations may be made from this data, but additional information from other regions supports a Benthic Assemblage 3 position. (From A. J. Boucot, 1975, *Evolution and extinction rate controls*, Figures 12–13. Copyright © 1975 by Elsevier Scientific Publishing Company.)

the *Striispirifer* Community from Benthic Assemblage 4.

In addition to assigning communities to a Benthic Assemblage, it is necessary to study them in terms of as many environmental factors as possible. Figure 207 outlines a Community Framework employed for this

purpose. A Community Framework is a graphic means of arraying the multiplicity of community data available from a particular biogeographic unit for a particular time interval in terms of environmental variables. By using a Community Framework one avoids the difficulty of trying to keep an overwhelming

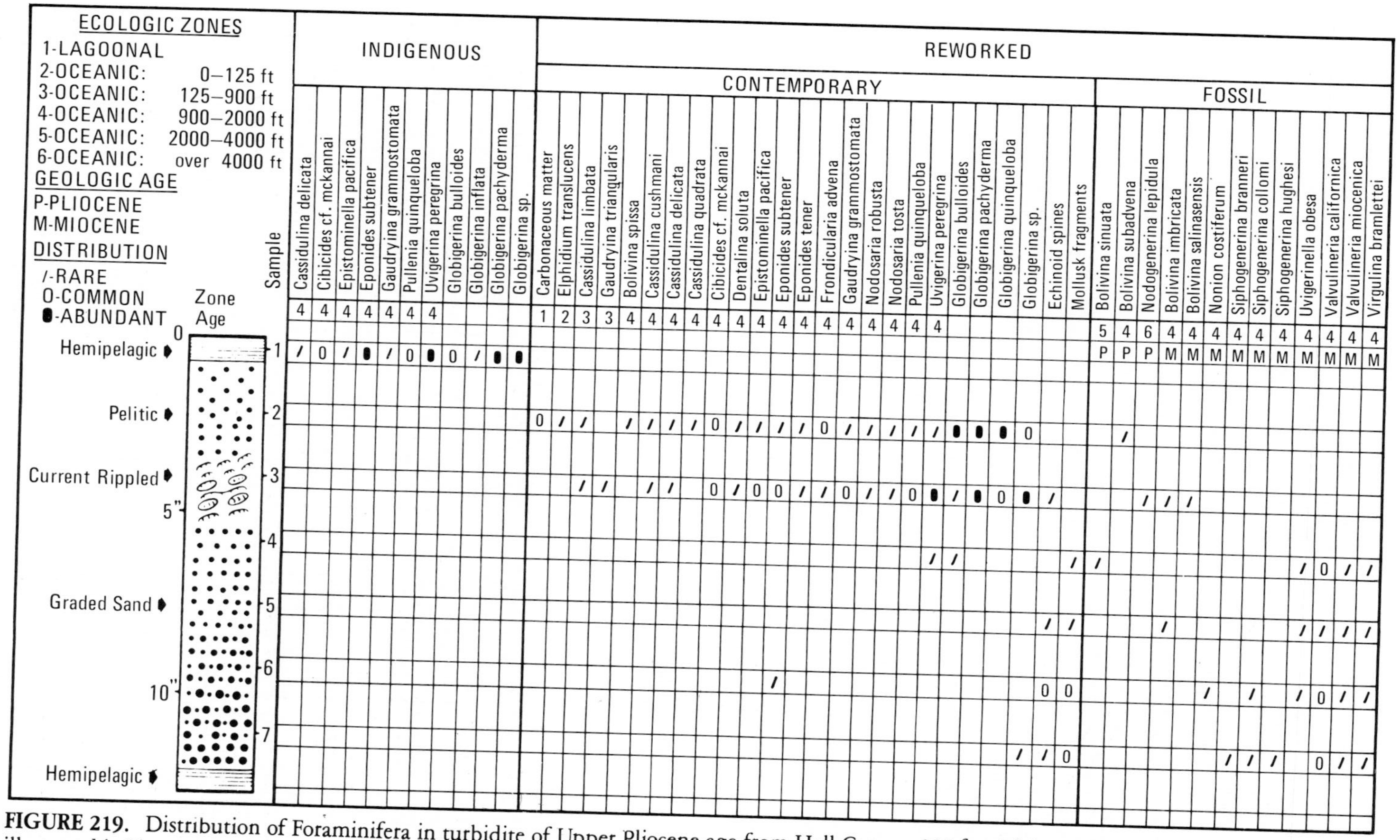

FIGURE 219. Distribution of Foraminifera in turbidite of Upper Pliocene age from Hall Canyon 300 feet higher stratigraphically than the sample illustrated in Figure 218. (From M. L. Natland, 1963. *Journal of Paleontology*, 37, pp. 946–951, Text-figure 3. Copyright © 1963 by the Society of Economic Paleontologists and Mineralogists.)

Another example is described by Stenzel and Turner (1944), who detail the interpretation of a vertebrate–invertebrate mixture containing elements from the brackish water–riverine freshwater and dry land fauna (oysters, rays, a sea cow, crocodilian, catfish, river snail, beaver-like rodent, river mussel, gar-pike, rhino, horse, camel). Stenzel and Turner explain their data in terms of a brackish-water depositional environment assisted by a little riverine transport. Wells (1976) and Fitch (1976) detail an Eocene example from Tonga that includes benthic shells derived from the shallow littoral to the bathyal environment, plus bathypelagic fishes and bathyal demersal fishes; the evidence for long-distance transport of contemporaneous materials is high, but it is not high for the mixing of materials of disparate ages. Spjeldnaes (1975, pp. 306–307) describes the transport and redeposition of Middle Ordovician to Lower Silurian silicified fossils in the Miocene and Pliocene of Denmark. There is not much possibility of confusing Lower Paleozoic fossils with those of the Neogene. Boekschoten (1966) discusses an excellent example of submarine Eocene beds being eroded off the Dutch coast in which the exhumed fossils are mixed on modern beaches with recent species. Wendt (1970) discusses examples from the Mesozoic of the Alps. Muller (1959) provides an excellent account of reworked Tertiary pollen present in the Recent sediment of the Orinoco delta. The citation of these examples should alert the geologist, but not so much as to dismiss animal community evidence of the past from serious consideration in the interpretation of environments. Geologists have been able to combine successfully both physical and biotic information and to decide, with a high degree of confidence in most instances, whether significant net transport has occurred.

Transported mixtures involving taxa of the same age are commonly more difficult to recognize than are transported mixtures of widely different ages. Reineck and Singh's (1973) evidence showing that the intertidal snail *Hydrobia* has been transported, in some cases, far into the North Sea and deposited amongst deeper water shells is an example. Wilson (1967) comments that: "On death the shells (*sic. Hydrobia*) become buoyant and are easily transported by onshore flood currents"; flotation is an uncommon phenomenon with most benthic marine invertebrates. Bimler (1976) provides another example of a gastropod that can float—in this case while alive.

Antia (1979), while reviewing bonebed characteristics, has described the mixture of marine and nonmarine organisms present in the Early Cretaceous Wealden Beds of England. Included are both freshwater organisms and terrestrial ones, as well as fully marine and brackish types. Antia also describes an Early Pleistocene English bonebed (The Suffolk Bonebed) that includes material eroded from Eocene through Pleistocene strata that were subsequently redeposited together. Both of these examples, while emphasizing vertebrate materials, are good examples of both contemporary mixing of remains derived from varied environments, and of noncontemporary remains redeposited together.

Wilson (1967) has described a modern intertidal-estuarine situation in which there has been considerable lateral transport, some of it far enough to mix thoroughly constituents of different communities in a postmortem manner unrelated to the biocoenosis.

Brenner and Davies (1973) reviewed, with examples, the concept of shallow water swells having the capability for concentrating benthic shells into lag aggregates associated with ripples in a manner that would certainly alter community structure.

Fleming (1953) has reviewed a number of New Zealand Cenozoic occurrences in which community ecology of the living shells plus geologic information is used to conclusively show whether or not significant transport has occurred. Fleming's examples are particularly good illustrations of how qualitative information can provide a reasonable solution to the question if some knowledge of the ecology of the living organisms is available. McAlester *et al.* (1964) provide an excellent discussion of how to recognize a transported mixture of materials using criteria similar to those used by Fleming. Middlemiss (1962) supplies another good example (Figure 220).

Kauffman (1969) has described a Tertiary *Thyasira* sample in which the joint occurrence of shallow-water and deep-water materials is logical in terms of the chaotic orientation of the materials due to turbulent transport conditions of both living and dead shells and other debris.

As a general rule, transport is a more important factor in the intertidal than in the subtidal zone, particularly in such places as intertidal gulleys of macrotidal regions, (see Kristensen, 1957, for an account of cockles being washed both up and down in the intertidal region of the macrotidal region of the macrotidal area), but the amount of net transport from the intertidal appears to be small. Haas' (1940) example of transport in the Recent has been quoted earlier (pp. 50–51).

Examples of deep water, even abyssal sedimentation of shallow-water materials, even abyssal sedimentation of shallow-water materials with the aid of turbidity currents are on record (see Lagaaij, 1973, for a Late Pleistocene shallow-water bryozoan assemblage displaced into deep water), but such things are unusual.

Transport of both living and reworked older materials has progressively more serious consequences as size decreases. Murray (1973) summarizes some of the information for smaller foraminifera. A large part of the problem is that settling velocity decreases very rapidly once a certain lower size limit is reached, in addition to the lower competent velocity needed to place the fossil in motion. Windle (1979) discusses a problem involving mixtures of Jurassic and Carboniferous spores. Dr. Jane Gray pointed out to me what is possibly the ultimate in redeposited-microfossil tall

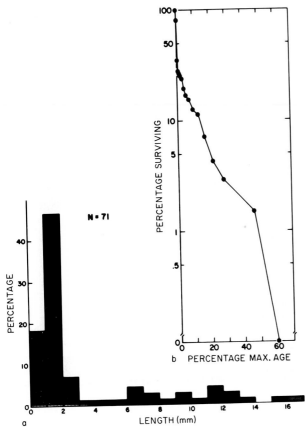

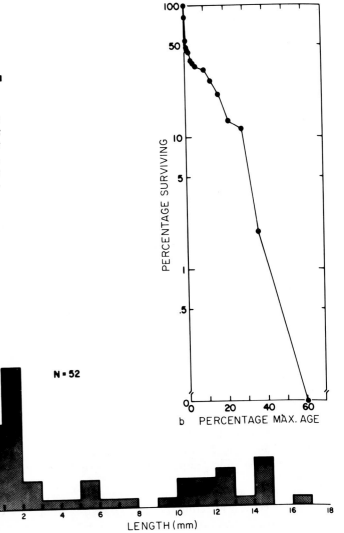

FIGURE 230. (*a*) Size–frequency distribution of *Onniella meeki* from the Upper Ordovician; (*b*) The survivorship curve suggests high infant and older mortality rates with a lower rate in between. (From R. P. Richards and R. K. Bambach, 1975, *Journal of Paleontology, 49*, pp. 775–798, Text–figure 9. Copyright © 1975 by the Society of Economic Paleontologists and Mineralogists.)

FIGURE 231. (*a*) Size–frequency distribution for *Onniella meeki* from the Upper Ordovician; (*b*) The survivorship curve suggests high infant and older mortality rates and a lower rate in between. (From R. P. Richards and R. K. Bambach, 1975, *Journal of Paleontology, 49*, pp. 775–798, Text–figure 8. Copyright © 1975 by the Society of Economic Paleontologists and Mineralogists.)

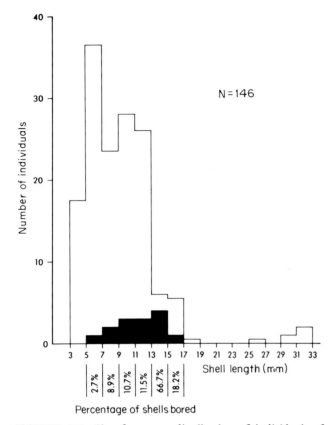

FIGURE 233. Bored specimen of *Discomyorthis musculosa;* internal view of pedicle valve interior (specimen was measured by P. M. Sheehan; photo courtesy of him). This is an Early Devonian brachiopod from Gaspé.

FIGURE 232. Size–frequency distribution of individuals of *Discomyorthis musculosa* from collection 2i. The size–frequency distribution of bored shells is indicated in black at the base of the histogram. The percent of shells bored in each size interval is given at the bottom of the figure. (From P. M. Sheehan and P. Lesperance, 1978, *Journal of Paleontology,* 52, Figure 2. Copyright © 1978 by the Society of Economic Paleontologists and Mineralogists.)

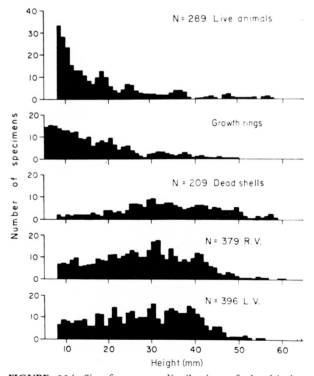

FIGURE 234. Size–frequency distribution of the bivalve *Codakia orbicularis,* living, dead, and disarticulated specimens plus growth rings, to show changes in population structure. The distributions imply a low infant mortality rate. (From G. Y. Craig, 1967, *Journal of Geology,* 75, pp. 34–45, Figure 6. Copyright © 1967 by Journal of Geology.)

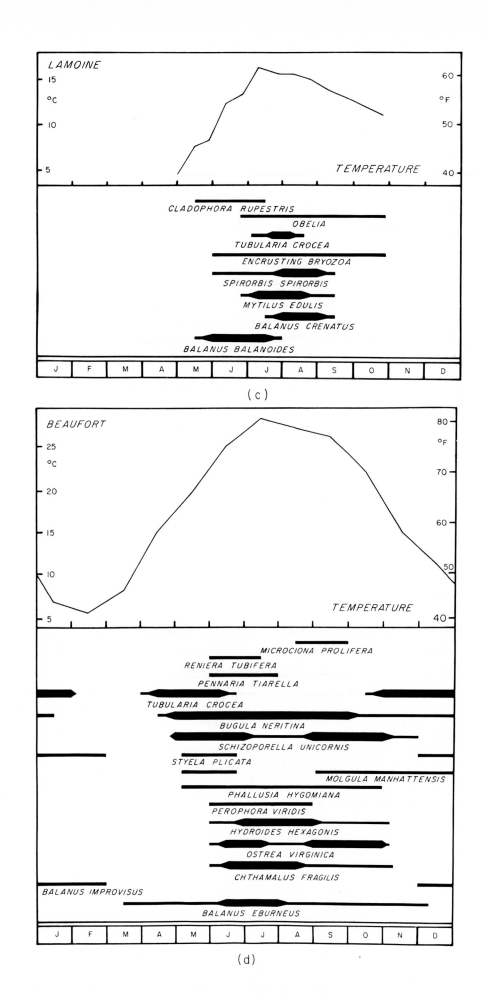

(c)

(d)

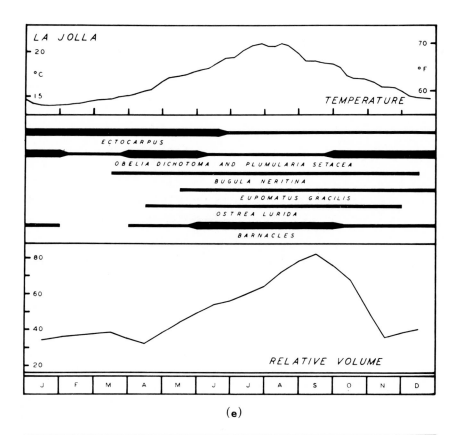

(e)

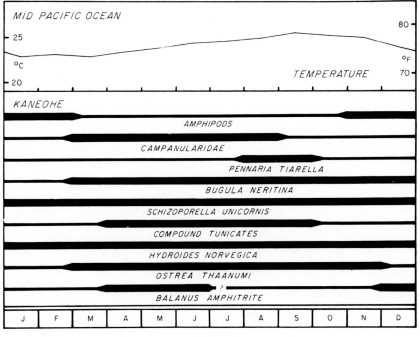

(f)

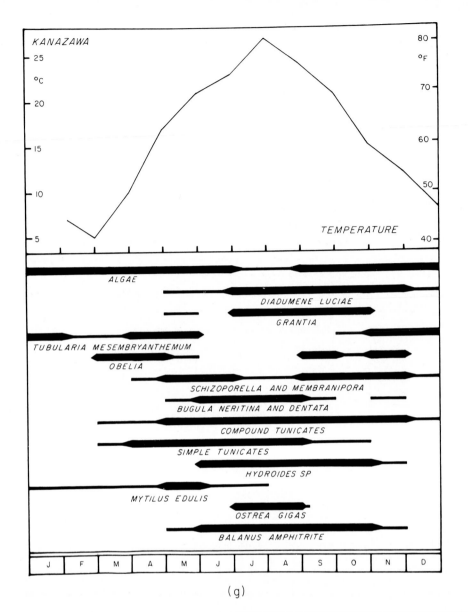

KANAZAWA

25 °C
20
15
10
5

80 °F
70
60
50
40

TEMPERATURE

ALGAE

DIADUMENE LUCIAE

GRANTIA

TUBULARIA MESEMBRYANTHEMUM

OBELIA

SCHIZOPORELLA AND MEMBRANIPORA

BUGULA NERITINA AND DENTATA

COMPOUND TUNICATES

SIMPLE TUNICATES

HYDROIDES SP

MYTILUS EDULIS

OSTREA GIGAS

BALANUS AMPHITRITE

| J | F | M | A | M | J | J | A | S | O | N | D |

(g)

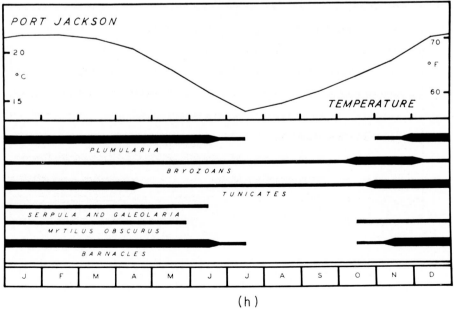

PORT JACKSON

20 °C
15

70 °F
60

TEMPERATURE

PLUMULARIA

BRYOZOANS

TUNICATES

SERPULA AND GALEOLARIA

MYTILUS OBSCURUS

BARNACLES

| J | F | M | A | M | J | J | A | S | O | N | D |

(h)

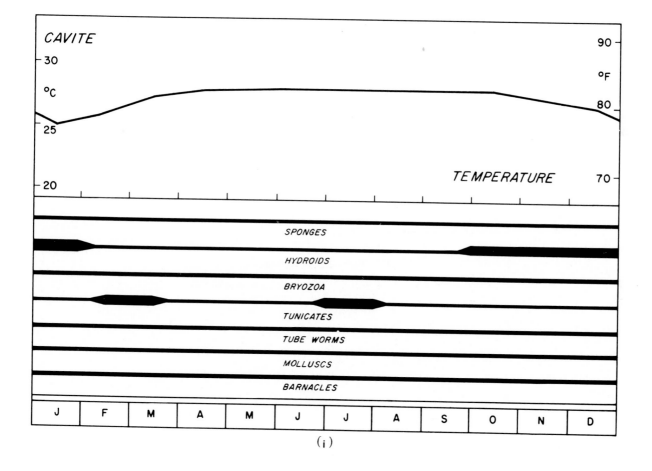

(i)

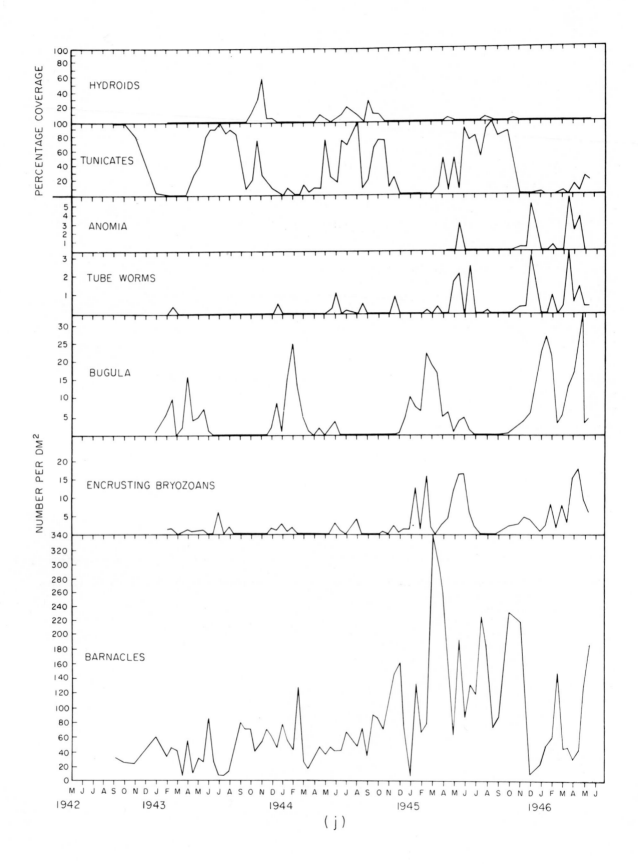

(j)

In addition to seasonal aspects of recruitment, there are a variety of irregular, local disturbances that take place on a year-to-year scale. Minor shifts in surface currents can disrupt the distribution of planktic larvae. Minor shifts in the upwelling pattern of cold waters can markedly displace larvae from places where they would normally be expected. Sindermann (1970) points out how some of the cyclic ups and downs in shellfish abundance may be intimately related to cycles in the incidence of diseases—including parasites. Disease outbreaks can be responsible for population crashes, followed either by slow buildups to another peak, or by rapid buildups. In any event, there are a number of factors involved in the recruitment process, some of them irregular through time, that can heavily perturb "normal" recruitment processes. There is no reason to suspect that these factors will affect all species in the same manner. One must expect that community composition will change over a period of time due to the vagaries of recruitment differentially affecting the constituent species. Over a geologically lengthy interval, however, these factors will tend to cancel out and produce a relatively monotonous product—unless one

looks at the short time intervals where the vagaries of recruitment actually become manifest.

Seed (1976; see Table 4.20) has tabulated data for the mussel *Mytilus edulis* indicating how variable settlement of the larvae (i.e., recruitment) may be, in terms of time, for a single species. It is doubtful if all the cooccurring species accompanying *M. edulis* in many localities have identical recruitment patterns. This type of situation inevitably complicates the observed community patterns from place to place.

ARTICULATION RATIO

I have (Boucot, 1953) emphasized the utility of studying the ratio of articulated to disarticulated bivalved shells in an effort to understand the amount of current activity suffered by a fauna. Broadhurst (1964) reported on a set of Carboniferous nonmarine bivalves in which the condition of almost total articulation is in agreement with other data suggesting little current activity and transport.

Worsley's (1971) data (Tables 4.21–4.22) from shales and shell beds shows how an articulation ratio may be

TABLE 4.20
Observed Settlement Periods of *Mytilus edulis*[a]

Locality	J	F	M	A	M	J	J	A	S	O	N	D	Comments
United Kingdom General N.E. Coast													Especially on zoophytes; maximum spatfall end of summer
Yorkshire													Some spatfall almost any time of year
E. Coast													Spatfall starts in spring and continues until early June
Morecambe Bay Morecambe Bay													Especially after denudations by storms and scour
River Exe													Small spat on *Enteromorpha*
Conway Menai Straits													Some spatfall in practically every month
Montrose													
Linne Mhuirich Millport													

(Continued)

TABLE 4.20 *(Continued)*

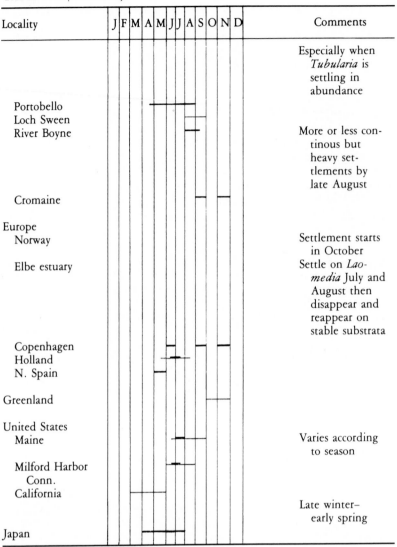

Locality	J	F	M	A	M	J	J	A	S	O	N	D	Comments
													Especially when *Tubularia* is settling in abundance
Portobello													
Loch Sween													
River Boyne													More or less continous but heavy settlements by late August
Cromaine													
Europe													
Norway													Settlement starts in October
Elbe estuary													Settle on *Laomedia* July and August then disappear and reappear on stable substrata
Copenhagen													
Holland													
N. Spain													
Greenland													
United States													
Maine													Varies according to season
Milford Harbor Conn.													
California													Late winter–early spring
Japan													

[a]From R. Seed, 1976, *In* B. L. Bayne, Ed., *Marine Mussels: Their Ecology and Physiology*, Table 2.4. Copyright © 1976 by Cambridge University Press.

used as a measure of total transport, although this does not necessarily indicate net transport unless studies similar to that of Watkins (1975) involving map relations are also done (also see Middlemiss, 1962; Figure 220). Watkins (1974a) has demonstrated how mapping of various parameters such as the disarticulation ratio, number of opposing valves remaining in the vicinity, breakage ratio, and so on, may be employed for such purposes (see also Boucot *et al.*, 1958; Figures 260 and 261). Watkins (1975) has provided additional data (Figure 262), as has Kern (1973, Figure 9).

I have also (Boucot, 1953) emphasized that different species have different tendencies toward disarticulation (i.e., disarticulation rates, under the same set of

physical conditions). Data from Boucot *et al.* (1958), as well as Amsden and Ventress (1963), and Amsden (1975) provides further evidence (see Figures 260–263, Table 4.23). Sheehan (1978) has made further contributions to the foregoing data on disarticulation tendencies shown by shells possessing different articulating structures.

OPPOSITE VALVE RATIO

I have earlier (Boucot *et al.*, 1958; Figures 260 and 261) used examples of ratios of brachial-to-pedicle valves in Devonian brachiopods in an effort to understand both amount of current activity and amount of

TABLE 4.21
Mode of Occurrence of Common Lower Silurian Brachiopods in a Norwegian Shale Example[a,b]

Brachiopod	Percentage of population	Pedicles: brachials	Articulation ratio (%)
6bα			
Leangella triangularis			
(= *Leptaena transversalis*			
var. *minor* of Kiær, 1908)	75	7.9 : 1.0	84
Dicoelosia osloensis			
(= *Bilobites biloba* of			
Kiær, 1908)	12	8.5 : 1.0	86
Skenidioides lewisii			
(= *Skenidium Lewisii*			
of Kiær, 1908)	6	7.0 : 1.0	89
6bβ			
Protatrypa malmoeyensis			
(= *Atrypa reticularis*			
of Kiær, 1908)	64	1.2 : 1.0	94
Eoplectodonta duplicata			
(= *Leptaena transversalis*			
var. *major* of Kiær, 1908)	14	6.1 : 1.0	81

[a] From D. Worsley, 1971, *Norsk Geol. Tidsskr., 51*, pp. 161–167, Table 1. Copyright © 1971 by Universitetsforlaget.
[b] A high percentage of articulated specimens are encountered, suggesting little disturbance.

TABLE 4.22
Mode of Occurrence of Common Lower Silurian Brachiopods in a Norwegian Siltstone Example[a,b]

Brachiopod	Pedicles: brachials	Articulation ratio (%)
Leangella triangularis	2.0 : 1.0	24
Dicoelosia osloensis	2.8 : 1.0	8
Eoplectodonta duplicata	0.6 : 1.0	6

[a] From D. Worsley, 1971, *Norsk Geol. Tidsskr. 51*, pp. 161–167, Table 2. Copyright © 1971 by Universitetsforlaget.
[b] A low percentage of articulated specimens are encountered, suggesting much disturbance.

net transport away from a growth site (also see Figure 262; p. 311; Tables 4.21 and 4.22, p. 307). Total transport, as opposed to net transport distance, will tend to disarticulate bivalves regardless of net transport distance, whereas the opposite valve ratio provides a better measure of net transport distance. Boyd and Newell (1972) provide interesting data on opposite valve ratios in Permian bivalves (Figure 264), which are thought to have suffered little net transport. Craig (1967) graphed size–frequency distributions for modern living and dead bivalves from relatively undisturbed environments (Figure 265), showing conclusively that similar numbers of right and left valves of the appropriate size, regardless of whether the species disarticulates easily or with difficulty after death, provide good evidence for *in situ* material. Conversely, he finds in common with Lever (1958), that bivalve shells thrown up on a beach are subject to right–left drift sorting (as also discussed by Boucot *et al.* [1958]).

Kornicker *et al.* (1963) have shown how opposite valve ratios of similar mirror image left–right valves in a bivalve do not always yield consistent departure from a 1:1 ratio as the organism undergoes transport (Figures 266 and 267). They also point out that bored valves may have higher or lower competent velocities than unbored shells. Schmidt and Warme (1969) provide data on left–right bivalve distributions, with information on the frequency of boring (Figure 268), and Veevers (1959), a sample (Figure 269) in which the similar form of the brachial–pedicle valve curves suggests little net transport despite the high level of disarticulation.

Lever (1961) has shown that bored bivalve shells have a lower transport velocity than do similar-sized unbored shells of the same species. His experimental work has shown that the presence of a hole in a shell, whether generated by boring or present naturally (as a pedicle foramen in some brachiopods) creates a very different hydrodynamic situation. It is very difficult to predict the hydrodynamic behavior of seashells; flume studies should be undertaken whenever possible.

TABLE 4.23

Different Tendencies toward Disarticulation[a] and the Ratio of Pedicle to Brachial Valves Shown by Lower Devonian Brachiopod Taxa in a Sample from Oklahoma[b]

Taxa	Arbuckle region				Sequoyah County			
	Articulated shells	Pedicle valves	Brachial valves	Pedicle/Brachial ratio	Articulated shells	Pedicle valves	Brachial valves	Pedicle/Brachial ratio
Rhipidomelloides musculosus	3	44	1	44.0	0	25	9	2.8
Levenea sp.	0	14	14	1.0	0	7	18	0.4
Platyorthis? sp.	0	0	4		0	0	3	
Leptostrophia magnifica	0	120	0		0	12	0	
Leptaena ventricosa	0	45	5	9.0	0	27	2	13.2
Pholidostrophia? sp.	0	0	0		0	4	2	2.0
Strophodonta sp.	0	0	0		0	4	0	
Strophonella sp.	0	0	0		0	1	1	1.0
Chonetes? sp.	0	0	0		0	1	0	
Chonostrophia complanata	0	0	0		0	23	0	
Anoplia nucleata	0	19	1	19.0	0	2	0	
Schuchertella becraftensis	0	0	0		0	4	4	1.0
Costellirostra peculiaris	29	30	12	2.5	9	5	4	1.2
Plethorhyncha cf. P. barrandi	0	22	13	1.7	0	0	0	
Plethorhyncha? welleri	1	16	8	2.0	0	9	3	3.0
Camarotoechia? cf. C. dryope	1	0	0		0	1	3	0.3
"Camarotoechia" sp.	0	0	0		1	0	3	
Spinoplasia oklahomensis	1	21	2	10.5	0	0	0	
Hysterolites (A.) murchisoni	1	179	12	15.0	0	87	19	4.5
Costispirifer arenosus	0	4	3	1.3	0	36	7	5.1
Kozlowskiellina new species	1	3	1	3.0	0	0	(?)5	
Eospirifer new species	0	3	0		0	7	2	3.5
Meristella carinata	1	376	66	5.6	0	15	11	1.4
"Meristella" vascularia?	0	2	0		0	18	4	4.5
Cyrtina rostrata	1	39	24	1.6	0	2	11	0.2
Trematospira sp.	0	2	0		0	0	0	
Rhynchospirina? sp.	1	1	0		0	0	0	
Renselaeria cf. R. elongata	1	71	47	1.5	0	7	15	0.5
Renselaeria sp.	0	1	0		1	0	0	
Prionothyris perovalis	4	17	6	2.8	0	0	0	
Oriskania simuata	6	3	27	0.1	0	0	0	
Beachia new species	0	0	0		0	12	2	6.0

[a]That is, disarticulation rate differences.
[b]From T. W. Amsden and W. P. S. Ventress, 1963, *Oklahoma Geological Survey Bulletin 94*, pp. 1–238, Table 1. Copyright © 1963 by the Oklahoma Geological Survey.

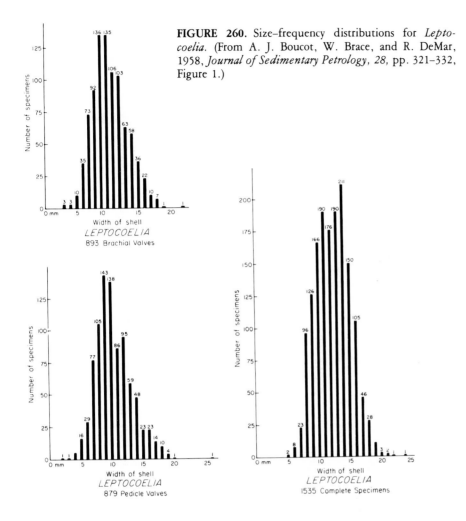

FIGURE 260. Size–frequency distributions for *Lepto-coelia*. (From A. J. Boucot, W. Brace, and R. DeMar, 1958, *Journal of Sedimentary Petrology*, 28, pp. 321–332, Figure 1.)

TABLE 4.24
Valve Characteristics and Flume Velocities Needed to Overturn Shells[a]

Characteristic	Continental Shelf			Wave-Washed Zone of Sand Beaches			
	Placopecten magellanicus	*Arctica islandica*	*Spisula solidissima*	*Spisula solidissima*	*Mytilus edulis*	*Arca ovalis*	*Arca campechensis*
Length (cm)	13.0	8.5	15.6	14.7	7.6	5.7	4.6
Height (cm)	12.5	7.5	10.9	10.2	3.2	4.0	3.7
Half thickness (cm)	1.9	2.2	2.9	2.8	1.3	1.8	1.6
Weight (gm)	42.9	48.4	121.2	110.0	6.1	17.4	7.5
Angle of balance (°)	11	15	15	15	26	19	16
Flume velocity to overturn (cm/sec)	22	44	46	46	23	29	35

[a]From K. O. Emery, 1968 *Journal of Sedimentary Petrology*, 38, pp. 1264–1269, Table 2. Copyright © 1968 by the Society of Economic Paleontologists and Mineralogists.

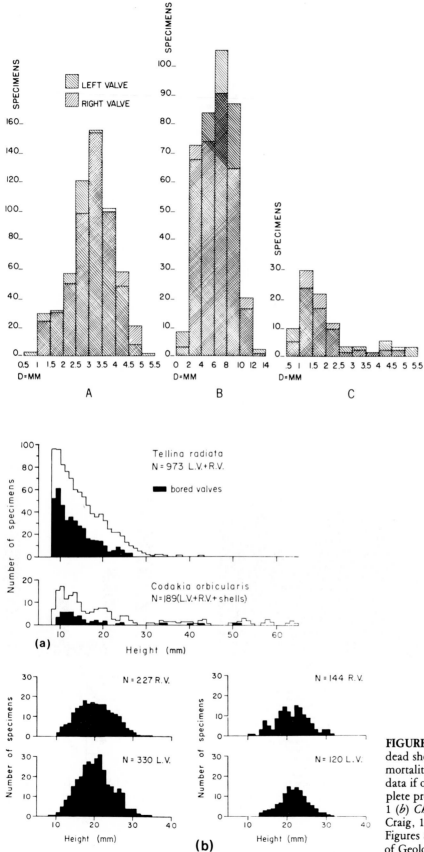

FIGURE 264. Size–frequency distribution of left and right valves for some Permian bivalves. Note the relatively close correspondence consistent with little net transport. (*a*) *Scaphellina bradyi;* (*b*) *Oriocrassatella elongata;* (*c*) *Kaibabella sp.* (From D. W. Boyd and N. D. Newell, 1973, *Journal of Paleontology,* *46,* pp. 1–14, Text–figure 2. Copyright © 1972 by the Society of Economic Paleontologists and Mineralogists.)

FIGURE 265. Size–frequency distributions of dead shells showing a variety of types. Different mortality rates could be interpreted from these data if one assumes little net transport and complete preservation. (*a*) *Chione cancellata,* locality 1 (*b*) *Chione cancellata,* locality 6. (From G. Y. Craig, 1967, *Journal of Geology,* *75,* pp. 34–45, Figures 8–10. Copyright © 1967 by the Journal of Geology.)

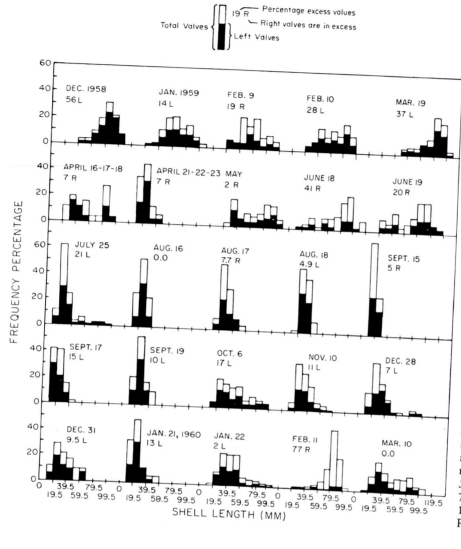

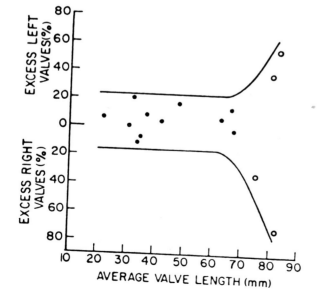

FIGURE 266. Size–frequency distributions of a transported bivalve (*Dinocardium robustum*) on a Gulf Coast Beach showing a variety of curves. Note that these transported assemblages could be interpreted in a variety of ways. (From L. S. Kornicker *et al.*, 1963, *Journal of Sedimentary Petrology, 33,* pp. 703–712, Figure 7. Copyright © 1963 by the Society of Economic Paleontologists and Mineralogists.)

FIGURE 267. Right–left bivalve relation on a Gulf Coast beach as a function of size. It is suggested that sorting occurs only above a certain size limit in this environment. Morphology of the shells is mirror image. (From L. S. Kornicker *et al.*, 1963, *Journal of Sedimentary Petrology, 33,* pp. 703–712, Figure 4. Copyright © 1963 by the Society of Economic Paleontologists and Mineralogists.)

varied fates of the many transported taxa provides a rational explanation for many of the puzzling mixed assemblages that turn up occasionally in the fossil record. Despite the fact that fossil assemblages of inordinately high taxic diversity immediately alert the paleontologist to look for something bizarre, many would hesitate to imagine a tale of the type Haas actually observed and put down on paper (see Chapter 3, pp. 50–51). One should not be carried away by this story and overuse such a possibility. Shaw's account (*in* Boucot, 1975) of the minimal effect of another Florida hurricane on the benthic shells of at least one area stresses that there are many types of storms in terms of disturbing the benthos.

Engle (1948) provides graphic descriptions of hurricanes working in shallow waters; they apparently are capable of blanketing oyster-covered bottoms enough to smother and bury the oysters completely. He makes clear, however, that movement of oysters is not the important component of the environmental shift brought on by the hurricane.

Rees *et al.* (1977) provide further data on the effects of storms on the transport of shallow-water shelly benthos; the effects may in some cases be significant in the subtidal as well as in the beach environments.

TRANSPORT BY SHORE-ICE

Spjeldnaes (1978) has reviewed the extensive literature on the transport by shore-ice of both living shells and fossils. Both mechanisms may seriously perturb the fossil record. The occurrence, for example, of polar benthic foraminifera in much lower latitude oceanic deposits, where the shore-ice melted, can lead to entirely erroneous paleoecological conclusions. The occurrence of living *Mytilus* on Arctic Ocean ice blocks is further evidence, summarized by Spjeldnaes, for serious transport possibilities. Spjeldnaes's work emphasizes the Pleistocene to present, but it is obvious that the problem of ice transport should also be considered in more ancient intervals subject to very cold conditions. The possibility that similar transport is occassioned by plant debris originating in the shallow-water region but rotting elsewhere so as to release its shelly epifauna should also be given serious consideration in some instances.

VERTICAL MOVEMENTS AND BIOTURBATION

Many students of paleoecology have considered the problem of transport in the horizontal direction. But there is also reliable evidence that shells are moved in a vertical plane by bioturbating organisms. In the deep sea, the work of Berger and Heath (1968) has shown the significance of vertical movement, and Clifton

(1971) reports that shells are moved upwards by bioturbating organisms (see Chapter 5, pp. 357–399). Thus, problems are induced by vertical mixing of communities that had no contact with each other in life. One cannot assume that all shells found in a particular layer of sediment actually lived together in that stratum. However, there is no evidence that there is much vertical transport for distances of more than a few tens of centimeters, although van Straaten's (1952, 1956) *Hydrobia* bed (Figure 289) example involves a depth of several hundred centimeters, and tidal gulleys (Schäfer, 1972; Figure 290) provide similar possibilities for vertical transport.

In addition to vertical movement due to bioturbation, infaunal elements suddenly covered by sediment, as during storm activity, may have differing capabilities for extricating themselves. Some with limited ability to move suddenly in a vertical direction through a newly deposited sediment blanket merely die in place; others are able to extricate themselves. Again, this complicates the definition of infaunal communities. MacGinitie and MacGinitie (1949) discuss a specific example, and Kranz (1974a,b) reviews several aspects of the question. Trewin and Welsh (1976) present an example in addition to van Straaten's in which the burrowing activity of invertebrates tends to form a shell stratum some distance below the sediment surface from which the shells were derived (see Figure 295).

STRATIGRAPHIC LEAKS

Additional problems are posed by stratigraphic leaks; evidence indicating cracks in the sea floor where younger materials (normally microfossils not macrofossils are involved) have slipped down into older beds, permitting the paleontologist to erroneously conclude that the fossils extracted from a rock sample actually lived together. There are some examples (Seddon, 1970) of conodonts of varying younger ages having been moved into intimate association with considerably older beds. The concept of stratigraphic leaks should be

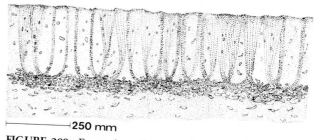

⊢————————┤ **250 mm**

FIGURE 289. Formation of a *Hydrobia* bed by the activities of *Arenicola*. (From W. Schäfer, 1952, *Aktuo-Paläontologie* Abb. 269. after van Straaten, 1952, 1956. Copyright © 1962 by Verlag Waldemar Kramer.)

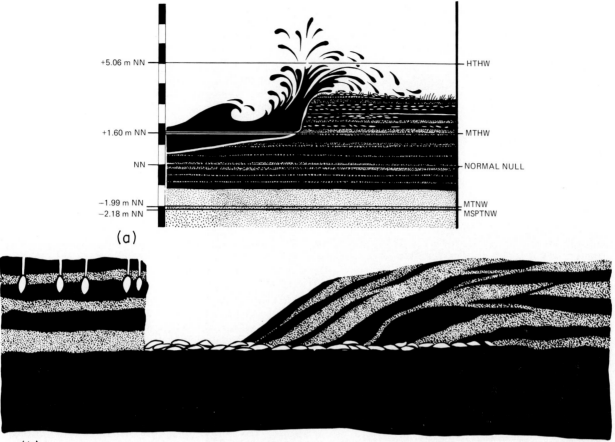

FIGURE 290. Shell bed formation by the activities of storm waves exhuming shells and throwing them up on the surface (*a*) and of a migrating intertidal gulley (*b*) in exhuming shells and depositing them in the bottom of the migrating gulley. (From W. Schäfer, 1952, *Aktuo-Paläontologie* Abb. 267–268; (*b*) after van Straaten, 1950 and 1954. Copyright © 1962 by Verlag Waldemar Kramer.)

used with caution, not as an excuse for retaining a rigid notion of the "correct" stratigraphic ranges of the fossils in question. Small, filled karstic features also have the potential for generating stratigraphic leaks if they remain unnoticed by the geologist.

There is abundant evidence in the geologic record of Neptunean dikes—some of the dike filling does contain younger fossils that, if the dikes remain undetected, can be mixed with older fossils occurring to either side of the dike material (Chlupac *et al.*, 1979, Figure 15).

COMMUNITY MIXING BY ENVIRONMENTAL SHIFT

Taxic associations inferred to represent mixed communities (Table 3.18, p. 166) normally restricted to separate environments may be interpreted in a variety of ways. There are examples of rough-water *Costistricklandia* Community shells associated with a high-

diversity, quiet-water *Striispirifer* Community that are at first puzzling (Figures 291 and 292). An abundant epifauna present on the disarticulated *Costistricklandia* valves, whose coequal numbers of brachial and pedicle valves suggest little net movement, is consistent with rough-water *Costistricklandia* Community conditions having given way in time to quiet-water conditions favorable to the *Striispirifer* Community, without any influx of sediment. The attached epifaunal organisms (Figures 291 and 292) on fragmentary and broken *Costistricklandia* shells provide additional supportive evidence for this interpretation. The ball–like tabulate colony (Figure 292, D–F) is most easily interpreted as an upright *Costistricklandia* fragment, on which the tabulate coral colony developed centrifically, rather than as an oncolite growth. Such growth is inconsistent with occasional, random rolling. We have here a community mixture in which there has been little transport and movement, merely a change from turbulence to quiet-water conditions.

Another type of environmental shift is one that

FIGURE 291. See Table 3.18, page 166 for locality, stratigraphic, and abundance information. (*a*)–(*d*) *Costistricklandia* fragment encrusted by a branching tabulate coral (*Syringopora*). (*e*)–(*f*) *Costistricklandia* pedicle valve beak region encrusted by a tabulate coral (possibly Alveolites). (*g*) *Costistricklandia* fragment encrusted by a bryozoan. (*h*)–(*l*) *Costistricklandia* pedicle valve encrusted by *Heliolites* cf. *H. megastoma* Northrop, 1939, and *Aulopora* sp. (*j* only), (*l*)–(*k*) Tubular growth on the interior of a *Costistricklandia* valve (Auloporoid? coral, possibly *Aulopora* sp.) (*m*)–(*n*) Obverse sides of a *Costistricklandia* shell acting as a hard substrate for a stromatoporoid and rugose corals. (Corals identified by W. A. Oliver, Jr., U.S. Geological Survey.) (× .75)

determines the success or failure of a taxon in a particular area (also see Chapter 4, pp. 188–200). A number of observers (Baxter, 1962; Coe, 1953, 1955, 1956) have commented on the irregular, fluctuating abundances of certain intertidal bivalves such as *Donax, Tivela,* and others. In the fossil record, such swings in the abundance of one taxon relative to a group of others might be puzzling, but should be kept in mind. Whether or not the times of high abundance represented different communities than times of low abundance or even times of total absence, is not answerable until more is known about environmental control over success and

failure of fluctuating (in terms of abundance) taxa relative to associated nonfluctuating taxa. Still another type of environmental shift is detailed by Powell (1937), in which the shell debris from a higher elevation community is transported into deeper water en masse, and a shelly substrate is formed for a deeper water, quiet-condition community. The *Costistricklandia-Striispirifer* example cited might be of this type, although we have no supporting evidence for such a conclusion except that rough versus quiet water is one of the variables.

McKnight (1968) cites an example from New

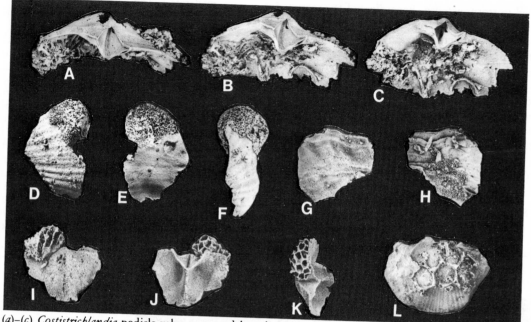

FIGURE 292. (*a*)–(*c*) *Costistricklandia* pedicle valve encrusted by a branching tabulate coral (*Romingeria*-like auloporoid coral) (*d*)–(*f*) *Costistricklandia* fragment encrusted by a spherical tabulate coral colony (possibly a favositid). (*g*)–(*h*) Obverse side of a *Costistricklandia* fragment practically encrusted (probably by an encrusting bryozoan). (*i*)–(*k*)*Costistricklandia* fragment encrusted by a favositid colony (*Palaeofavosites* sp.) (*l*) *Costistricklandia* valve encrusted by a branching tabulate coral (*Aulopora* sp.) (Corals identified by W. A. Oliver, U.S. Geological Survey.) (× .75)

Zealand where an ancient infaunal association appears to have been exhumed and then encrusted with living materials: Modern epifauna settled down in its midst, presenting a confusing situation to the unwary paleontologist.

MOVEMENT AND DESTRUCTION AS EVIDENCED BY SHELL BREAKAGE AND ROUNDING

Movement, with or without net transport, has the potential for breaking shells and for rounding and breaking massive biogenic materials such as aggregates of colonial organisms (stromatoporoids, corals, bryozoans, etc.) and calcareous algae. Potential for cleavage of shelly material along parting planes, which is predetermined by shell structure, such as the two-layered septae of certain brachiopods (pentameroids) is also present. All these factors can provide data relevant to whether or not a particular fossil assemblage consists of *in situ* material in terms of life and death assemblages. Most shells are thin, so breakage and cleavage are common evidence for movement, but rounding occurs on those portions of the shell that are relatively massive (beak regions of brachiopods, bivalve umbones, gastropod columellae). Certain growth forms are resistant to breakage and to rounding, including such items as pelmatozoan columnals (rounded to begin with and of equant form) and solitary small- to medium-size corals. It is common to find assemblages of material that have suffered considerable movement, consisting of both broken and subrounded materials, to have appropriate median diameters depending on their hydrodynamic properties. The ball mill effect, encountered in the intertidal region in areas of steep beach slope, rapidly breaks and rounds, as well as sorts, available shelly materials, including items collected from the intertidal and the nearby subtidal (Haas, 1940, see Chapter 3, pp. 50–51).

Intertidal ball-milling is more rapid and destructive in its action that is subtidal movement and transport. Ball-milling appears to be more common as an intertidal than as a subtidal phenomenon. Evidences of turbidity-flow transport correlate with a low level of breakage and rounding in many cases, although there are some cases where extensive breakage is evident.

In general, evidence of breakage and rounding correlates with an artificially high taxic diversity, indicating that more than one community has been amalgamated during movement and transport.

Ball-milling of concavo-convex shells affects the convex external structures disproportionately; one must not be misled by the presence of well preserved internal beak-region structures into thinking that significant movement and/or transport has not occurred.

Boyd and Newell (1972) record a Permian situation in which obviously rounded bellerophontids that were thought to have suffered transport were mixed with thoroughly unrounded material thought to be essentially *in situ*. Brenner and Davies (1973) describe a variety of badly broken shell materials from the Jurassic of Wyoming and Montana where the breakage was

Newman's (1960) example suggests that the low abundance of balanoid barnacles in the tropical–subtropical intertidal reflects the habits of parrot-fishes and others that nibble the carbonate rocks clean, and possibly accounts for the rarity of seaweeds (Stephenson and Searles, 1960). This may be an example of a Keystone group.

It is easy to define dominance in terms of the presence of a particular taxon above a certain absolute number, or above a percentage of the biota. But it also may be significant to heed the percentage of samples in which a particular taxon is present as opposed to those in which it is absent, regardless of absolute number or fixed percentage. There are no "rules" for defining dominance in either of these approaches to the problem. The concept of dominance involves relative abundance in individual samples, as well as regional abundance. A species found as a numerical dominant at a few sites may still form the basis for the definition of a community, albeit rare, whereas a species found at a large percentage of the sampled localities but in modest abundance, might also be considered to be a dominant.

SHELLY BIOMASS

For the marine ecologist and for the terrestrial ecologist, the concepts of biomass, productivity, standing crop, etc., are routine. For the paleoecologist, they are essentially unknown and uninvestigated. The well known relation in the modern sea between increasing biomass and approaching shoreline in progressively shallower water from the ocean has not been really examined in the fossil record. It is difficult for the paleontologist to distinguish standing crop from rate of production, although so little thought has been given to the problem that the question cannot be dismissed as impossible of solution.

The paleoecologist, although lacking the capability for measuring biomass, does have the means for measuring shelly biomass. Whether or not shelly biomass is a true reflection of biomass; whether different biomass levels existing in the past were a direct function of the preserved shelly biomass, and so on, are questions that probably will not be answered easily. It is unfortunate that so little data is available from the modern environment relating shelly biomass to total biomass (see Lawrence, 1968, Stanton, 1974, 1976;

Warme *et al.*, 1976, for exceptions). The relation of modern shelly biomass to total biomass from environment to environment, much less from community to community is virtually unknown. There must be trends in the modern environment worthy of study. Separating rate of sedimentation from organic productivity complicates the questions of shelly biomass. There is a wide spectrum of sedimentation rate from fractions of a mm/year to truckloads per brief time interval.

Despite an aura of indifferent ignorance on the part of both ecologists and paleoecologists in the past, we have a qualitative impression that shelly biomass decreases markedly as one leaves the intertidal–shallow subtidal regions, as it probably did in the past. Until time interval by time interval has been sampled, we cannot say much more about variation in shelly biomass.

Futterer and Paul (1976) provide data that suggests shelly productivity is a decreasing function of departure from shoreline (Figure 299).

COMMUNITY GROUPS—ANALOGOUS AND HOMOLOGOUS COMMUNITIES

I have previously (1975, pp. 227–232) defined and illustrated the terms Community Group, Analogous Community, and Homologous Community. A brief review of the three terms is useful here. Petersen (1914) defined what we now refer to as species-level *Petersen Animal Communities*. These species-level communities are regularly recurring associations of the same species with similar abundance levels for each species involved. Thorson (1958) noted that the genera present in similar regularly recurring associations collected from widely distant regions were sometimes the same, but that the species involved were commonly different. For these similar associations occurring on distant pieces of bottom belonging to biogeographically different units, Thorson coined the term *Parallel Community*. The term *Community Group,* unites the Petersen Animal Community, species-level association, the Thorson Parallel Community concept, and the evolutionary change in communities that occurs through geologic time. A Community Group is defined as a time sequence of Petersen Animal Communities, which may or may not have biogeographically separate Thorson Parallel Community branches consisting of sequences of parallel Petersen Animal Communities. The Petersen and Thorson concepts, although entirely adequate for the present needed to be extended to account for

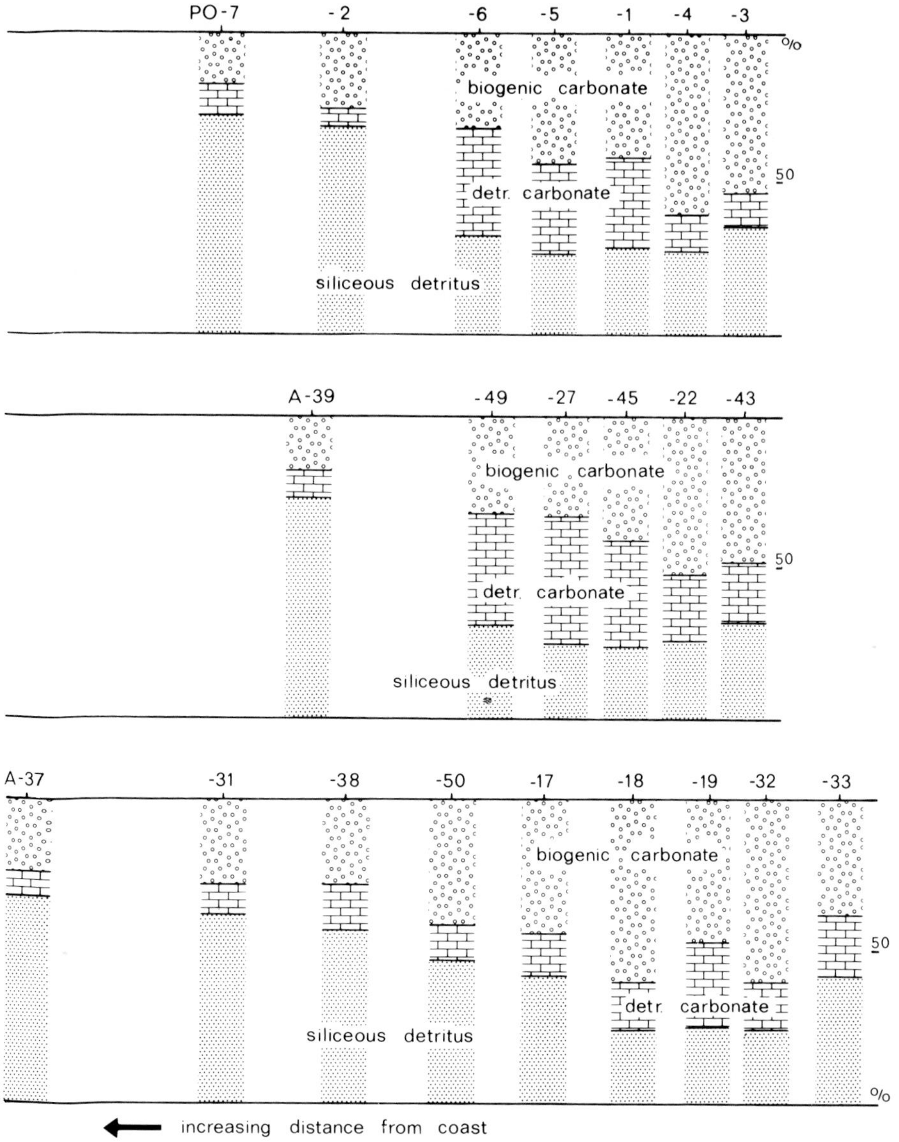

FIGURE 299. Quantitative distribution of sediment components off the Istrian coast. Note how the biogenic component, and carbonate in general, decreases with distance from shoreline; this relation may reflect decreased productivity away from the shoreline. (From D. Futterer and J. Paul, 1976, *Senckenbergiana Maritima*, 8, pp. 1–21, Figure 9.)

the facts of biogeographic change and organic evolution.

An *Analogous Community* is defined as one whose functional attributes are similar to another, although the Darwinian relations of the two are negligible. Thus, one can discuss a Middle Paleozoic set of level-bottom, shallow, subtidal filter-feeding shells as forming a community Analogous to a Cenozoic association of shells despite the fact that completely unrelated taxa are involved. What welds them together is their functional similarity and ecologic position in the geographies of two time intervals, or of the same time interval.

A *Homologous Community* is defined as a community whose contents have direct, Darwinian continuity with another community—regardless of function. Some Homologous Communities will function in

a similar fashion to that of the ancestral unit; others will not.

A Homologous Community may have been involved in enough functional change that it will no longer be thought of as belonging to the same Community Group as its ancestor.

Thorson (1957, p. 504) discussed the term "Iso-community" in such a way as to make it clear that it is very close to the term analogous community. Baldi (1973) uses the term *isocoenose* in a manner suggesting that he includes both the iso-community, analogous community, plus homologous community concepts within it. In any event, it is important to keep in mind that similarly functioning, similar appearing, taxonomically closely related or taxonomically unrelated communities occur in the fossil record and are clues to the presence of similar environments.

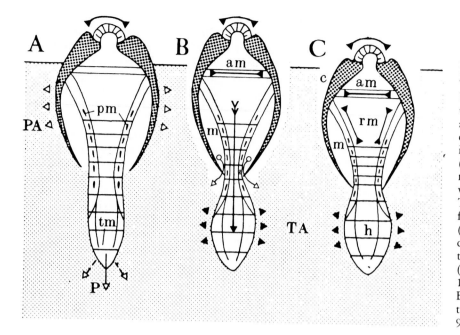

FIGURE 308. Generalized bivalve burrowing shown as three stages. (*a*) The penetration anchor (PA) is formed by the shells as the ligament forces them apart. Protractor muscles (pm) and transverse muscles (tm) in the foot contract, narrowing the foot and forcing it to penetrate (P) the sediment. (*b*) Adductor muscles (am) contract rapidly forcing water into the sediment ventrally out of the mantle cavity (m). The haemocoel of the foot (h) is inflated forming the terminal anchor (TA). (*c*) Retractor muscles (rm) then contract drawing the shell down into the sediments after the anchored foot. (From E. R. Trueman and A. Ansell, 1969, Figure 12, after Trueman, 1968, Burrowing habit and the early evolution of body cavities, *Nature, 218,* pp. 96–98.)

3. Shell adductor muscles rapidly draw the two shells together, ejecting water from the ventral margin of the shell serving as an hydraulic excavating device.

4. Anterior and posterior retractor muscles constrict, pulling the shell down after the anchored foot.

5. Shell adductors relax, and the ligament causes the gape to increase forming a penetration anchor.

6. Static period prior to resumption of the cycle.

In animals that burrow this way the foot with its inflatable hemocoele is a highly modified burrowing organ and terminal anchor. The retractor muscles of the foot provide the power for movement of the body, and the shell provides both anchorage and burrowing assistance. The ventrally directed jet of water is controlled by the mantle folds (Trueman, 1968) and represents an adaptation that allows for alteration of the physical properties of the sediment. Stanley (1970) has discussed the possible effects of the radial ribbing and other ornamentation found on the shells of burrowers. Some ornamentation may be friction elements arising as penetration-anchor specializations. Stanley also attributes several forms of ornamentation to an enhancement of penetration ability, serving to bore or saw through the sediment as the shell rotates during burrowing. A shell may be a hindrance to burrowing when compared to soft bodied animals, yet it offers several obvious advantages in addition to protection from predation. The animal has little exposure to the unpredictable sediment environment of its burrow and thus may have relatively little trouble with toxicity, mechanical irritation, or clogging of vital organs. The shell is much harder than soft tissue and can aid in penetration of hard substrates. The firm shell and the elastic ligament protect the animal from some external

mechanical forces and allow the penetration anchor to be formed passively.

Burrowing by Lingulid Brachiopods. According to Rudwick (1970), the absence of a siphon-like structure in most brachiopods prevented adoption of an infaunal habit by most forms with the exception of the lingulids. The hard bivalve test with a long muscular external peduncle places lingulids in the same burrowing category as the bivalve mollusks. Observations on burrowing have been reviewed by Trueman and Ansell (1969), but were not particularly detailed. According to Morse (1902), burrowing takes place by a combination of wormlike movement of the contractile peduncle and excavation with the dorsal valve.

Complete Rigid External Covering with Numerous Flexible Appendages

The greatest number and diversity of extant species in this category are the crustacean arthropods with their chitinous exoskeleton and specialized, hinged legs. To these animals can be added the thecate, hard echinoderms. Although very dissimilar in structure, both have lost the ability to burrow by inflation or extension of the body. Hinged legs, tube-feet and mobile spines have become the major penetration and anchoring organs. Burrowing is commonly accomplished by shovelling, and in the case of the Crustacea may involve complex behavior patterns.

Burrowing by an Irregular Urchin. The irregular urchins are highly specialized for burrowing in a variety of sediments (see Durham, 1966, for a discussion of phylogeny). In a series of papers, Nichols (1959a,b;

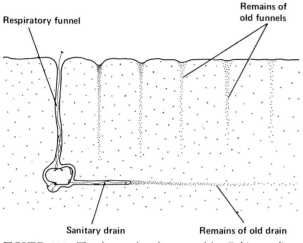

FIGURE 309. The burrowing heart urchin *Echinocardium cordatum* in its burrow maintaining its respiratory funnel with the use of elongated mucus producing podia and maintaining the sanitary drain with similar podia. (From D. Nichols, 1962, Differential selection in populations of a hearturchin, *Systematics Assoc. Publ. 4*, Text–figure 1.)

1962) has described excellently the burrowing habits and functional morphology of several forms of spatangids (heart urchins) found in Britain. The general pattern stressing specialization of tube-feet and mobile spines (Figure 309) is given in this statement by Nichols (1962):

> When the urchin is burrowing it lies in a snug cavity whose walls are plastered with mucus secreted by the animal's spines. A group of highly extensible tube-feet in the dorsal part of the anterior ambulacrum excavate and maintain a funnel-shaped tube connecting the burrow with the surface of the substratum. The tube-feet wipe mucus on to the walls of this funnel to prevent it from falling in for a time, its purpose being to provide a supply of fresh sea-water to the burrow, carrying food and oxygen to the animal. When adult, the urchin may be anything up to 6 inches below the surface of the substratum.
>
> As the animal moves forward the short spines at the front scrape the anterior wall of the burrow, and the sand removed in this way is passed by long ventrolateral spines to the rear of the burrow, where it falls in behind the advancing animal, except that a single tube remains as a backward prolongation of the burrow for a few centimeters. This originates just below the anus, and is made and maintained by a tuft of sub-anal spines and a group of from 4 to 6 sub-anal tube-feet; the spines move in such a way as to prevent the sand from falling in close behind the animal in the sub-anal region and the tube-feet extend along the passage thus formed, far beyond the ends of the spines, to wipe mucus on to the passage walls to prevent their immediate collapse. The purpose of this tube is to provide a soak-away for the products of their respiration and for fecal matter [p. 107].

Burrowing by Crustaceans. Crustaceans are morphologically diverse and capable of considerable behavioral complexity. Limb specialization is a major theme in crustacean evolution, and many of the limbs are capable of multiple functions. An equally great diversity of burrowing mechanisms may be expected in this group, and very little can be said concerning general patterns. Most orders contain some species which are capable of burrowing. The burrowers of the superfamily Thalassinoidea: Axiidae, Thalassinidae, and Callianassidae, which rapidly excavate deep and elaborate burrow systems are of particular interest because of their fossil record and abundance. Burrowing beach crabs such as the fiddler crab, *Uca*, are major extant, intertidal excavators of burrows. Gammarid amphipods such as *Corophium*, which are very abundant on some beaches, make smaller, U-shaped burrows. Marshall and Orr (1960) have reviewed the specialization of limbs associated with feeding, but did not deal directly with burrowing.

Burrowing by a beach amphipod Corophium volutator. Meadows and Reid (1966) provide one of several detailed accounts of the burrowing and feeding of a beach amphipod. The following account is based upon their description. Burrowing is initiated by rapid fanning of the pleiopods, which possibly serves to increase the water content of the sediment immediately under the animal facilitating penetration. The second antennae are modified into the shape of large legs. These are thrust into the sediment first and then the large gnathopods are used to excavate the hole. The body is pushed into the sediment by active fanning of the pleiopods that generate a current aimed to the rear. This current may help to carry away excavated material. A semicircular burrow is formed bringing the head and antennae to the sediment surface a few centimeters away from where the animal began burrowing head first (Figure 310). Once the burrow is formed the animal gathers light flocculent material from the surface of the sediment, rolls it into balls, and then plasters this material (and presumably some mucus) onto the walls of the burrow. Once the burrow is stabilized by the organic plaster, the setose gnathopods are used to filter material out of the water current generated by the pleiopods and funneled through the burrow (Hart, 1930).

Burrowing by a thalassioid crustacean. The burrows of some thallassioid shrimps are too deep to be studied easily. However, species of the related genera *Upogebia* and *Callianassa* dig less deeply and are widely distributed in some dense populations. The description of *Upogebia litoralis'* burrowing behavior given by Ott *et al.* (1976) is of particular interest for its ecologic and sedimentologic considerations. The description given here is based largely upon their work. The animals initiate burrowing by picking at the sediment surface

TABLE 5.4
Feeding Strategy Classification of Benthic Marine Polychaetes, Excluding Predominantly
Carnivorous Species and the Family Sphaerodoridae[a]

FST—Filtering, sessile, tentaculate
  Sabellariidae
  Sabellinae
  Serpulidae
  Spirorbidae
FSP—Filtering, sessile, pumping
  Chaetopteridae (except *Phyllochaetopterus*)
FDT—Filtering, discretely motile, tentaculate
  Fabriciinae
  Owenia
FDP—Filtering, discretely motile, pumping
  Arenicolidae
FDJ-P—Filtering, discretely motile, jawed and pumping
  Neanthes diversicolor
  Nereis zonata
  Platynereis
  Rhamphobrachium
F-SST-P—Filtering and surface deposit feeding, sessile, tentaculate, and pumping
  Phyllochaetopterus
SST—Surface deposit feeding, sessile, tentaculate
  Ampharetidae
  Boccardia
  Dodecaceria
  Polydora
  Pygospio
  Terebellidae (with exceptions below)
  Tharyx (some species only)
  Trichobranchidae
SS-DJ—Surface deposit feeding, sessile or discretely motile, jawed Onuphidae (except *Hyalinoecia* and *Onuphis conchylega)*
SDT—Surface deposit feeding, Discretely motile, Tentaculate
  Acrocirridae (with exceptions below)
  Apistobranchidae
  Artacaminae
  Cirratulidae (except some *Tharyx*)
  Flabelligeridae

  Longosomidae
  Magelonidae
  Myriowenia
  Nicolea
  Polycirrinae
  Sabellongidae
  Spionidae (with exceptions above)
  Trochochaetidae
SMJ—Surface deposit feeding, Motile, Jawed
  Dorvilleidae (except *Meiodorvillea*)
  Eunicidae
  Hesionidae (some without jaws)
  Hyalinoecia
  Nereidae (with exceptions above)
  Onuphis conchylega
BSE—Burrowing, Sessile, Eversible proboscis
  Fauveliopsis glabra
  Maldanidae
  Myriochele
BMJ—Burrowing, Motile, Jawed
  Lumbrineridae
  Meiodorvillea
  Nephtys incisa
  Nephtys picta
BMX—Burrowing, Motile, various other modes (X)
  Aphroditidae
  Bogueidae
  Capitellidae
  Cossuridae
  Fauveliopsis (except *F. glabra*)
  Flabelligella (except palpate species)
  Lacydoniidae
  Opheliidae
  Orbiniidae
  Paraonidae
  Pectinariidae
  Scalibregmidae
  Sternaspidae

[a]From P. A. Jumars and K. Fauchald, 1977, *In* B. C. Coull, Ed., *Ecology of Marine Benthos,* pp. 1–21. Univ. South Carolina Press.

The Burrow of *Scoloplos robustus*

Scoloplos robustus is a smaller and more mobile, peristaltic burrowing, deposit feeder whose autecology has been described by Myers (1977a). It shares with other worms the habit of building a system of temporary and more permanent burrows (Frey and Howard, 1972; Myers, 1977a). According to Myers, the first burrow type is a smooth, 2 mm diameter tunnel of irregular plan reinforced with mucus and connected to the sediment surface by a mucus-lined vertical shaft 3–4 cm long. The second burrow is also a tunnel of similarly irregular plan and dimension, reinforced poorly or not at all. The worm feeds to the side of its burrow system, thrusting the prostomium into the sediment. The prostomium is withdrawn and the pharynx everted into the cavity thus formed. The worm

appears to feed selectively on the sediment, ingesting very little of that tested. This feeding continues for about 5 minutes, after which the worm withdraws and the feeding cavity collapses. Fecal material was, at least in part, egested to the sediment surface forming small fecal mounds. Water currents were maintained by the worm within its burrow.

The degree of sediment reworking by these two different types of worm-burrowing is quite different. Myers (1977a) reported that populations of *Scoloplos* did not alter appreciably artificially layered sediments. This is because the worm burrows through the sediment by pushing it out of the way laterally and not by excavating. Feeding is restricted to selected particles, minimizing the disruption due to defecation at the sediment surface. On the other hand, conveyor belt

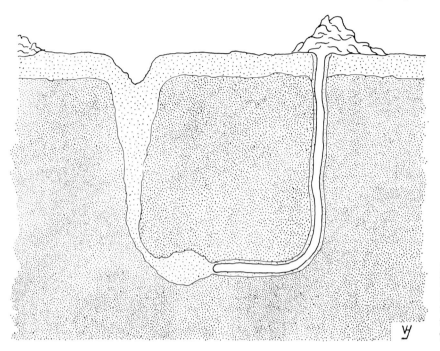

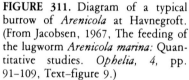

FIGURE 311. Diagram of a typical burrow of *Arenicola* at Havnegroft. (From Jacobsen, 1967, The feeding of the lugworm *Arenicola marina:* Quantitative studies. *Ophelia,* 4, pp. 91–109, Text-figure 9.)

feeders such as *Arenicola* do destroy the structure of the sediment overlying the actively feeding worm.

Many different trace fossils may be attributable to polychaetes or other burrowing vermiform animals. Among these are *Arenicolites* Salter, *Chondrites* von Sternberg, *Diplocraterion* Torrell, and others. Additional examples of the burrows of extant worms can be found in Schaefer (1972).

Arthropod Burrows

Burrows and other traces attributable to multilimbed arthropods are distributed from the lower boundary of the Cambrian to the Recent. From the Cambrian through the mid-Upper Devonian, nearshore trilobites were common benthic animals. Their sediment–water interface trace, *Cruziana,* and their burrow trace, *Rusophycus,* are relatively common trace fossils. They have approximately the same stratigraphic distribution as the trilobite body fossils, unlike other burrow forms that seem to have been made by a phylogenetically diverse group of organisms over a considerable time range. Crustacea of the extinct Eocarida were present by the Devonian. The shrimplike bodies of these forms suggest that they were nekto-benthonic (Glaessner, 1969), and it is reasonable to assume that some may have been burrowing forms. Extant crustacea of the superorder Eucarida appear to be related to this extinct group. Modern burrowing decapod crustacea such as *Hippa* and *Thalassina* are in the superorder Eucarida.

Within the decapoda, the lobsterlike infraorder Astacura were present as early as the Permian–Triassic, and possibly included burrowing forms. The champion burrowing shrimp, Thalassinoidea, were present as

early as the Lower Jurassic. Crablike decapods, with their abdomens greatly reduced and tucked under the carapace, are a more recent addition to the benthic fauna. Many of this group are first reported from the Tertiary. Remains of thalassinid shrimp have been found within their preserved burrows, but definitely identified decapod burrows have only been found in Holocene sediments.

Burrows Attributable to Thalassinid Shrimp

Trace fossils resembling the modern burrows of mud shrimp, Thalassininae, have been placed in the two ichnogenera *Thalassinoides* Ehrenberg and *Ophiomorpha* Lundgren. Both ichnogenera are three-dimensional branching burrow systems with dichotomously branching tunnels. The main distinguishing feature is in the structure of the burrow lining. In *Ophiomorpha* the burrow is lined with pellets, giving the outer surface of a preserved burrow a mammillary texture (Figure 312). *Thalassinoides* lacks this texture (Figure 313), but intergrading forms have been reported (Müller, 1970). The literature on both of these forms has been reviewed in Hantzschel (1975). More recently, Beikirch and Feldman (1980) have reported the second finding of a fossilized shrimp found within its burrow.

Its large populations, conspicuous burrows, and trace-fossil record have resulted in a fair amount of study directed at the ecology and biology of extant thalassinid shrimp (e.g., Buchanan, 1963b; MacGinitie and MacGinitie 1930, 1934; Pohl, 1946; Thompson and Pritchard, 1969) and many more observations have been made upon its burrowing ability during field studies in areas heavily populated by these shrimp.

	Group 1 Trace fossils indicative of early Cambrian age	Group 2 Trace fossils not useful in delineating the basal Cambrian boundary	Group 3 Trace fossils known only from the late Precambrian
Dwelling burrows	Diplocraterion Monocraterion Laevicyclus Dolopichnus Bergaueria	Skolithos	
Feeding burrows	Phycodes Rhyzocorallium Teichichnus Arthrophycus Syringomorpha Zoophycos Dictyodora	Planolites unbranched horizontal burrows unbranched horizontal backfilled burrows	Archaeichnium
Trails or horizontal feeding burrows	Plagiognus Psammichnites Cochlichnus Belorhaphe Helminthopsis Astropolithon Dactyloidites Oldhamia	Scolicia Curyolithus Didymaulichnus Torrowangea Helminthoidichnites unbranched horizontal trails	Bunyerichnus Buchholzbrunnichnus
Trilobite or arthropod traces	Rusophycus Cruziana Diplichnites Proptichnites Dimorphichnus Monomorphichnus		

FIGURE 324. Trace fossils diagnostic of Cambrian age (Group 1); trace fossils that span the Precambrian–Cambrian transition (Group 2); and trace fossils known only from the Precambrian (Group 3). (From Alpert, 1977, page 6.)

the burrowing of wormlike forms. It has been hypothesized that major evolutionary advances of the metazoa that occurred during the Precambrian (Clark, 1964) were actually results of adaptations for improved burrowing ability. According to this concept, the fluid filled coelom surrounded by a sheath of muscle tissue arose as a hydrostatic burrowing organ.

The appearance of various thecal materials (e.g., chitin sheaths, calcite plates, and massive carbonate shells) during and after the Cambrian may also reflect selection for improved survival of infaunal organisms. These theca serve, to some extent, as a portable burrow lining. They prevent clogging of respiratory surfaces by sediments, protect against the local chemical environment, and resist sediment weight. This last function would reduce the energy required to prevent compression by the surrounding sediment. With the appearance of theca to assist in burrowing and maintaining the burrow, mechanical restrictions upon the soft tissue

hydrostatic burrowing organs must have decreased. There followed the appearance of increasingly elaborate and specialized body forms typical of the Paleozoic marine environment.

The annelid worms are a group whose diversity may reflect different early adaptations to mode of burrowing. Dales (1967) suggested that the ancestral polychaete worm was a small, metameric burrowing animal. It was a detritus feeder with a proboscis that was used both as a terminal anchor in burrowing and simple setae along the body to act as penetration anchors. Subsequently, there was great elaboration of the anterior end. In some forms, it is still a burrowing–feeding proboscis. In others, such as Magelona (Jones, 1968), there are large anterior feeding tentacles, and the proboscis is primarily a burrowing organ. In others, such as the sabellids, the oral tentacles have become highly elaborate for suspension feeding, precluding burrowing by the anterior end of the worm.

Lewis (1968) has described the behavior of two sabellids that form the penetration anchor with the posterior end.

When discussing the diversity of polychaete parapodia and setae, Fauchald (1974) speculated on the possible role that selection for enhanced burrowing ability might have had in the early evolution of the taxa. These structures might have arisen in a Precambrian ocean with a very soupy sedimentary environment. In such an environment, hydrostatic burrowing alone would not be successful as the sediment lacks sufficient consistancy to support penetration anchors. The setae and supportive tissue that had served as the penetration anchor enlarged until they might facilitate undulatory swimming through the soft sediment.

ECOLOGICAL ASPECTS OF BIOTURBATION

Modern synecological marine benthic studies have centered around concepts of resource utilization in hard-bottom multispecies communities. In many ways the fauna of hard bottoms is better suited for detailed research in this area. Since the hard-bottom fauna is largely sessile, it remains available for study day and night, high tide and low. The permanence of the substrate allows precise measurement and description of the physical environment. Rocky shores tend to be located in esthetically pleasing settings. Mud bottoms, on the other hand, are inhabited by a more motile fauna; the animals tending to withdraw into the sediments at low tide or to be obscured by turbidity when submerged. The physical properties of the mud environment are not easily measured, especially with respect to spatial distribution. As a result, hard bottom faunas have been more carefully studied with respect to spatial patterns. Mud bottom faunas are more commonly studied after spatial pattern has been largely destroyed by removal from the sediment. Thus we have little knowledge about the spatial relationships between individuals and traces within the sediment mixed layer. How do different components of the fauna react to the presence of a large burrow containing oxygen-rich water and an organically rich lining? How do faunal components respond to the overall effects of bioturbation rather than to the presence of a particular type of burrow? As pointed out by Frey (1975), ichnology is a field in which the past may be the key to the present. With trace fossils there is the possibility of determining the types of behavior that can be found within the sediment, but which could not be detected in the modern environment with most benthic sampling techniques. Conveniently, preservational processes have solidified, stained, and even sectioned burrow systems, allowing for the leisurely, dry study of the system's details.

In this section, we will quickly review some of the information concerning the ways in which bioturbation and related processes affect the ecology of the soft bottom. This is one area in which the future offers exciting research possibilities.

Biodeposition and Pelletization

The feeding activities of both suspension feeders and sediment feeders have three major impacts upon the nature of the soft-sediment environment. While passing through the gut, the chemical processes by which organics are removed may alter the organic and inorganic chemistry of the sediments. The manipulation of sediment during ingestion and in the gut of the organisms will change the apparent grain-size structure of the sediment. This is the process of pelletization, by which the sediment is bound by mucus into fecal pellets or strings. The filtering or passive feeding activity of suspension feeders will result in the deposition of particles that are smaller than those that would be deposited by strictly physical processes (Rhoads, 1974). This is the process of biodeposition.

When marine sediments are examined under a microscope prior to the extensive disruptions needed for particle size analysis, much of the sand-sized material is seen to consist of ellipsoidal and cylindrical fecal pellets. As discussed by Rhoads (1974), in some environments the top centimeter of the sediment may be almost entirely bound up into fecal pellets. Suspension-feeding animals actively remove particles from the water and bind them into pellets or pseudofeces (fecal-like material not actually ingested by the animal but formed during fine sorting of the filtered material). Relatively little work has been done on this aspect of biological sediment processing, and the works by Lund (1957) and Haven and Morales-Alamo (1966) remain the principle references (see also Tables 5.5 and 5.6). In a Texas coastal estuary, Lund found that the common oyster Crassostrea virginica removed from suspension and deposited .7 tons of sediment per acre per day. Haven and Morales-Alamo reviewed the subject of biological deposition and found sedimentation rates ranging from 18.4 mg dry sediment per day per individual barnacle (Balanus eburneus) up to 125 mg dry sediment per day per individual for the bivalve Modiolus demissus.

The chemical alteration of sediments during passage through animal digestive tracts has not been extensively studied. It is logical to expect that some organics are utilized as a food source (see pp. 136–142), but it has been suggested for over 100 years that some alteration of the inorganics may also take place. In tropical regions there are often dense populations of large epifaunal, sediment-feeding holothuroids, usually of the genera Holothuria and Stichopus, feeding on sand and gravel-sized coral or other carbonate rubble. In Bermuda, Crozier (1918) studied the possibility that the

flysch deposits are presumably deposited in bathyal depths, and our knowledge of the behavior of deep-sea infauna is quite limited, or based upon knowledge of shallow-water forms. Third, these trace fossils may be, in part, the result of postdepositional diagenesis, requiring geologically long time periods to develop. If, however, at least some of these trace fossils are indeed biogenic, they may be of some importance in studying the manner in which benthic organisms exploit available nutrient resources within the soft bottom.

Much of Seilacher's interpretation of flysch trace fossils has been centered around a resource-utilization model. Given simple behavioral patterns, what foraging patterns would be produced over a restricted area of the bottom with a minimum of wasted time and effort? Employing computer-graphic techniques and relatively simple foraging algorithms, actual trace meanders have been reproduced (Papentin and Röder, 1975; Raup and Seilacher, 1969). According to Seilacher (1977b) the increasing morphological elaborateness of some traces in flysch deposits through time reflects an increased efficiency of foraging (Figure 325) under conditions of food limitations and stable environmental conditions.

Kitchell (1978a, 1979) has tested one prediction of the Seilacher–Papentin model through a very detailed examination of forage traces seen in deep-sea-bottom photographs from abyssal depths in the Arctic and Antarctic. On the basis of fossil material and simulations, it was suggested that there is an increase in the occurrence of more structured, nonrandom traces with depth. Kitchell (1978a) could find no such consistent trend. It was pointed out (Kitchell, 1979) that the deep-sea fauna must also respond to selective pressure due to predation, and that this might lead to foraging traces unlike those predicted by a foraging efficiency model. Thus, while the Seilacher–Papentin model is probably too simple to accurately predict foraging strategy under different nutrient and stability regimes, it is valuable in that it points to the possibility that some members of the deep-sea megafauna have been in that environment following the same feeding strategy since the Cretaceous.

An extremely interesting function of some deep-sea graphoglyptid burrows has been proposed by Seilacher (1977a,b). According to Seilacher, the burrows of many foraging sediment feeders are filled behind the animal, as there is no purpose in maintaining a more permanent tunnel. In contrast, other graphoglyptid burrows appear to have remained open in the sediment, but were reinforced with some substance, possibly like mucus (Figure 326). These open burrows

FIGURE 325. Backfilled burrows of sediment-feeding worms from the muddy and silty parts of flysch sections, reflecting the evolutionary increase in fitness by improved foraging programs and sensory monitoring. They also show the size decrease in established lineages. (For details see Seilacher, 1974, Figure 2, from Seilacher, 1977b, Evolution of trace fossil communities, Text–figure 4. *In*, A. Hallam, Ed., *Patterns of Evolution as Illustrated by the Fossil Record.* Copyright © 1977 by Elsevier, Amsterdam.)

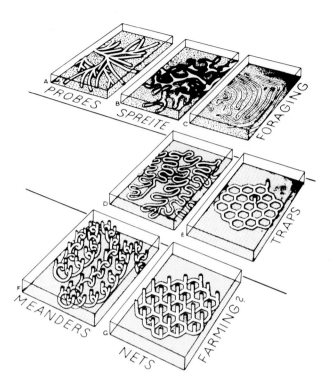

FIGURE 326. In contrast to the burrows of foraging sediment feeders, which are typically backfilled and tend to cover a given surface area as completely as possible (*a–c*) graphoglyptid burrow systems (*d–g*) are designed to leave gaps and remain unfilled, allowing repeated passages. Passive trapping of mobile and dissolved food within the sediment would explain the simpler forms (*d–e*) active farming the forms with multiple openings (*f*) (*g*). (*a*) *Chondrites* with terminal backfilling; (*b*) *Phycosiphon* with "spreite" structures (white) inside loops; (*c*) *Helminthoida labyrinthica* with separate backfilling of rim zone (white) and central tunnel. (*d*) *Cosmorhaphe* (*e*) Simple *Paleodictyon;* (*f*) *Helicolithus* with multiple openings; (*g*) *Paleodictyon* with multiple openings. (From Seilacher, 1977b, Text–figure 8.)

formed nets within the sediments through which the burrowing organisms could move repeatedly. These elaborate systems may have served as traps for meiofauna moving through the sediment or the extensive surface area of the walls farmed for bacteria and fungi.

Whatever the function of these burrow systems, if they had been restricted to the deep sea, they would have indicated the presence of unique feeding strategies in that environment.

INTENSIVE BIOTURBATION

INTRODUCTION

In previous sections, bioturbation has been treated as a process producing discrete burrows or other structures that can be used to determine animal behavior in ancient environments. Until recently, this has been the major use of trace fossils in paleontology. Now there is increasing interest in the blurring and erasing effect of bioturbation upon the resolution of paleoenvironmental information recorded in the marine micropaleontological record. An integration of the traditional concept and this newer view of bioturbation has yet to come about, although both could benefit from a synthesis. The paleoecologist would benefit from a study of sediment-mixing models in which bioturbation was viewed as an agent with the potential to homogenize all features at the top of the sediment column. The worker attempting to attribute sediment gradients to intensive bioturbation would benefit from a knowledge of the actual burrowing and sediment-feeding behavior of organisms.

The mixed layer is central to all sediment-mixing models regardless of the distribution being described (i.e., chemical gradients, particle distributions, minor textural structures, etc.). This layer is a vertical zone of unspecified horizontal extent bounded by the sediment–water interface above. The bottom of the mixed layer is self-defined as that layer below which there is no postdepositional mixing of sediments. By definition, the mixed layer is that part of the benthic environment of concern to ecologists; it is that part subject to bioturbation.

Of particular importance in extracting paleoenvironmental information from marine sediments is the knowledge that processes in the mixed layer can be the major factor in determining the content and vertical distribution of fossil material in the sedimentary record. This is especially true of trace fossils and shelly community successions. Preserved sediment features and some interstitial gradients found below the mixed layer have been altered by the extensive burrowing of the sediments. Only when there is evidence that biological activity within the mixed layer was stopped instantaneously can trace fossil assemblages be safely used as indicators of penecontemporaneous benthic conditions. In fact, trace fossils at bedding planes may well fit into this category. However, trace fossils found in relatively homogeneous sediment or rock columns represent that which escaped erasure within the mixing layer.

Efforts to describe the dynamics of the mixed layer have been made in response to the observed distribution of some sedimentary feature. These fall into two general categories: (a) descriptive, often informal, models used to explain distribution of minor internal features in sediment cores; and (b) formal mathematical models used to describe the observed gradient of small particulate materials such as microtektites, microfossils, and interstitial chemicals. With the current popularity of mathematical models and quantifiable sedimentary parameters, descriptive models are receiving little attention. They are, however, of value to the ecologist and the paleoecologist who may have small interest in more easily modeled gradients. An informal model used to explain the distribution of minor internal structures (Moore and Scrutton, 1957) is of special interest to the paleoecologist for two reasons. First, it actually is concerned with trace fossils (although not explicitly), and second, it served as the basis for later more sophisticated models.

MINOR INTERNAL STRUCTURES IN UNCONSOLIDATED PRESERVED SEDIMENTS

A model of the dynamic processes that produce burrow mottling in cores of unconsolidated sediments is important because it places emphasis upon the preserved trace fossil record as the product of several factors that must be considered in any attempt at interpretation. Descriptive models of these processes have been proposed on the basis of different sets of cores by Moore and Scrutton (1957) and Berger et al. (1979). Although based upon very different environments of deposition (the nearshore Gulf of Mexico in the former case and the abyssal Pacific in the latter), both models recognize that preservation is a function of sedimentation rate, burrowing rate, and rate of burrow destruction. There are five main points that can be found in the combination of the two models:

1. Burrowing and other bioturbation activity is most intense in the upper few centimeters of sediment where a truly homogenized mixed layer may be formed.
2. The extent to which any given horizon of the sediment is disturbed by bioturbation is a function of how long that layer remains susceptable to animal penetration.
3. The length of time a horizon is exposed to bioturbation is determined by the burial rate due to sedimentation.
4. Intensive bioturbation (such as is found in the mixed layer) can erase burrows, preventing their preservation. Thus, the mixed layer acts as a filter with respect to information preservation (Berger's description).
5. Even if a burrow survives structural erasure within the mixed layer, it may go undetected in the preserved sediments unless there is a textural or color contrast between it and the surrounding matrix.

These observations suggest that large burrows that penetrate past the contemporary mixed layer have the greatest likelihood of preservation. This is due to the decreased probability of erasure in the more intensively stirred upper sediments and to enhanced possibility of a contrast with the matrix.

The enhanced contrast of deeper burrows can be expected for two reasons. First, color contrast is commonly associated with a local chemical anomaly. This is best seen in modern sediments where a burrow serves as a corridor of oxygen penetration into sediments in which reducing conditions predominate, causing formation of a color halo around the burrow. A shallow burrow that does not cross marked chemical gradients produces no such color-changing reactions. Second, a deep burrow is more likely to cross textural discontinuities caused by changing sedimentation or dissolution patterns than is a shallow burrow. If the deep burrow is filled by surface sediment or with fecal pellets from the surface, the textural differences will enhance recognition in the record. An extreme example is the alpine flysch trace fossils that seem to have been cast by the coarser turbidite sediments.

Moore and Scrutton's (1957) Example.

In a large series of cores taken from the Mississippi River Delta and Eastern Texas in the Gulf of Mexico, Moore and Scrutton (1957) found that the minor internal structure of the sediment underwent a fairly regular change seaward. Nearshore, the sediment was coarse and vertically homogeneous; at the seaward end of the transect the sediment was fine and homogeneous. In between there was a seaward increase in fine material and a progression from regularly layered structures to progressively mottled and finally homogeneous sediment. It was found in aquarium experiments that bioturbation could progressively homogenize layered sediment (Figure 327, A–E) and could produce a mottled texture in homogeneous sediments if there was selective textural and size sorting.

The pattern of minor structure for one of their transects was explained as the result of an interaction between erasing agents and intensity of bioturbation (Figure 328). The original model proposed by Moore

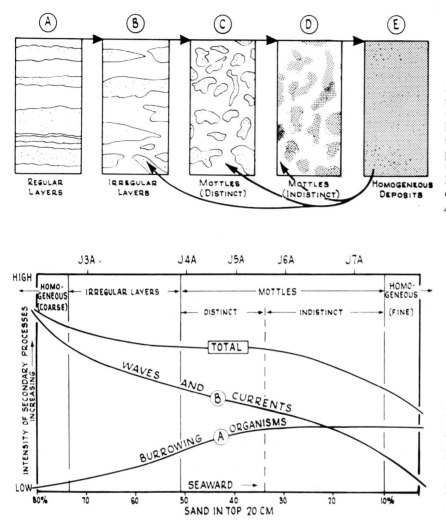

FIGURE 327. Progressive (a–e) alterations of two extreme primary structures; distinct regular layers (a) and homogenous deposits (e), by bioturbation as suggested in aquarium experiments and field observations. (From Moore and Scrutton, 1957, Minor internal structures of some recent unconsolidated sediments. *Bull. AAPG, 41,* pp. 2723–2751.)

FIGURE 328. The interrelation between bioturbation activity as suggested by faunal density (line *a*), physical erasing agents (line *b*), the decrease in sand content seaward, and the internal minor structures seen in a series of cores collected off the East Texas Coast. (From Moore and Scrutton, 1957.)

and Scrutton has received some criticism as underestimating the nearshore intensity of bioturbation. Howard (1975) pointed out that the apparent rarity of bioturbation nearshore may be attributable to the poor preservational characteristics of coarse sediments. He also warned against the general application of the scheme, stressing that sedimentological, physical, and biological factors are not uniform throughout the ocean. Hill and Hunter (1976) also pointed out that some of the homogeneous layers considered to be primary structure by Moore and Scrutton had actually been homogenized by bioturbation, as revealed by shell orientation. However, the basic observations of Moore and Scrutton still provide a good general descriptive model. The intensity of mottling seen in vertical section is determined by the rate of bioturbation, the rate of sedimentation, the rate of physical and biological erasure, and a mechanism for preservation Without knowledge of all of these factors, the degree of mottling cannot be used as an indicator of the intensity of benthic activity in paleoenvironments.

Berger, Ekdale, and Bryant's (1979) Example

A series of 14 box cores taken below 3496 m in the Pacific formed the basis of Berger, Ekdale, and Bryant's (1979) descriptive model. The top 5–8 cm of sediment appeared to be homogenized by intensive burrowing. Below this mixed layer was a transition zone stretching 20–35 cm deeper, where burrows were progressively less common, yet easily seen because of color contrast with the matrix. Deeper still was a historical layer with no active burrowing and a marked loss of the color contrast found in the transition layer. Since the 45 cm of the sediment record recovered by the cores spanned several thousand years, alternate models were suggested. The historical model supposes that the vertical pattern of burrowing reflects actual changes through time of burrowing rate and burrow preservation, due to changes in sediment properties. The dissolution of carbonates from the sediments is thought to be a prime factor in altering these sediment properties. The dynamic model supposes that the vertical pattern is the result of the steady-state between burrow formation, erasure by the mixed layer, preservation of a few deep burrows, heightened diagenetic color contrast below the mixed layer, and then diminished diagenetic color contrast associated with long-term burial in the historic layers. While Berger *et al.* (1979) do not select a preferred model, they do stress that historical interpretation of burrowing intensity from burrow count in vertical sections is of dubious validity, in that it ignores the importance of the mixed layer in destroying most evidence of bioturbation.

CONTINUOUS GRADIENTS IN THE SEDIMENT

As early as 1924, Solowiew recognized that infauna could completely homogenize the upper few centimeters of sediment (also see comments by Dapples, 1942; Richter, 1927). However, wider recognition of the importance of bioturbation came about only after extensive sedimentological and micropaleontological studies were conducted upon deep-sea cores. Of greatest importance in this development were studies on microfossils (Bramlette and Bradley, 1942; McIntyre *et al.*, 1967) and on microtektites and ash (Glass, 1969; Ruddiman and Glover, 1972). In both instances, the sedimentological record showed relatively gradual gradients of particle concentration where sudden appearance or disappearance had been expected.

In the case of the microfossil, a coccolithophore, the Plio-Pleistocene boundary in the North Atlantic deep-sea cores, appeared to be smeared vertically. An indicator species did not vanish abruptly from the core; rather, its numbers decreased exponentially. Such a record could be due to an exponential decrease of the indicator in the surface plankton caused by gradual climatic change. Or, it could have been due to a more rapid faunal change, the record of which was subsequently smeared by bioturbation following deposition. Twenty-six years after the original observation, Berger and Heath (1968) formally demonstrated that the observed exponential gradient was consistant with the pattern that bioturbation could produce.

Data on microtektite distribution provided a stronger confirmation of the record-smearing properties of bioturbation. Whereas the vertical record of a microfossil might be influenced by numerous unrecognized ecological phenomena in the overlying water, microtektite input could be considered more convincingly as an instantaneous, or "spike," event. In areas of the ocean bottom where lateral input from erosion can be discounted, bioturbation must be the major factor acting to smear the input spike into a gradual distribution gradient. Glass's data (1969) was later used by Guinasso and Schinck (1975) in the development of a numerical mixing model.

There are a growing number of mathematical mixing models. However, most of the most recent ones draw upon either the Berger-Heath or the Guinasso-Schinck formulation. The former is relatively simple and is concerned only with the distributions that remain in the sediment record once it is buried below a completely homogenized mixed layer. The latter is more mathematically complex but also more general, in that it can be used to model distributions within the active mixed layer.

The Berger-Heath model is based upon the assumption that in deep-sea sediments biological mixing is so intense relative to deposition rate, that for the geological time scales being considered, it was sufficient to consider the mixed layer as completely homogenized. By this model, any material deposited at the surface of the mixed layer was instantaneously distributed (on a geological time scale) throughout the mixed layer. As sediment was added to the top of the mixed layer by sedimentation, a preserved layer was dropped from the bottom of the mixed layer. The Berger-Heath model attempts to describe distributions in the preserved layer; there is no consideration of exactly how the mixed layer is homogenized.

The effect of a progressive mixed layer of fixed thickness, and with a constant rate of sedimentation, can be seen in Figure 329. When a particular sediment component, such as a layer of ash, falls on the sediment surface, it is immediately mixed downward by bioturbation and begins to be incorporated into the preserved layer at a horizon determined by the vertical extent of the mixed layer. Thus, it appears in the fossil record at a lower horizon than the one in which it was deposited. When the deposition from the water column ceases, the intensive mixing of the mixed layer retains an exponentially decreasing amount of material, mixing it with new sediments. The particular material thus continues to be preserved in horizons well above its actual horizon of extinction.

In the deep sea, sedimentation rates are commonly less than 1 cm per 1000 years. If one assumes a mixed layer just 10 cm thick, it is easy to realize that events that occur over a few thousand years will be extremely distorted in the geological record. Berger *et al.* (1977) have devised an "unmixing model," and have processed oxygen isotope data obtained from preserved foraminiferal tests (Figure 330). Although the unprocessed isotope data suggested gradual changes in the extent of glacial melting, the processed data suggested the possibility that there had been a rapid melting of glacial cover that may have created a low salinity surface layer over the entire ocean.

Guinasso and Schinck (1975) developed a mathematically complex model that described the effect of mixing both in the mixed layer and upon preserved layers. The model is more general than the Berger-Heath model, and may be used to predict vertical gradients in the sediment of solutions, radioisotopes, and so on, as well as particulate material. However, in spite of the claimed applicability within

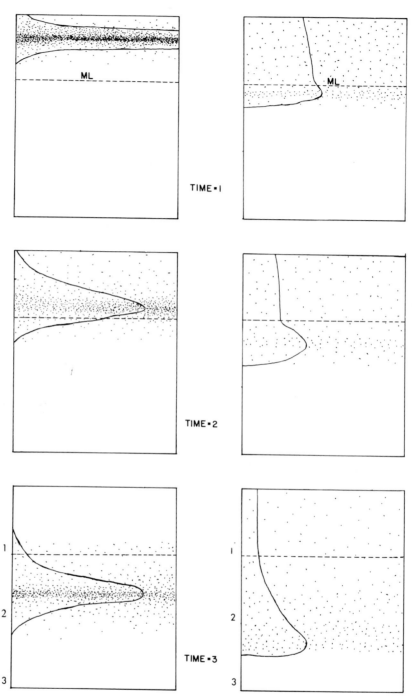

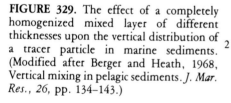

FIGURE 329. The effect of a completely homogenized mixed layer of different thicknesses upon the vertical distribution of a tracer particle in marine sediments. (Modified after Berger and Heath, 1968, Vertical mixing in pelagic sediments. *J. Mar. Res.*, 26, pp. 134–143.)

the mixed layer, it has not been established whether there is sufficient biological realism in the model to make it a valuable tool in studying sedimentary ecology. Whether the mathematical analog of actual biological mixing is a reasonable facsimile of the actual processes in operation in many sedimentary environments must be carefully considered.

The movement of sediment within the confines of the mixed layer is modeled (Guinasso and Schinck, 1975; Schinck and Guinasso, 1977a,b) to resemble a

fluid in turbulent motion between an upper and a lower boundary. By analogy to the movement of a solute or colloidal material through a turbulent fluid, particles within the mixed layer may diffuse through that layer by two mechanisms, molecular and eddy diffusion. Movement through the mixed layer by molecular diffusion is restricted to the solutes and colloidal material in the interstitial fluid. This type of movement does not require any biological movement of the sediment, although bioturbation can enhance molec-

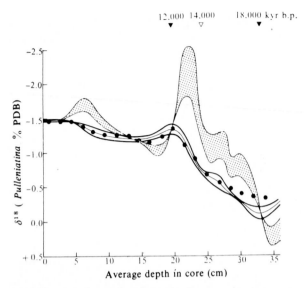

FIGURE 330. Observed and unmixed oxygen isotope curves determined from core collected specimens of the pelagic foraminifera *Pulleniatina obliquiloculata*. The two boldest lines represent data based on four cores each. The thin line between is the average of these two observed curves. The hatched envelope represents an unmixed signal assuming a mixed layer of 4 cm for the lower edge and a mixed layer 8 cm for the upper edge. The observed data would suggest gradual change while the unmixed signal suggests more drastic change. (From W. H. Berger, R. F. Johnson, and J. S. Killinley, 1977, "Unmixing," of the deep-sea record and the deglacial melt water spike. *Nature 269,* pp. 661–663.)

ular diffusion by opening the fabric of the sediment and by increasing the size of the fluid-filled voids between sediment grains. Particulate material and material in the interstitial water move by eddylike diffusion when the sediment containing them is disturbed by processes such as bioturbation.

Limitations of Mixing Models

The most commonly used mixing models appear to describe animal sediment interactions in very unrealistic terms. The two potentially most serious problems arise in the use of a uniform mixing rate through the entire mixed layer and the assumed diffusionlike nature of the modeled mixing. The possible need to employ models that treat mixing rate as a function of depth within the mixed layer has been recognized (Berger *et al.,* 1979; Jumars, 1978), and initial investigations have been made in such a direction (Schinck and Guinasso, 1977a). Therefore, this limitation may be overcome in short order.

Any model that treats bioturbation as a diffusionlike process over relatively short time spans should be used with extreme caution when distributions within the mixed layer are being studied. The burrowing and sediment-feeding activities of the better studied organisms are not random. These animals, such as

Arenicola, move sediment about in a distinct manner analogous to convection. Deep sediment is brought to the surface and surface sediment is brought downward. If there is a dense population of a conveyor-belt species feeding at the bottom of the mixed layer, the sediment will be mixed in a convective manner and not in a manner analogous to eddy diffusion (Figure 331). According to the eddy-diffusion analogy, a tracer material placed on the surface of the sediment would be seen to move downward into the mixed layer, diffusing until a uniform concentration was reached. If conveyor-belt species are active, the tracer would progress downward as a fairly discrete layer until it reached the bottom of the mixed layer, at which point it would be expelled, rising to the surface to appear as a second layer.

The amount of bioturbation that can be treated as convective and as diffusive is not known; nor is it known how these separate types of mixing might change vertically through the mixed layer. Available evidence does suggest that diffusivelike bioturbation is most intense and active near the top of the mixed layer, whereas convectivelike mixing might be equally intense across the full width of the mixed layer. Infaunal organisms tend to be most abundant nearer the sediment–water interface, and are probably more active near the surface. Shallow burrows may be more subject to destruction, forcing frequent reburrowing and resulting eddylike diffusion. The feeding and shallow burrowing of epifaunal animals are also of the diffusivelike type of bioturbation, and are obviously most intense near the sediment surface.

Some of the often cited conveyor-belt species, such as the holothuroid *Molpadia oolitica* and the polychaete worm, *Arenicola,* are relatively large animals in comparison to most infauna. The numerical abundance of these animals, their life position, and their size cause them to be the principal agents setting the width of the mixed layer. The convective effect of their feeding can extend, with uniform intensity, the full width of the layer they create, whereas few diffusionlike bioturbators may be active deep in the mixed layer.

The Guinasso–Schinck model has been applied most extensively to deep-sea sediments, whereas observations on conveyor-belt species are most extensive in shallow water. It is not known how much effect the lack of a convection term in the model will have. It is not known whether there are many conveyor-belt species in the deep sea. At least within the Holothuria, deep-sea species with morphologies and sizes comparable to the shallow-water conveyor-belt species do exist. This is easily seen for common deep-sea species of *Molpadia* and *Protankyra,* which resemble the shallow-water burrower *Leptosynapata* in many ways. However, the vertical profiles of radionuclides found by Noshkin and Bowen (1973) in deep-sea sediments suggest that diffusive movement of particles predominates.

The actual vertical profiles of tracer materials that have been studied in the field can provide some

FIGURE 331. Appearance of *Hydrobia* bed in vertical section from a Tidal Flat, Nes, Ameland. The grading from finest material at the bottom to largest shell material is obvious. (From L. M. J. U. Van Straaten, 1956, Composition of shell beds formed in tidal flat environment in the Netherlands and in the Bay of Arcachon (France) *Geologie en Mijnbouw Nieuwe Serie, 18e Jaargang nummer 7,* Juli 1956, text photos A and B, Plate 2.)

evidence, although it is scant, as to whether convectionlike bioturbation is a major cause of movement within the mixed layer in particular environments. If the tracer has moved downward in the sediment as a relatively distinct horizon at a rate too fast to be attributed to sedimentation, then convection is to be suspected. However, if the tracer level is spreading mainly by diffusionlike bioturbation, it should not have a distinct horizon of highest concentration, and

the downward movement (see Guinasso and Schinck, 1975, for full details) should be explainable in terms of the known sedimentation rate. In the deep sea, the radionuclide profiles of Noshkin and Bowen (1973) probably reflect movement only within the mixed layer, where sedimentation has been negligible for the time period concerned. In this case, the tracer isotopes seem to have moved into the sediment by a diffusionlike bioturbation process and not by convection. In shallow water, the possibility of higher sedimentation and physical stirring makes interpretation more difficult. Aller and Dodge (1976) have interpreted the cycling of a dyed sediment tracer in Jamaican sands as largely due to the convection type of bioturbation. In a somewhat similar study by Haven and Morales-Alamo (1966), diffusion and convection may both have been at work, but since the sedimentation rate was not considered, this conclusion is doubtful. Other studies have not reported data in a sufficiently quantitative manner to allow for critical reexamination. It appears that convection-type bioturbation may be the dominant mechanism in some environments. In shallow environments dominated by conveyor-belt feeders such as *Arenicola*, or deep excavators such as Thalassinids mixing models that lack a convection term should not be used. In the deep sea, nonconvection models should be applied with caution until the distribution and prominence of conveyor-belt species is known.

RATES OF BIOTURBATION AND THICKNESS OF THE MIXED LAYER

The scales and units by which the rate of bioturbation and the thickness of the mixed layer are measured depend greatly upon the type of bioturbation being envisioned. When the behavior of a particular animal is being considered, rates are expressed, typically, as volume or weight of sediment displaced per unit time. Mixed-layer depth is simply the maximal depth of burrowing, in centimeters. When a mixing model is being used without reference to the actual biology of the sediment, the model dictates scales and units. Typically, bioturbation rate is given as a diffusionlike coefficient (with the units of square centimeters per 1000 years in the Guinasso–Schink model). The mixed layer may be measured in centimeters, but its thickness is often taken to be a standard unit. In this manner, depth in the sediment is in terms of mixed-layer units, and sedimentation rate is in terms of mixed layer units per 1000 years.

Actual measurement of the movement of sediment by organisms has usually focused most closely upon a single type of bioturbatory activity. Thus the amount of sediment expelled by a relatively sedentary bivalve may have been carefully documented, but the amount of sediment displaced as that animal readjusts its position relative to the sediment–water interface remains unstudied. The observations needed to classify the type of bioturbation have seldom been made. Thus the data of Table 5.7 do not include a judgment as to which type of bioturbation the reported measurement is dealing with. In general, small animals are considered to be diffusion-type bioturbators as they can only move sediment very short distances relative to the width of the mixed layer. Mobile epifauna are also considered dif-

fusive because they stir only the top centimeter or so of the sediment and can have little direct effect on deeper convection in the mixed layer. Among burrowers, those that produce a basically vertical shaft to the surface by excavation are considered convection bioturbators, whereas those that burrow most extensively on a horizontal plane are considered diffusive.

If there is a large conveyor-belt species dominating the infauna, the mixed layer may be a relatively well defined zone within the sediments that can be measured as the maximum depth of burrowing by that dominant species. The values given in Table 5.7 show that the actual depth of such a convection-dominated mixed layer will vary considerably, depending upon the most abundant deep-burrowing conveyor-belt species. Warme (1971a) reported a thalassinid-controlled mixed layer approximately 75 cm thick in a southern California coastal lagoon. The relatively permanent galleries of the deep-burrowing Crustacea allow them to penetrate deeper and generate the thickest mixed layers.

If the bottom fauna lacks an abundant convection-type bioturbator, the mixed layer may lack a well defined lower boundary. In such a case, the bottom of the mixed layer may need to be defined statistically by the distribution of animals or animal burrows in the sediment; a method to accomplish this purpose has not yet been proposed. The maximal depth of penetration of infauna into the sediment will be limited by the availability of vital resources and the nature of the deeper sediments. As these parameters vary greatly from one benthic environment to another, maximal penetration will vary also; however, with spectacular exceptions, such as the deep-burrowing thalassinid shrimp, most animals probably live in a layer less that 1 m thick.

Since diffusionlike bioturbators shift sediment only short distances, the extent to which they mix a sediment will be related to their abundance within the sediment. Thus, even though some fauna may live as deep as 1 m, a mixed layer in the sense implied by the models will exist only where there is a population sufficiently dense and active to homogenize each level in the biologically occupied level before it is more deeply buried.

Most faunal surveys that made use of grab- or corer-type samplers have sieved the faunas from the sediment sample with little attention to the vertical distribution within the sediment. Therefore, more is known about the horizontal distribution of multispecies assemblages than is known concerning vertical changes in assemblage structure. There is, however, general agreement that the great majority of species present in any sediment are encountered in the top 5 cm or less (with many exceptions), although there may still be animals many centimeters deeper (Jumars, 1978; McIntyre, 1961, 1964; Molander, 1928; Moore, 1931; Muus, 1967; Thiel, 1966). As we have pointed out, the abun-

TABLE 5.7
Bioturbation Rates Reported in the Literature[a]

Species	Taxon	Life habit	Burrow depth (cm)	Sediment transport (gr/day)	Typical faunal densities per m²	Reference
Macoma baltica	Bivalvia Tellinacea	Infaunal, mobile deposit feeder	4–6 cm	.043	100–1000	(Bubnova, 1972)
Yoldia limatula	Bivalvia Nuculacea	Infaunal, mobile deposit feeder	2–4 cm	.28	10–100	(Rhoads, 1963)
Nucula annulata	Bivalvia Nuculacea	Infaunal, mobile deposit feeder	1–2 cm	.0033	100–1000	(Young, 1971)
Pectinaria gouldi	Polychaeta	Infaunal, mobile deposit feeder	1–6 cm	1.600	10	(Gordon, 1966)
Arenicola marina	Polychaeta Maldanidae	Infaunal, mobile deposit feeder	15–60 cm	4.500	10–100	(Jacobsen, 1967)
Clymenella torquata	Polychaeta Maldanidae	Infaunal, sedentary deposit feeder	20–30 cm	.9	100–1000	(Rhoads, 1967)
Amphitrite ornata	Polychaeta Turribelidae	Infaunal, sedentary surface deposit feeder	30 cm	4.5	1–10	(Rhoads, 1967; Aller, 1977)
Leptosynapta tenuis	Holothuroidea Apodida	Infaunal, mobile deposit feeder	10 cm	5–14	10–100	(Meyers, 1973)
Parasticbopus parvimensis	Holothuroidea Aspidochirotida	Epifaunal, mobile deposit feeder	0	22.	.4	(Yingst, 1974)
Holothuria atra	Holothuroidea Aspidochiro-tida	Epifaunal, mobile deposit feeder	0	86.5	.44	(Yamanouchi, 1941)
Holothuria vitiensis	Holothuroidea Aspidochiro-tida	Epifaunal, mobile surface deposit feeder	0	73.0	.072	(Yamanouchi, 1941)
Holothuria arenicola	Holothuroidea Aspidochiro-tida	Infaunal, mobile deposit feeder	15–20 cm	105	1.2	(Mosher, 1980)
Paracaudina chilensis	Holothuroidea Molpadonea	Infaunal, mobile deposit feeder	20 cm	158	1–10	(Yamanouchi, 1927)

[a]From Aller, 1977.

dance of these relatively small animals near the sediment–water interface can be used as evidence supporting the contention that diffusivelike mixing decreases downward into the sediment.

The best estimates of the thickness of the active mixed layer in the modern ocean is based upon a compilation of C[14] core top ages by Peng *et al.* (1977). Recognizing that the radiocarbon age of the top of a sediment core is affected by the amount of older sediment carried up by bioturbation, the depth of mixing necessary to produce the measured age was determined assuming a Berger–Heath type of homogenized mixed layer. The compiled data suggested a deep sea mixed layer between 6 and 18 cm thick. Before this C[14] determined thickness was obtained, 10 cm typically was used in computations to convert from mixed layer units to absolute thickness.

If the areas of the bottom where mixing processes are being studied lack large convection-type bioturbating species, then an assumed mixed-layer depth of 10 cm is a reasonable first approximation. Although the actual variation in mixed-layer thickness over the ocean under different ecological conditions is not known, the consistancy of estimates from radionuclides can be taken to suggest that the thickness is relatively uniform. Such uniformity seems reasonable on biological grounds, since there is no strong evidence of great variation is size range among deep-sea organisms living in different environments. Although the absolute number of large and small animals may vary over the ocean bottom, the time scales under consideration are so great (relative to ecological phenomena) that only a few large organisms always may control the depth of the mixed layer. However, this does not imply that mixing rate is uniform over the same areas where thickness is uniform.

It has been suggested that the depth of burrows in the sediment decreases with increasing bathymetric depth and that this pattern may allow burrow assemblages to be used as paleobathymetric indicators (Seilacher, 1967). Part of this argument is based upon logical assumptions concerning physiology. Intertidally, thermal and salinity fluctuations are less within the bottom than near the interface. Thus, a deep-burrowing animal escapes much environmental fluctuation. Progressing deeper into the ocean, conditions become more stable, and physiological need for burrowing disappears. In spite of the argument's logic, it cannot prove the absence of deep burrows in deep-sea sediment. The sampling gear commonly employed in deep-sea infaunal study is not capable of collecting large organisms that burrow much deeper than .5 m. The possibility that populations of deep-burrowing organisms do exist in the deep sea and create thick mixed layers remains open.

Rates of bioturbation derived from application of mixing models to observed gradients in the sedimentary record are not directly comparable to rates obtained from burrowing observations due to the units required by the diffusion models (centimeters squared per unit time). On the basis of the relatively few measurements made, these values tend to fall within fairly well defined limits. Nozaki *et al.* (1977); Peng *et al.* (1979); and Turekian *et al.* (1978) found values ranging from $.2 \times 10^{-6}$ cm^2/sec in shallow water to 14×10^{-9} in abyssal ooze. Such computed rates, however, are dependant upon estimations of the mixed-layer thickness and the sedimentation rate in the environment of concern. The decrease with depth in the sea is consistent with current views on decreased animal activity in the deep sea relative to the more productive shallow bottoms.

AN IDEAL CASE: INTENSE CONVECTION-TYPE BIOTURBATION, A WELL DEFINED MIXED LAYER, AND A SIZE-SELECTIVE BIOTURBATOR

Van Straaten (1952a,b; 1956) presents the most interesting and clearcut account of biogenic bedding (Rhoads and Stanley, 1965) caused by a distinct, intensively mixed layer in the sediments in the Dutch Wadden Sea. Twenty to 30 cm below the sediment surface for great expanses of the Wadden Sea, there is a distinct shell bed composed mostly of the shells of the gastropod *Hydrobia ulvae* overlain by sand. Several deeper beds can often be found in sections only 2 m deep, and the uppermost parallels surface contours. The following argument (slightly modified) that these beds were formed by dense populations of the conveyor-belt burrower *Arenicola marina* is put forward:

1. The beds are so extensive that they must have been generated by processes simultaneously operating over large areas. *Arenicola* is practically ubiquitous on those flats.
2. The depth of the shallowest bed coincides with the horizontal feeding portion of the *Arenicola* burrow.
3. The bed parallels the surface as does the worm population.
4. In the soft-mud bottoms of tidal channels both the *Arenicola* population and the uppermost *Hydrobia* bed are generally absent.
5. The depth of the bed tends to coincide with a sediment color change indicating the predominance of reducing conditions in deeper sediment. This would be expected if the mixed layer was intensely mixed and had a distinct lower boundary.
6. Cases where laminated sediments are found on top of *Hydrobia* beds are quite rare and probably represent postdepositional erosion and redeposition. Typically, the beds are overlain by bioturbated sediments and underlain by laminated sediments.

7. The sand covering the beds is usually that in which *Arenicola* is typically found.
8. The beds themselves grade from fine material on the bottom to coarser material on top (see Figure 331). This may be due to that fact that fine material could cycle through the mixed layer faster and thus would tend to reach the bottom of the mixed layer before large shell debris did.
9. In bottoms that can be interpreted as young, no *Hydrobia* is found in spite of the presence of *Arenicola*. This is consistent with the view that it takes considerable time to form the beds.
10. In spite of the depth of the uppermost *Hydrobia* bed, it contains recent shell debris such as *Crepidula fornicata*, which established populations in Holland about 1925.
11. The morphology of the shell debris suggests that the shells have been vertically concentrated but not laterally transported to build up the high concentration.
12. The quantitative composition of the *Hydrobia* bed is similar to that of the much sparser live material, and it does not resemble the composition seen in shell beds formed by other processes.

These beds are the best example in the literature of biogenic bedding, which was first suggested as a structuring mechanism by Schäfer (1952). Most animals are selective to some extent with respect to the way in which they react to different sediment components. Even the "nonselective deposit feeders" have simple mechanical limits to what they can swallow or shove aside while digging. The importance of size selection within an intensively mixed layer lies in the fact that not all particles will be spread about in a random fashion. If the particular particle being selected against is a microfossil, then information content of the record will be distorted. Presumably, all the material in the *Hydrobia* bed represents sediment components rejected by the dense populations of *Arenicola* and left to ac-

cumulate at the bottom of the mixed layer while other sediment components were convectively cycled.

Even when there is selective mixing, if that mixing is continuous through time and the input of the selected particle is continuous, there should still be an homogenized fossil layer.

The fact that several distinct *Hydrobia* beds could be found in the section strongly suggests that the mixed layer has not been a constant feature through time. In typical form, Van Straaten (1952b) listed the possibilities that could lead to the discontinuities. We have modified them to suit our terminology:

1. The existence of an *Arenicola*-controlled mixed layer may fluctuate in space and time with major changes in worm population density or size structure.
2. Since the shell bed is formed of biogenic material, the animals contributing this rejected coarse fraction may have fluctuated in space and time.
3. Physical disturbance at some time may have exceeded greatly convection due to *Arenicola*. Under this condition the mixed layer would have been predominantly diffusion-type. Shell material would have tended to remain scattered through the mixed layer when transferred to the preserved layers through the less well defined bottom boundary; no marked shell bed would have been formed.
4. The sedimentation rate (the rate at which the mixed layer was swept upward) may have varied greatly through time. *Hydrobia* beds represent times in which sedimentation was slow enough to allow the mixed layer to be cycled many times and for many gastropods to make shells and die. When rapid sedimentation carries the mixed layer rapidly upward, some combination of a failure to fully cycle the mixed layer and a lack of sufficient shell material would prevent the formation of a distinct shell bed.

There is no uniform formula or factor that may be applied, except the use of a measure of common sense guided by experience.

With microfossils, most of the data is a variant of counts per volume of sample. In some cases, the investigator starts with a weight that is converted to a volume based on reasonable specific gravities. Macro- and microfossil data are seldom run on the same samples.

SAMPLE SIZE

There is no fixed sample size suited to all purposes. Sander's (1968) rarefaction method is a practical way of estimating the probability that most of the available taxa have been encountered. However, based on practical consideration, each investigator must decide when the law of diminishing returns has begun to apply.

SAMPLE DENSITY

Sample density is another major factor that cannot be fixed in absolute terms. It is often regulated by practical considerations that are not involved in the formal logic designing sampling programs. Clearly, one cannot sample every fossiliferous layer in the world and then subject the fossils to careful statistical analysis. One normally works with a variety of samples obtained by oneself or by collaborators, and with data from the literature, often of very uneven quality. It is foolish to reject any data not collected by oneself, but it is equally foolish to assume that all available data is of equal validity. One must learn by one's mistakes, and be willing to repeat sampling experiments as understanding increases.

One problem unique to macrofossil sampling is experienced when the sample is collected by hand from the exposure, as contrasted with that prepared from a field-collected bulk sample. Ager (1963; Figure 332) presents a good example of this bias; a bias that should be avoided when possible. However, it is important to realize that hand collecting enables one to obtain many of the larger, far less common, species that are commonly missing in a reasonably-sized bulk sample. This method should be employed if a really comprehensive sample is desired. The hand-collected sample seldom contains specimens under a few millimeters in size, and is obtained by ranging over a fairly wide area. The bulk sample contains the minute specimens in the in-

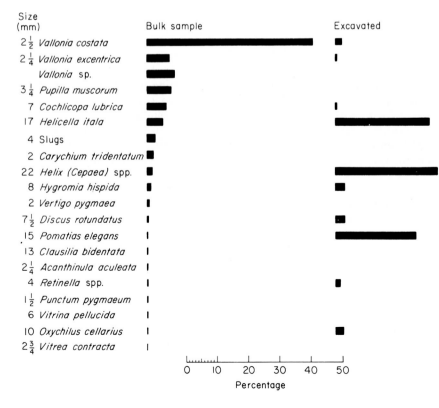

FIGURE 332. Comparison of samples obtained by bulk sampling with sieving, and hand picking of Quaternary mollusks on the Isle of Wight. Note the radically different biases involved. (From D. V. Ager, 1963, *Principles of Paleoecology* Figure 12.2. Copyright © 1963 by McGraw–Hill.)

dividual stratum in greater numbers and variety, and may also contain species missing from the hand-collected sample, particularly if the latter specimens were not more than a millimeter or so in maximum dimension. Serious collectors never fail to involve at least one 7-year-old with eyes close to the ground to ensure better sampling than can be achieved with bifocals and arthritic limbs.

NUMBER OF SPECIMENS

Counts of the number of individuals belonging to each species in a community are critical if one is to effectively define and classify communities. In dealing with fossils of marine invertebrates, benthic invertebrates, in particular, this is seldom a critical problem for most major taxa, as many have no more than one, or at most two, skeletal elements. However, when dealing with vertebrates, as in the nonmarine environments, this is a most difficult, time-consuming, and critical piece of work (see Estes and Berberian, 1970, for an excellent introduction and set of examples; also see Behrensmeyer, 1975b). The question of disarticulation ratios and their significance in connection with estimating amounts of movement has been reviewed previously. Disparate numbers of right and left valves, pedicle and brachial valves, pygidia and cephala versus thoracic segments, and so on, must be rationally considered. It is a common experience, when dealing with disarticulated materials, to find that one element is consistently less abundant than another. Some fixed procedure must be followed in normalizing the data obtained from collections of disarticulated materials. However, there are some benthic invertebrate taxa that call for special, more time-consuming estimating procedures. In particular, such groups as varied types of branching bryozoans are commonly preserved as broken fragments. Echinoderms are usually preserved as piles of disarticulated skeletal elements. It is possible to weigh disarticulated and badly scattered skeletal elements in order to arrive at an estimate of relative abundance, or to point count sawn slabs of the sample. An additional possibility is to find the volume displaced by the specimens belonging to each group. All of these estimating procedures create serious sampling problems that result in relative abundance numbers that are far from perfect. The results may be weighted in a number of ways. Ultimately, a measure of common sense must be introduced in interpreting the estimates. Wherever possible, it is useful to study modern samples aggregated in various ways in an attempt to develop guidelines for interpreting the fossil samples. Little work of this type has been carried out with marine-invertebrate materials.

PERCENTAGE OF SPECIMENS BELONGING TO DIFFERENT MAJOR TAXA

In most faunas the species of one major taxon dominate. Thus we have bivalve-dominated communities, trilobite-dominated communities, coral-dominated communities, brachiopod-dominated communities, and so on. However, as pointed out by Rohr (personal communication, 1977) on the basis of his experience with Lower Paleozoic brachiopod-dominated communities from which he is studying gastropod paleontology, the same community may be recognized and delineated in terms of the species belonging to the dominant major taxon or to the minor, rare taxa. But the minor, rare taxa present a difficult sampling problem in terms of obtaining a reasonable representation of the dominant, rare, and moderately abundant species belonging to their group. One must collect, Rohr observes, many more specimens in order to obtain adequate representation of the minor group to characterize the recognized community with a far smaller sample of the major group. It is encouraging to recognize that even the minor groups, in this sense, may be employed for purposes of community recognition and characterization. This suggests that the minor groups played a significant, integral role in the past, just as they do at present, consistent with their numerical importance.

PREPARATION AND ROCK TYPES

The nature of the sample is greatly influenced by the preparation techniques employed, and also by the rock types in which the fossils occur. In order to compare samples obtained by means of different preparation techniques from different rock types, a variety of "factors" must be employed in normalizing the raw data. A faunal list with a particular head count is difficult to employ for comparative purposes unless accompanied by information on rock type and preparation techniques.

In general, marine-invertebrate macrofossil[1] occurrences may be classified into (a) those enclosed by a poorly consolidated matrix that lends itself to easy disaggregation; (b) those in which the fossils have been replaced by silica in a soluble matrix that permits freeing of the fossils by mineral acid treatment; (c) those in which the fossils occur in a well cemented, acid-resistant matrix lending itself to study of casts and molds of the fossils after careful splitting of the rock parallel to the bedding planes; and (d) fossils that occur in a matrix of the same composition, where recourse must be had to splitting the fossils one-by-one out of the matrix. Each of these four major preservational categories is characterized by its own sampling characteristics, and each yields preferentially a higher percentage of certain fossil types. Raw data from each of these four major preservational types cannot be directly compared at a high confidence level to any of the others.

SILICIFIED FOSSILS IN A SOLUBLE MATRIX

The extraction of silicified fossils from an acid-soluble matrix is most satisfying. A minimum of labor is required to extract specimens that preserve structures with exquisite fidelity. However, there are many sampling problems involved. First is the problem that a block of limestone or dolomite containing silicified fossils will dissolve on all sides after being placed in acid. If nothing is done about this, a large percentage of the specimens in the bottom portion of the block will be crushed by the weight of the undissolved material lying above. This problem can be eliminated by coating the bottom of the block with any of a number of acid-resistant materials such as latex, or with various plastics (dissolved in appropriate solvents and allowed to dry before placing the block in the acid vat).

Second is the question of differential silicification. Experience has taught that not all fossil groups are as prone to silicification as are others. The behavior of different fossil groups varies from outcrop to outcrop, from formation to formation, from region to region, and so on. We know very little about the process of silicification under natural conditions, but we know enough to realize that there probably is more than one causal factor leading to silicification. For the paleontologist, it is enough to know that these various factors lead to different results capable of badly biasing the

sample. For example, in my own experience I have found localities where articulate brachiopods are commonly silicified but trilobites and bryozoans are not, and I have found other localities where the reverse is true. Also, it is clear that in any major group the representatives of some groups are more commonly silicified than others; namely, atrypaceans are more commonly silicified than are chonetids. Tetracorals, tabulate corals, and stromatoporoids are commonly silicified in comparison with brachiopods or trilobites in dolomites.

Third, a consideration of picking. Picking is the process of using tweezers (never one's fingers) of the right type (those with very delicate tines made of fairly flexible material; never rigid steel blades) to sort the silicified fossils from the mound of debris left after solution of the block. The block should, of course, have been placed on a fine-meshed plastic screen (plastic window-screening material will do nicely) to ensure that small specimens did not fall through the holes in the supporting plate. The picking and sorting of the fossils should be done systematically by working from one side of the screen to the other, which ensures that every piece of silicified material is examined. If one's eyesight is middle-aged and failing, watchmakers' or jewelers' eye assistants should be worn. In any event, the common practice is to pick out all of the "good" specimens and to leave behind the useless "scraps." In terms of sampling, this commonly means that a large percentage of the original specimens are left behind. There is no "factor" that can be employed universally to estimate the percentage of specimens left behind as scrap. In some cases, shattering and jointing of the rock during its geologic history may ensure that well over 50 or even 90% of the original shells are left behind as an unidentifiable hash after acid treatment. In other cases, well over 90% of the original specimens are left behind in pristine condition after acid treatment. In other words, a factor that takes preservation into account must be made for each rock type at each locality. There is also the fact that silicification preserves the smaller specimens preferentially relative to the larger specimens, and the extraction procedures serve to exaggerate this original disparity, which is made worse because small species almost always occur in far greater population densities than do larger species.

It is easy to see that since localities yielding silicified fossils are very important in our attempts to gather information of ecologic value, they must be considered very carefully when one is trying to arrive at conclusions.

FOSSILS IN AN EASILY DISAGGREGATED MATRIX

Fossils preserved in an easily disaggregated matrix; that is, one in which cementation has not gone very far or one in which the cement is easily removed, are a joy

---

[1]The reader should refer to Kummel and Raup (1965) to gain some appreciation of the biases introduced into different type microfossil samples due to different preparation techniques. Think for a moment of the solution of conodonts in formic acid, employing samples measured in hundreds of grams, as contrasted with solution in mineral acids for acritarch and chitinozoan samples weighing only a few grams.

to work with. All that need be done is carefully sieve the sample after having "softened" the matrix in water, preferably warm, or in various other liquid media such as kerosene or gasoline (see Kummel and Raup, 1965, for a lengthy discussion of most of the major preparational techniques). Unfortunately, such easily disaggregated matrices are increasingly rare as one goes back in time. The Pleistocene samples are almost always easily disaggregated, whereas those of the Cambrian are almost never of this type; the other time intervals are spread between these extremes in a sequential manner. In general, many Cenozoic matrices may be easily disaggregated, a moderate number of Mesozoic matrices may be easily disaggregated, and very few Paleozoic matrices may be easily disaggregated.

Disaggregation and subsequent sieving ensures that a very fine sample-size distribution, from smallest to largest, will be obtained.

FOSSILS IN AN ACID-RESISTANT MATRIX

Fossils preserved in an acid-resistant matrix are the common mode away from the Platform Carbonate Suite of sediments. Fossils preserved in this manner present a wide spectrum of extraction problems ranging from the moderately well cemented matrix common in many Cenozoic beds to the chlorite grade metamorphic rocks cut by slatey cleavage that are all too characteristic of many of the older noncarbonate situations.

Preparation techniques employed with these rocks involve the use of metal tools for splitting the blocks parallel to the bedding planes, then soaking the pieces in acid to remove the carbonate of the fossil in order that good, clean casts and molds may be made available for study after washing. The actual tools may range from the mechanical rocksplitter, which does a superior job in the laboratory, to the plain old hammer and chisel, which suffices in the field for the initial splitting. The chief sampling problem encountered has to do with the large percentage of specimens that are broken by the splitting procedure and, most importantly, the fact that layers of rock may be split only so thin and no thinner. Therefore, a significant number of the smaller specimens, relative to the larger specimens, remain unexposed. A chlorite-grade metamorphic rock will result in far greater destruction of specimens during preparation than is commonly the case with an unmetamorphosed equivalent.

FOSSILS IN AN ACID-NONRESISTANT MATRIX

For the bulk of carbonate rocks containing unsilicified fossils, and for many rocks having carbonate cement, it is necessary to extract the individual fossils with metal tools: rock splitters, hammer, and chisel. Acid cannot be used to clean away obscuring material as it can in the case of the acid-resistant matrix. This situation leads to a prepared sample that is highly biased toward larger specimens and toward flatter specimens that parallel bedding planes. However, in really massive rocks, limestones or not, the flatter specimens may be selected against as compared with the spherical specimens. In some situations, the blocks can be heated in a furnace for a lengthy time interval in order to weather them artificially and thus make the matrix break down and disaggregate more easily. In any event, these situations result in a heavy bias against the smaller specimens.

A FAUNAL LIST VERSUS A COMMUNITY LISTING

The literature of stratigraphic geology from the beginning of the nineteenth century to the present is replete with faunal lists. These faunal lists are compilations of taxa obtained from specific, spot localities, from individual "time" units, from individual rock units, or from individual "areas." In many instances, these lists incorporate taxa from more than a single community. For example, a single rock unit, a formation or member, commonly yields fossils from more than one layer at one locality. Therefore, there is a strong possibility that more than one community is involved in the resulting faunal list. At a single locality, such as a roadcut, quarry, natural stream bank exposure, and so on, it has been almost the rule in the past to collect fossils from more than one individual bed or layer of fossiliferous rock. The amalgamated fossil collection, even if representing only a meter or two of strata, may include many communities if environmental conditions have fluctuated with time. There are, of course, situations in which it is clear that a single locality, formation, or area represents a remarkably stable environment through considerable time. In this minority of cases, it is reasonable to treat a faunal list as a community listing. But in most instances this will not do. However, faunal lists do provide clues as to what communities are present in a region. After one becomes familiar with the communities present in a specific area, the previously published faunal lists may then be profitably considered in terms of whether or not they include more than one community. Therefore, one may not with safety merely compile faunal lists from the literature and consider them to be communities.

COPING WITH CORE MATERIAL FROM BOREHOLES

The previous discussion has tacitly assumed that adequate collections will be available from surface exposures. However, if coenologic studies are to have broad impact it is essential that they deal with available subsurface material derived from boreholes so that adequate areal coverage may be obtained. The restricted, small samples available from such cores are of course, discouraging. However, Rubel (1970) has provided a sampling program and procedure employing the results of both coeval surface and subsurface material that goes a long way toward employing subsurface core material as advantageously as possible. His approach should be used wherever possible to take full advantage of available core material.

SUMMARY

The foregoing considerations are merely a brief outline of the problem. In reality, no two samples are ever precisely the same. Therefore, one must give very careful consideration to the various sample biases introduced at various stages of the preparation process. It is clear that not only is absolute population density biased in the course of working with different type matrices, but also that the relative abundances of different taxa, and the relative abundances of different growth stages, are also biased. Thus it is most important not to take too narrow a view in defining communities based on the available diversity data. One should be realistic about the nature of the data and how it was obtained.

Finally new and innovative preparation techniques can seriously alter our concept of what is present in a sample. For example, Meijer (1969) has pointed out how noncalcareous algae may be recovered from limestones. His technique, although not widely applied up to the present, presents us with the possibility of learning a great deal about the nature of organisms present during the past. In general, the paleontologist has made use of bulk macerations for the recovery of various acid-resistant cuticles, both plant and animal. When bulk macerations are used more routinely we will undoubtedly have to modify seriously concepts regarding the nature of many communities of the past.

CONCLUSION: THE ACTUAL DEFINING AND RECOGNIZING OF THE COMMUNITY

The paleontologist will always be faced with the problem of defining the community from the fossil collections on hand. There are, of course, any number of clustering techniques and useful statistical methods for determining the degrees of similarity between collections of objects. In actual practice, however, it is almost as easy for the paleontologist to make a table on which one ordinate is devoted to a listing of taxa present and the other to collection numbers from which the taxa were obtained. In the cubbyholes, it is then useful to indicate both absolute number and percentage for each taxon for each collection—with a convenient diagonal line separating the two numbers. Then the sorting-out process begins. The table may be sliced so that a slip of paper remains intact for each collection. The various slips can then be sorted in accord with their obvious affinities. The term "obvious" is employed here particularly as regards the dominant taxa. It is also employed to call attention to the fact that the paleontologist is far more than a statistician who has been presented with a mass of raw data. The paleontologist has presumably collected the fossils, prepared the fossils, sorted the fossils, and finally counted the fossils—he or she may even have described the fossils.

The paleontologist, therefore, is uncommonly familiar with the collections—more than a little life blood and time have gone into the work. The paleontologist will clearly have some well based notions as to which collections are most like each other, unless the mind has been allowed to wander during the course of the lengthy preparation, collecting, and identification process. Therefore, the paleontologist will almost certainly be able to arrange the slips of paper in a preliminary manner that accords with nascent thoughts about similarities between collections. The arranging of the slips will test these similarities. The paleontologist may even resort to such childish but useful practices as coloring in some of the squares representing the dominant taxa using a variety of colors if a diverse fauna is being handled. The paleontologist will almost certainly be aware of various lithologic, stratigraphic, and geographic correlatives of the taxonomic associations, and will govern the arranging of the slips accordingly. In those situations where the total number of collections does not exceed a few hundred, such a simple, graphic approach should prove satisfactory. If, however, thousands of collections are involved, it will clearly be necessary to resort to more sophisticated, formal statistical procedures. But, such time-consuming formal procedures should not be employed initially unless the amount of the data makes it mandatory. There is no reason for paleontologists to be ashamed of their memories, with their innate capabilities, nor to be unwilling to admit when the volume of data exceeds the storage capacity of those memories. The alert paleontologist will have a number of hunches based on lengthy contact with fossil collections; these should be used to advantage rather than discarded in favor of one program or another. **After** having employed such a basic analysis, the results can be checked with a formal analysis if the situation seems to require one.

Bibliography

Aarseth, I., Bjerkli, K., Bjorklund, K., Boe, D., Holm, J., Lorentzen-Styr, T., Myhre, L., Uglnad, E., and Thiede, J. (1975). Late Quaternary sediments from Korsfjorden, Western Norway. *Sarsia 58:* 43–66.

Abel, O. (1916). "Paläobiologie der Cephalopoden aus der gruppen der Dibranchiaten." G. Fischer, Jena: 281 p.

Abel, O. (1935). "Vorzeitliche Lebensspuren." G. Fischer, Jena: 644 p.

Abele, L. G. (1976). Comparative species richness in fluctuating and constant environments: coral-associated decapod crustaceans. *Science 192:* 461–463.

Adams, S. M. (1976). The ecology of eelgrass, *Zostera marine* (L.), fish communities. I. Structural analysis. *J. Exp. Mar. Biol. Ecol. 22:* 269–291.

Addicott, W. O. (1966). Late Pleistocene marine paleoecology and zoogeography in central California. *U. S. Geol. Surv. Prof. Paper 523-C:* 21 p.

Addicott, W. O. (1969). Tertiary climatic change in the marginal Northeastern Pacific Ocean. *Science 165:* 583–586.

Addicott, W. O. (1970). Latitudinal gradients in Tertiary molluscan faunas of the Pacific Coast. *Palaeogeogr., Palaeoclimatol., Palaeoecol. 8:* 287–312.

Adegoke, O. S., and Tevesz, M. J. S. (1974). Gastropod predation patterns in the Eocene of Nigeria. *Lethaia 7:* 17–24.

Adey, W. H. (1978). Coral reef morphogenesis: A multidimensional model. *Science 202:* 831–836.

Ager, D. V. (1961). The epifauna of a Devonian spiriferid. *Quart. J. Geol. Soc. London 71:* 1–10.

Ager, D. V. (1963). Principles of Paleoecology. McGraw–Hill, New York: 371 p.

Ager, D. V. (1979). Paleoecology. *In:* "The Encyclopedia of Paleontology, Encyclopedia of Earth Sciences Series, V. VII." (R. W. Fairbridge and D. Jablonski, Eds.) Pp. 530–540.

Alderman, D. J., and Gareth Jones, E. B. (1967). Shell disease of *Ostrea edulis* L. *Nature 216:* 797–798.

Aldinger, H. (1968). Ecology of algal–sponge-reefs in the Upper Jurassic of the Schwabische Alb, Germany. *In:* "Recent Developments in Carbonate Sedimentology in Central Europe." (G. Muller and G. M. Friedman, Eds.) Springer–Verlag, Inc., New York: pp. 250–253.

Alexander, M. (1965). Biodegradation: Problems of molecular recalcitrance and microbial fallibility. *In:* "Advances in Applied Microbiology." (W. W. Umbreit, Ed.,) Academic Press, New York: pp. 35–79.

Alexander, R. R. (1975). Phenotypic lability of the brachiopod *Rafinesquina alternata* (Ordovician) and its correlation with the sedimentologic regime. *J. Paleontol. 49:* 607–618.

Alexander, R. R. (1977). Growth, morphology and ecology of Paleozoic and Mesozoic opportunistic species of brachiopods from Idaho–Utah. *J. Paleontol. 51:* 1133–1149.

Alexander, R. R. (submitted for publication). Predation-scars preserved in Chesterian brachiopods: Probable culprits and evolutionary consequences for the Articulates: *J. Paleo.*

Alexandersson, E. T. (1972). Micritization of carbonate particles: Processes of precipitation and dissolution in modern shallow-marine sediments. *Bull. Geol. Inst. Univ. Uppsala, N.S. 3:* 201–236.

Alexandersson, E. T. (1974). Carbonate cementation in coralline algal nodules in the Skagerrak, North Sea: Biochemical precipitation in undersaturated waters. *J. Paleontol. 44:* 7–26.

Alexandersson, E. T. (1975). Etch patterns on calcareous sediment grains: Petrographic evidence of marine dissolution on carbonate minerals. *Science 189:* 47–48.

Allan, R. S. (1937). On a neglected factor in brachiopod migration. *Rec. Canterbury Mus. N. Z. 4:* 157–165.

Allen, J. A. (1953). Observations on the epifauna of the deep water muds of the Clyde Sea area, with special reference to *Chlamys septemradiata* (Muller). *J. Animal Ecol. 22:* 240–260.

Allen, J. A. (1969). Observations on size composition and breeding of Northumberland populations of *Zirphaea crispata* (Pholadidiae: Bivalvia). *Mar. Biol. 3:* 269–275.

Allen, P., and Keith, M. L. (1965). Carbon isotope ratios and paleosalinities of Purbeck–Wealden carbonates. *Nature 208:* 1278–1280.

Allen, P., Keith, M. L., Tan, F. C., and Deines, P. (1973). Isotopic ratios and Wealden environments. *Palaeontology, 16:* 607–621.

Aller, R. C. (1977). The influence of macrobenthos on chemical diagenesis of marine sediments. Doctoral dissertation, Yale University, New Haven, Conn.: 600 pp.

Aller, R. C. (1978). Experimental studies of changes produced by deposit feeders on pore water, sediment, and overlying water chemistry. *Am. J. Sci. 278:* 1185–1234.

Aller, R. C., and Dodge, R. E. (1974). Animal–sediment relations in a tropical lagoon, Discovery Bay, Jamaica. *J. Mar. Res. 32:* 209–232.

Alpert, S. A. (1977). Trace fossils and the basal Cambrian boundary. *In:* Trace fossils 2. (T. P. Crimes, and T. C. Harper, Eds.) *Geol. J. Spec. Publ. No. 9:* 1–8.

Alverson, D. L., Pruter, A. T., and Ronholt, L. L. (1964). "A Study of Demersal Fishes and Fisheries of the Northeastern Pacific Ocean." H. R. MacMillan Lectures in Fisheries, Inst. Fish. Univ. of B. C.: 190 p.

Alvim, P. T. (1964). Tree growth periodicity in tropical climates. *In:* "Formation of Wood of Forest Trees. (M. Zimmermann, Ed.) Academic Press, New York: pp. 479–496.

Amsden, T. W. (In press.) Early Late Silurian biofacies in South–Central Oklahoma as determined by point-counting. *Okla. Geol. Surv. publication.*

Amsden, T. W. (1975). Hunton Group in the Anadarko Basin of Oklahoma. *Okla. Geol. Surv. Bull. 121:* 214 p.

Amsden, T. W., and Ventress, E. P. S. (1963). Early Devonian brachiopods of Oklahoma. *Okla. Geol. Surv. Bull. 94:* 238 p.

Anderson, A. E., Jonas, E. C., and Odum, H. T. (1958). Alteration of clay minerals by digestive processes of marine organisms. *Science 127:* 190–191.

Anderson, D. Q. (1939). Distribution of organic matter in marine sediments and its availability to further decomposition. *J. Mar. Res. 2:* 225–235.

Andrews, P. B. (1964). Serpulid reefs, Baffin Bay, Southeast Texas. *Gulf Coast Assoc. Geol. Soc., Field Trip Guidebook Ann. Mtg.:* pp. 102–120.

Anonymous. (1969–1971). Fossil ostracods and fishes from Brazil. *In:* "Report on the British Museum (Natural History) 1969–71": pp. 42–44.

Ansell, A. D., and Nair, N. B. (1969). A comparative study of bivalves which bore mainly by mechanical means. *Am. Zool. 9:* 857–868.

Ansell, A. D., Sivadas, P., Narayanan, B., and Trevaillion, A. (1972). The ecology of two sandy beaches in southwest India. II. Notes of *Emerita holothuisi. Mar. Biol. 17:* 311–317.

Ansell, A. D., and Trueman, E. R. (1968). The mechanism of burrowing in the anemone *Peachia hastata* Gosse. *J. Exp. Mar. Biol. Ecol. 2:* 124–134.

Antia, D. D. J. (1977). A comparison of diversity and trophic nuclei of live and dead molluscan faunas from the Essex Chenier Plain, England *Paleobiol. 3:* 404–414.

Antia, D. D. J. (1979). Bone-beds: A review of their classification, occurrence, genesis, diagenesis, geochemistry, palaeoecology, weathering and microbiotas. *Mercian Geologist 7:* 93–174.

Arakawa, K. Y. (1970). Scatological studies of the Bivalvia (Mollusca). *Adv. Mar. Biol. 8:* 307–436.

Arkell, W. J. (1956). "Jurassic Geology of the World." Hafner Press, New York: 806 p.

Arnold, J. M., and Arnold, K. O. (1969). Some aspects of hole-boring predation by *Octopus vulgaris. Am. Zool. 9:* 991–996.

Arntz, W. E., Brunswig, D., and Sarnthein, M. (1976). Zonierung von Mollusken und Schill im Rinnensystem der Kieler Bucht (Westliche Ostsee). *Senck. marit. 8:* 189–269.

Asgaard, U. (1968). Brachiopod palaeoecology in Middle Danian limestone at Fakse, Denmark. *Lethaia 1:* 103–121.

Asgaard, U. (1976). *Cyclaster danicus,* a shallow burrowing non-marsupiate echinoid. *Lethaia 9:* 363–375.

Ausich, W. I., and Gurrola, R. A. (1979). Two boring organisms in a Lower Mississippian community of southern Indiana. *J. Paleontol. 53:* 335–344.

Axelrod, D. I. (1958). Evolution of the Madro–Tertiary Geoflora. *Botan. Rev. 24:* 433–509.

Ayyakkannu, K., and Chandramohan, D. (1971). Occurrence and distribution of phosphate solubilizing bacteria and phosphatase in marine sediments at Porto Nova. *Mar. Biol. 11:* 201–205.

Bacescu, M. C. (1972). Animals. *In:* "Marine ecology 1(3)" (O. Kinne, Ed.): Wiley–Interscience, New York: pp. 1291–1322.

Bader, R. G. (1954). The role of organic matter in determining the distribution of pelecypods in marine sediments. *J. Mar. Res. 13:* 32–47.

Bader, R. S. (1955). Variability and evolutionary role in the oreodonts. *Evolution 9:* 119–140.

Baer, J. G. (1951). "Ecology of Animal Parasites." Univ. Illinois Press, 224 p.

Bainbridge, R. (1961). Migration. *In:* "Physiology of Crustacea, 2." (T. Waterman, Ed.) Academic Press, New York: pp. 431–463.

Baird, G. C., and Fürsich, F. T. (1975). Taphonomy and biologic progression associated with submarine erosion surfaces from the German Lias. *N. Jb. Geol. Paläontol., Monatsh. 6:* 321–338.

Bakus, G. J. (1964). The effects of fish-grazing on invertebrate evolution in shallow tropical waters. *Occ. Paper Allan Hancock Foundation No. 27:* 29 p.

Bakus, G. J. (1969). Energetics and feeding in shallow marine waters. *Int. Rev. Gen. and Exper. Zool. 4:* 275–369.

Bakus, G. J. (1974). Toxicity in holothurians: A geographical pattern. *Biotropica 6:* 229–236.

Baldi, T. (1973). "Mollusc Fauna of the Hungarian Upper Oligocene (Egerian)." Akademiai Kiado, Budapest: 511 p.

Bandel, K. (1974). Faecal pellets of Amphineura and Prosobranchia (Mollusca) from the Caribbean, Coast of Colombia, South America. *Senck. Marit. 6:* 1–31.

Bandy, O. L. (1964). General correlation of foraminiferal structure with environment. *In:* "Approaches to Paleoecology." (J. Imbrie, and D. Newell, Eds.) J. Wiley & Sons, New York: pp. 75–90.

Lower/Middle Devonian Boundary beds in the Barrandian area, Czechoslovakia. *Geol. et Palaeo.*, *13*, 125–156.

Christensen, D. J. (1973). Prey preference of *Stylochus ellipticus* in Chesapeake Bay. *Proc. Nat. Shellfisheries Assoc.* *63:* 35–38.

Christensen, W. K. (1976). Palaeobiogeography of Late Cretaceous belemnites of Europe. *Paläontol. Zeit. 50:* 113–129.

Christie, N. D. (1975). Relationship between sediment texture, species richness and volume of sediment sampled by grab. *Mar. Biol. 30:* 89–96.

Chuang, S. H. (1977). Larval development in *Discinisca* (Inarticulate brachiopod). *Am. Zool. 17:* 39–53.

Church, S. B. (1974). Lower Ordovician patch reefs in Western Utah. *Brigham Young Univ. Geol. Studies 21*(3): 41–62.

Clark, G. R., II. (1974a). Growth lines in invertebrate skeletons. *Ann. Rev. Earth & Planet. Sci. 2:* 77–99.

Clark, G. R., II. (1974b). Depth-related variations in morphology of *Argopecten gibbus*. *Geol. Soc. Am. Abstr. with Prog., Ann. Mtg. 6:* 689.

Clark, G. R., II (1975). Aberrant growth line formation in transplanted *Argopecten gibbus*. Geol. Soc. Am. Abstr. with Prog., Ann Mtg. 1028–1029.

Clark, G. R., II (1978). Byssate scallops in a Late Pennsylvanian lagoon. *Geol. Soc. Am. Abstr. with Prog., Ann. Mtg. 10*(7): 380.

Clark, R. B. (1964). "The Dynamics of Metazoan Evolution." Clarendon Press, Oxford: 313 p.

Clark, R. B., and Clark, M. E. (1960). The ligamentary system and the segmental musculature of *Nephtys*. *Quart. J. Micros. Sci. 101:* 149–176.

Clarke, J. M. (1908). The beginnings of dependent life. *N. Y. St. Mus. Bull. No. 121:* 146–169.

Clarke, J. M. (1921). Organic dependence and disease; their origin and significance. *N. Y. St. Mus. Bull. Nos. 221, 222:* 113 p.

Clarke, R. H. (1968). Burrow frequency in abyssal sediments. *Deep-sea Res. 15:* 397–400.

Clifton, H. E. (1971). Orientation of empty pelecypod shells and shell fragments in quiet water. *J. Sed. Pet. 41:* 671–682.

Clifton, H. E. (1976). Wave-formed sedimentary structures—a conceptual model. *In:* Beach and nearshore sedimentation. (R. A. Davis, Jr. and R. L. Ethington Eds.) *Soc. Econ. Paleontol. & Mineral. Spec. Publ. 24:* 126–148.

Clifton, H. E., and Boggs, Jr., S. (1970). Concave-up pelecypod (*Psephidia*) shells in shallow marine sand, Elk River Beds, Southwestern Oregon. *J. Sed. Pet. 40:* 888–897.

Clifton, H. E., and Hunter, R. E. (1972). The sand tilefish, *Malacanthus plumieri*, and the distribution of coarse debris near West Indian coral reefs. *Nat. Hist. Mus. Los Angeles Co., Sci. Bull. 14:* 87–92.

Clifton, R. L. (1942). Invertebrate faunas from the Blaine and the Dog Creek Formations of the Permian Leonard Series. *J. Paleontol. 16:* 685–699.

Clifton, R. L. (1944). Paleoecology and environments inferred for some marginal Middle Permian marine strata. *Am. Assoc. Pet. Geol. Bull. 28:* 1012–1031.

Cloud, P. E., Jr. (1941). Color patterns in Devonian terebratuloids. *Am. J. Sci. 239:* 905–907.

Cloud, P. E., Jr. (1959a). Paleoecology-retrospect and prospect. *J. Paleontol. 33:* 926–962.

Cloud, P. E., Jr. (1959b). Submarine topography and shoalwater ecology. *In:* "Geology of Saipan, Mariana Islands." Pt. 4. *U. S. Geol. Surv. Prof. Paper 280K:* 361–445.

Cloud, P. E., Jr. (1972). A working model of the primitive earth. *Am. J. Sci. 272:* 537–548.

Coates, A. G., and Kauffman, E. G. (1973). Stratigraphy, paleontology and paleoenvironment of a Cretaceous coral thicket, Lamy, New Mexico. *J. Paleontol. 47:* 953–968.

Coe, W. R. (1942). Influence of natural and experimental conditions in determining shape of shell and rate of growth in gastropods of the genus *Crepidula*. *J. Morphol. 71:* 35–52.

Coe, W. R. (1953). Resurgent populations of littoral marine invertebrates and their dependence on ocean currents and tidal currents. *Ecol. 34:* 225–229.

Coe, W. R. (1955). Ecology of the bean clam *Donax gouldi* on the coast of Southern California. *Ecol. 26:* 512–514.

Coe, W. R. (1956). Fluctuations in populations of littoral marine invertebrates. *J. Mar. Res. 15:* 212–232.

Coe, W. R. and Fitch, J. E. (1950). Population studies, local growth rates and reproduction of the Pismo clam (*Tivela stultorum*). *J. Mar. Res. 9:* 188–210.

Coe, W. R., and Fox, D. L. (1944). Biology of the California sea-mussel (*Mytilus californianus*). III. Environmental conditions and rate of growth. *Biol. Bull. M. B. L. 87:* 59–72.

Cohen, J. E. (1968). Alternate derivations of a species-abundance relation. *Am. Nat. 102:* 165–172.

Cole, H. A., and Hancock, D. A. (1955). *Odostomia* as a pest of oysters and mussels. *J. Mar. Biol. Assoc. U. K. 34:* 25–31.

Colman, J. S., and Segrove, F. (1955). The fauna living in Stoupe Beck Sands, Robin Hood's Bay (Yorkshire, North Riding). *J. Animal Ecol. 24:* 426–444.

Comfort, A. (1957). The duration of life in molluscs. *Proc. Malacol. Soc. 30:* 219–241.

Conaghan, P. J., Mountjoy, E. W., Edgecomb, D. R., Talent, J. A., and Owen, D. E. (1976). Nubrigyn algal reefs (Devonian), eastern Australia: Allochthonous blocks and megabreccias. *Geol. Soc. Am. Bull. 87:* 515–530.

Connell, J. H. (1956). Spatial distribution of two species of clams *Mya arenaria* Lamarck and *Petricola pholadiformis* Lamarck in an intertidal area. *Woods Hole Oceanogr. Inst. Contrib. No. 876,* v. *8:* 15–25.

Connell, J. H. (1963). Territorial behavior and dispersion in some marine invertebrates. *Res. Popul. Ecol. 5:* 87–101.

Connell, J. H. (1970). A predator–prey system in the marine intertidal region. *Ecol. Monogr. 40:* 49–78.

Connell, J. H., and Slayter, R. O. (1977). Mechanisms of succession in natural communities and their role in community stability and organization. *Am. Nat. 111:* 1119–1144.

Conover, J. T., and Sieburth, J. M. (1964). Effect of *Sargassum* distribution on its epibiota and antibacterial activity. *Botan. Mar. 6:* 147–157.

Constans, R. E., and Wise, Jr., S. W., (1975). Fluctuations in the carbonate compensation depth recorded in deep sea cores. *Geol. Soc. Am. Abstr. with Prog., Ann. Mtg.:* 1036.

Dow, R. L. (1969). Cyclic and geographic trends in seawater temperature and abundance of American lobster. *Science 164:* 1060–1063.

Dow, R. L. (1972). Fluctuations in Gulf of Maine sea temperatures and specific molluscan abundance. *J. Conseil Internat. L'Explor. de la Mer 34:* 532–534.

Dow, R. L. (1973). Fluctuations in marine species abundances during climatic cycles. *MTS Journal 7:* 38–40.

Dow, W. G. (1978). Petroleum source beds on continental slopes and rises. *Bull. Am. Assoc. Pet. Geol. 62:* 1584–1606.

Driscoll, E. G. (1967a). Experimental field study of shell abrasion. *J. Sed. Pet. 37:* 1117–1123.

Driscoll, E. G. (1967b). Attached epifauna–substrate relations. *Limnol. and Oceanogr. 12:* 633–641.

Druschits, V. V., and Zevina, G. B. (1969). Novie predsraviteli usonogikh rakov iz nizhnemelovikh otloshenii severnogo Kavkaza. *Pal. Zhur. 2:* 73–85.

DuBar, J. R., and Taylor, D. S. (1962). Paleoecology of the Choctawhatchee deposits, Jackson Bluff, Florida. *Trans. Gulf Coast Assoc. Geol. Soc. XII:* 349–376.

Dudley, E. C., and Vermeij. G. J. (1978). Predation in time and space: Drilling in the gastropod *Turritella. Paleobiol. 4:* 436–441.

Duff, K. L. (1975). Palaeoecology of a bituminous shale—the Lower Oxford Clay of central England. *Paleontology 18:* 443–482.

Duffus, J. H. (1969). Associations of marine mollusca and benthic algae in the Canary Island of Lanzarote. *Proc. Malacol. Soc. London 38:* 343–349.

Durham, J. W. (1966). Evolution among the Echinoidea. *Biol. Rev. 41:* 368–391.

Durham, J. W., and Zullo, V. A. (1961). The genus *Bankia* Gray (Pelecypoda) in the Oligocene of Washington. *The Veliger 4:* 1–4.

Dzik, J. (1979). Some terebratulid populations from the Lower Kimmeridgian of Poland and their relations to the biotic environment. *Acta Palaeontologica Polonica 24:* 473–492.

Eagar, R. M. C. (1977). Shape of shell in relation to weight of *Margaritifera margaritifera* (L.) (Bivalvia: Margaritiferidae). *J. Conchol. 29:* 207–218.

Eagar, R. M. C. (1978). Shape and function of the shell: A comparison of some living and fossil bivalve molluscs. *Biol. Rev. 53:* 169–210.

Easton, W. H., and Olson, E. A. (1976). Radiocarbon profile of Hanauma Reef, Oahu, Hawaii. *Geol. Soc. Am. Bull. 87:* 711–719.

Ebling, F. J., Kitching, J. A., Purchon, R. D., and Bassindale, R. (1948). The ecology of the Lough Ine Rapids with special reference to water currents. 2. The fauna of the *Saccorhiza* canopy. *J. Animal Ecol. 17:* 223–244.

Edwards, D. C. (1969a). Zonation by size as an adaptation for intertidal life in *Olivella biplicata. Am. Zool. 9:* 399–417.

Edwards, D. C. (1969b). Predators on *Olivella biplicata*, including a species–specific predator avoidance response. *The Veliger 11:* 326–333.

Edwards, D. C. (1975). Preferred prey of *Polinices duplicatus* in Cape Cod inlets. *Bull. Am. Malacol. Union, Inc. 40:* 17–20.

Edwards, D. C., and Huebner, J. D. (1978). Feeding and growth rates of *Polinices duplicatus* preying on *Mya arenaria* at Barnstable Harbor, Massachusetts. *Ecol. 58:* 1218–1236.

Ekdale, A. A. (1973). Relation of invertebrate death assemblages to living benthic communities in Recent carbonate sediments along eastern Yucatan coast. *Am. Assoc. Pet. Geol. Bull. 57:* 777.

Ekdale, A. A. (1974a). Marine molluscs from shallow-water environments (0 to 60 meters) off the northeast Yucatan coast. *Bull. Mar. Sci. 24:* 638–668.

Ekdale, A. A. (1974b). Recent marine molluscs from Northeast Quintana Roo, Mexico. *In:* Field seminar on water and carbonate rocks of the Yucatan Peninsula, Mexico. (A. E. Weidies Ed.) *Geol. Soc. Am. Guidebook No. 2, Field trip to Yucatan Peninsula:* 199–218.

Ekdale, A. A. (1977). Abyssal trace fossils in worldwide Deep Sea Drilling Project cores *In:* Trace fossils 2, (T. P. Crimes, and J. C. Harper Eds.) *Geol. J. Spec. Issue 9,* Seel House Press, Liverpool: pp. 163–182.

Ekdale, A. A. (1980). Graphoglyptid burrows in modern deep-sea sediments. *Science 207,* 304–306.

Ekman, S. (1946a). Uber die Festigkeit der marinen Sedimente als Faktor der Tierverbreitung. *Zool. Bid. Uppsala 25:* 1–20.

Ekman, S. (1946b). Zur verbreitungsgeschichte der warmwasserechinodermen im Stillen Ozean (Asteroidea, Ophiuroidea, Echinoidea). *Nova Acta Regiae Soc. Sci. Uppsala, Ser. 4, 14:* 5–42.

Elias, M. K. (1937). Depth of deposition of the Big Blue (Late Paleozoic) sediments in Kansas. *Geol. Soc. Am. Bull. 48:* 403–432.

Elias, R. J. (1980). Borings in solitary rugose corals of the Selkirk Member, Red River Formation (late Middle or Upper Ordovician), southern Manitoba. *Can. J. Earth Sci. 17:* 272–277.

Elles, G. L. (1939). Factors controlling graptolite successions and assemblages. *Geol. Mag. 76:* 181–187.

Elliot, G. F. (1963). A Palaeocene teredinid (Mollusca) from Iraq. *Palaeontology 6:* 315–317.

Elmhurst, R. (1922). Habits of *Echinus esculentus. Nature 110:* 667.

Emery, K. O. (1968). Positions of empty pelecypod valves on the continental shelf. *J. Sed. Pet. 38:* 1264–1269.

Emery, K. O., and Steveson, R. E. (1957). Estuaries and lagoons. *In:* Treatise on marine ecology and paleoecology. (J. Hedgpeth, Ed.) *Geol. Soc. Am. Mem. 67(1) Ecology:* 673–750.

Emiliani, C., Hudson, J. H., Shinn, E. A., and George, R. Y. (1978). Oxygen and carbon isotopic growth record in a reef coral from the Florida Keys and a deep-sea coral from Blake Plateau. *Science 202:* 627–629.

Endean, R. (1973). Population explosions of *Acanthaster planci* and associated destruction of hermatypic corals in the Indo–Pacific region. *In:* "Biology and Geology of Coral Reefs 2(1) Biology." (O. A. Jones, and R. Endean, Eds.) Academic Press, New York: pp. 389–438.

Engle, J. B. (1948). Investigations of Mississippi, Louisiana and Alabama following the hurricane of September 19, 1947. *U. S. Fish & Wildlife Serv., Spec. Rept. No. 59:* 71 p.

Enright, J. T. (1978). Migration and homing of marine invertebrates: A potpourri of strategies. *In:* "Animal Migration, Navigation, and Homing." (K. Schmidt-Koenig and W. T. Keaton, Eds.) Springer-Verlag, New York: pp. 440–446.

Erez, J. (1978). Vital effect on stable-isotope composition seen in Foraminifera and coral skeletons. *Nature 273:* 199–202.

Estcourt, I. N. (1967). Distributions and associations of benthic invertebrates in a sheltered water soft-bottom environment (Marlborough Sounds, New Zealand). *New Zealand J. Mar. Freshwater Res. 1:* 352–370.

Estcourt, I. N. (1968). A note on the fauna of a ripple-marked sandy sediment in Western Cook Strait, New Zealand. *New Zealand J. Mar. Freshwater Res. 2:* 654–658.

Estes, J. A., and Palmisano, J. F. (1974). Sea otters: Their role in structuring nearshore communities. *Science 185:* 1058–1060.

Estes, R., and Berberian, P. (1970). Paleoecology of a Late Cretaceous vertebrate community from Montana. *Breviora No. 343:* 35 p.

Ettensohn, F. R. (1976). Environmental and evolutionary significance of Paleozoic stemless crinoids. *Geol. Soc. Am. Abstr. with Prog., Ann. Mtg.:* 856.

Evans, J. W. (1968). The role of *Penitella penita* (Conrad, 1837) (Family Pholadidae) as eroders along the Pacific Coast of North America. *Ecology 143:* 156–159.

Ewing, M., and Davis, R. A. (1967). Lebensspuren photographed on the ocean floor. *In:* "Deep-sea Photography." (J. B. Hershey, Ed.) Johns Hopkins Press, Baltimore: pp. 259–268.

Faber, P., Vogel, K., and Winter, J. (1977). Beziehungen zwischen morphologischen Merkmalen der Brachopoden und Fazies, dargestellt an Beispielen des Mitteldevons der Eifel und Sudmarokkos. *N. Jb. Geol. Paläontol., Abh. 154:* 21–60.

Fager, E. W. (1964). Marine sediments: Effects of a tube-building polychaete. *Science 143:* 356–359.

Fager, E. W. (1968). A sand-bottom epifaunal community of invertebrates in shallow water. *Limnol. and Oceanogr. 13:* 448–464.

Farrow, G. E. (1971). Back-reef and lagoonal environments of Aldabra Atoll distinguished by their crustacean burrows. *Symp. Zool. Soc. London 28:* 455–500.

Farrow, G. E. (1974). On the ecology and sedimentation of the Cardium shellsands and transgressive shellbanks of Traigh Mhor, Island of Barra, Outer Hebrides. *Trans. Roy. Soc. Edinburgh 69:* 203–227.

Farrow, G. E., and Clokie, J. (1979). Molluscan grazing of sublittoral algal bored shell material and the production of carbonate mud. *Trans. Roy. Soc. Edinburgh 70B:* 139–148.

Farrow, G. E., Cucci, M., and Scoffin, T. P. (1978). Calcareous sediments on the nearshore continental shelf of western Scotland. *Proc. Roy. Soc. Edinburgh 76B:* 55–75.

Fauchald, K. (1974). Polychaete phylogeny: A problem in protostome evolution. *Syst. Zool. 23:* 493–506.

Fauchald, K. and Jumars, P. (1979). The diet of worms: a study of polychaete feeding guilds. *Ocean. and Mar. Biol. Ann. 17:* 193–284.

Feare, C. J. (1970). Aspects of the ecology of an exposed shore population of dogwhelks *Nucella lapillus* (L.) *Ocoelogia 5:* 1–18.

Feder, H., and Christensen, A. M. (1966). Aspects of asteroid biology. *In:* "Physiology of Echinodermata." (R. A. Boolootian, Ed.) Wiley–Interscience, New York: pp. 87–127.

Fell, H. B. (1966a). Ecology of crinoids *In:* "Physiology of Echinodermata." (R. A. Boolootian, Ed.) Wiley–Interscience, New York: pp. 49–62.

Fell, H. B. (1966b). The ecology of ophiuroids. *In:* "Physiology of Echinodermata." (R. A. Boolootian, Ed.) Wiley-Interscience, New York: pp. 129–143.

Fenchel, T. (1971). Vertical distribution of photosynthetic pigments and the penetration of light in marine sediments. *Oikos 22:* 172–182.

Fenchel, T. (1977). Competition, coexistence and character displacement in mud snails (Hydrobiidae). *In:* "Ecology of Marine Benthos." (B. C. Coull, Ed.) Univ. South Carolina Press: pp. 229–243.

Fenchel, T. (1978). The ecology of micro- and meiobenthos. *Ann. Rev. Ecol. Syst. 9:* 99–122.

Fenton, C. L., and Fenton, M. A. (1932). Orientation and injury in the genus *Atrypa*. *Am. Midl. Nat. 8:* 63–74.

Filatova, Z. A. (1957). General review of the bivalve mollusks of the Northern Seas of the USSR. *Trans. Inst. Ocean. 20,* Marine Biology: 1–44.

Fine, M. L. (1970). Faunal variation on pelagic *Sargassum*. *Mar. Biol. 7:* 112–122.

Finney, S. C. (1979). Mode of life of planktonic graptolites: Flotation structure in Ordovician *Dicellograptus* sp. *Paleobiology 5:* 31–39.

Fischler, K. J., and Walburg, C. H. (1962). Blue crab movement in coastal South Carolina: 1958–59. *Trans. Am. Fish. Soc. 91:* 275–278.

Fitch, J. E. (1976). Eocene fish fauna of Eua, Tonga, based upon additional otoliths. *U. S. Geol. Surv. Prof. Paper, 640-G:* G13–G16.

Fleming, C. A. (1951). The molluscan fauna of the fiords of Western Southland. *New Zealand J. Sci. Tech., Gen. Res. Sec. B 31:* 20–40.

Fleming, C. A. (1952). A Foveaux Strait oyster-bed. *New Zealand J. Sci. Tech. Sec. B 34:* 73–85.

Fleming, C. A. (1953). The geology of Wanganui Subdivision. *New Zealand Geol. Surv. Bull., N. S. 52:* 101–274.

Flügel, H. (1972). Animals. *In:* (O. Kinne, Ed.) *Mar. Ecol. 1(3):* 1407–1450. Wiley–Interscience, New York.

Flügel, H. W. (1979). Injuries and teratological phenomena in the rugose coral *Phaulactis*. *Geol. Fören. Stockholm Förhandl. 101:* 233–236.

Foerste, A. F. (1930). The color patterns of fossil cephalopods and brachiopods with notes on gastropods and pelecypods. *Mich. Univ. Mus. Paleontol. Contrib. 3(6):* 109–150.

Fogg, C. E. (1965). "Algal Cultures and Phytoplankton Ecology." Univ. Wisconsin Press, Madison: 126 p.

Forbes, A. T. (1973). An unusual abbreviated larval life in the estuarine burrowing prawn *Callianassa kraussi* (Crustacea: Decapoda: Thalassinidea). *Mar. Biol. 22:* 361–365.

Förster, R. (1969). Epökie, Entökie, Parasitismus und Regeneration bei fossilen Dekapoden *Mitt. Bayerischen Staatssammlung f. Paläont. u. hist. Geol. Heft 9:* 45–59.

Fortey, R. A. (1975). Early Ordovician trilobite communities. *In:* "Evolution and Morphology of the Trilobita, Trilobitoidea, and Merostomata. Fossils and Strata No. 4." (A. Martinsson, Ed.) Universitetsforlaget, Oslo: pp. 331–352.

Foster, B. A. (1969). Tolerance of high temperatures by some intertidal barnacles. *Mar. Biol. 4:* 326–332.

Foster, B. A. (1971). On the determination of the upper limits of intertidal distribution of barnacles (Crustacea: Cirripedia). *J. Animal Ecol. 40:* 33–48.

Foster, M. W. (1974). Recent Antarctic and Subantarctic brachiopods. *21st Antarctic Res. Ser. Am. Geophys. Union:* 189 p.

Foxon, G. E. H. (1936). Notes on the natural history of certain sand dwelling cumacea. *Ann. Mag. Nat. Hist. 10:* 377–393.

Frank, P. W. (1965). The biodemography of an intertidal snail population. *Ecology 46:* 831–844.

Frank, P. W. (1975). Latitudinal variation in the life history features of the black turban snail *Tegula funebralis* (Prosobranchia: Trochidae). *Mar. Biol. 31:* 181–192.

Frankel, E. (1977). Previous *Acanthaster* aggregations in the Great Barrier Reef. *Preprint, III Internat. Coral Reef Symp., Miami.*

Frankenberg, D., and Leiper, A. S. (1977). Seasonal cycles in benthic communities of the Georgia continental shelf. *In:* "Ecology of Marine Benthos." (B. C. Coull, Ed.) Univ. South Carolina Press, Columbia: pp. 383–397.

Frankenberg, D., and Smith, K. (1967). Coprophagy in marine animals. *Limnol. and Oceanogr. 12:* 443–450.

Franz, D. (1976). Benthic molluscan assemblages in relation to sediment gradients in northeastern Long Island Sound, Connecticut. *Malacol. 15:* 377–399.

Franz, D. (1977). Size and age-specific predation by *Lunatia heros* (Say, 1822) on the surf clam *Spisula solidissima* (Dillwyn, 1817) off western Long Island, New York. *The Veliger 20:* 144–150.

Franzen, C. (1974). Epizoans on Silurian–Devonian crinoids. *Lethaia 7:* 287–301.

Franzen, C. (1977). Crinoid holdfasts from the Silurian of Gotland. *Lethaia 10:* 219–234.

Frey, R. W. (1970). The lebensspuren of some common marine invertebrates near Beaufort, North Carolina. II. Anemone burrows. *J. Paleontol. 44:* 308–311.

Frey, R. W. (1972). Paleoecology and depositional environment of Fort Hays Limestone Member, Niobrara Chalk (Upper Cretaceous), West–Central Kansas. *Univ. Kansas Paleontol. Contrib. Art. 58* (Cretaceous 3): 72 p.

Frey, R. W. (1973). Concepts in the study of biogenic sedimentary structures. *J. Sed. Pet. 43:* 6–19.

Frey, R. W. (1975). The realm of ichnology, its strengths and limitations. *In:* "The Study of Trace Fossils." (R. W. Frey, Ed.) Springer–Verlag, New York: pp. 13–38.

Frey, R. W., and Howard, J. (1972). Radiographic study of sedimentary structures made by beach and offshore animals in aquaria. *Senck. Marit. 4:* 169–182.

Frey, R. W., and Howard, J. (1975). Endobenthic adaptations of juvenile Thalassinidean shrimp. *Bull. Geol. Soc. Denmark 24:* 283–297.

Frey, R. W., and Mayou, T. V. (1971). Decapod burrows in Holocene barrier island beaches and washover fans, Georgia. *Senck. Marit. 3:* 53–77.

Frey, R. W., Voorhies, M. R., and Howard, J. D. (1975). Estuaries of the Georgia coast, U.S.A.: Sedimentology and biology. VIII. Fossil and Recent skeletal remains in Georgia estuaries. *Senck. Marit. 7:* 257–295.

Fujiyama, I. (1967). A fossil scutellerid bug from marine deposit of Tottori, Japan. *Bull. Nat. Sci. Mus.* (Tokyo) *10:* 393–402.

Fürsich, F. T. (1975a). Trace fossils as environmental indicators in the Corallian of England and Normandy. *Lethaia 8:* 151–172.

Fürsich, F. T. (1975b). A. Hallam, Evolutionary size increase and longevity in Jurassic bivalves and ammonites. *Nature 258:* 493–496.

Fürsich, F. T. (1976). The use of macroinvertebrate associations in interpreting Corallian (Upper Jurassic) environments. *Palaeogeogr., Palaeoclimatol., Palaeoecol. 20:* 235–256.

Fürsich, F. T. (1977). Corallian (Upper Jurassic) marine benthic associations from England and Normandy. *Palaeontology 20:* 337–386.

Fürsich, F. T. (1978). The influence of faunal condensation and mixing on the preservation of fossil benthic communities. *Lethaia 11:* 243–250.

Fürsich, F. T. (1979). Genesis, environments, and ecology of Jurassic hardgrounds: *Neues Jahrb. Pal., Abh. 158:* 1–63.

Furst, M. J., Lowenstam, H. A., and Burnett, D. S. (1978). Paleosalinity determinations based on boron in sponge spicules. *Geol. Soc. Am. Abstr. with Prog., Cordilleran Sec. 10*(3): 106.

Futterer, D., and Paul, J. (1976). Recent and Pleistocene sediments off the Istrian coast (Northern Adriatic, Yugoslavia). *Senck. Marit. 8:* 1–21.

Futterer, E. (1978). Studien über die Einregelung, Anlagerung und Einbettung biogener Hartteile im Strömungskanal: *Neues Jahrb. Geol. Paläont. Abh. 156:* 87–131.

Gage, J. (1966). The life-histories of the bivalves *Montacuta substriata* and *M. ferruginosa*, "commensals" with spatangoids. *J. Mar. Biol. Assoc. U. K. 46:* 499–511.

Gage, J. (1977). Structure of the abyssal macrobenthos community in the Rockall Trough. *In:* "Biology of Benthic Organisms." (B. F. Keegan, P. O. Ceidigh, and P. J. S. Boadean, Eds.) Pergamon Press, New York: pp. 247–260.

Gage, J., and Coghill, G. G. (1977). Studies on the dispersion patterns of Scottish Sea Loch benthos from contiguous core transects. *In:* "Ecology of Marine Benthos." (B. C. Coull, Ed.) Univ. South Carolina Press, Columbia: pp. 319–337.

Gall, J. C. (1976). "Environmennements sedimentaires anciens et milieux de vie: Introduction a la paleoecologie." Doin, Paris: 228 p.

Galtsoff, P. S. (1942). Wasting disease causing mortality of sponges in the West Indies and Gulf of Mexico. *Proc. Eighth Am. Sci. Congr. 3:* 411–421.

Galtsoff, P. S., and Loosanoff, V. L. (1939). Natural history and methods of controlling the starfish (*Asterias forbesi*, Desor). *Bur. Fish. Bull. 31:* 75–132.

Gardner, J. (1945). Mollusca of the Tertiary formations of northeastern Mexico. *Geol. Soc. Am. Mem. 11:* 332 p.

Garrett, P. (1970). Phanerozoic stromatolites: Non-competitive ecological restriction by grazing and burrowing animals. *Science 169:* 171–173.

Garrett, P. (1977). Biological communities and their sedimentary record. *In:* "Sedimentation on the Modern Carbonate Tidal Flats of Northwest Andros Island, Bahamas." (L. A. Hardie, Ed.) Johns Hopkins Univ. Press, Baltimore: pp. 124–158.

Gekker, R. F., and Uspenskaya, E. A. (1966). Ob indikatornom znachenii sglazhennikh poverkhnostei izvestnyakov issverlennikh kamnetochtsami. *In:* "Organism i Sreda v Geologicheskom Proshlom." (R. F. Gekker, Ed.) Akad. Nauk, Otdel Biologii, Moskva: pp. 246–254.

George, J. D. (1964). Organic material available to the polychaete *Cirriformia tentaculata* (Montagu) living in an artificial mud flat. *Limnol. and Oceanogr. 9:* 453–455.

Gernant, R. E. (1970). Paleoecology of the Choptank Formation (Miocene) of Maryland and Virginia. *Md. Geol. Surv. Rept. Invest. No. 12:* 90 p.

Gessner, F., and Schramm, W. (1971). Plants. *In:* "Marine Ecology 1(2)." (O. Kinne, Ed.) Wiley–Interscience, New York: pp. 705–820.

Gignoux, M. (1955). "Stratigraphic Geology." Freeman, New York: 682 p.

Gill, E. D. (1973). Application of Recent hypotheses to changes of sealevel in Bass Strait, Australia. *Proc. Roy. Soc. Victoria 85:* 117–124.

Gill, E. D. (1975). Coast of Australia. *In:* G. D. Aitchison, A preliminary—appraisal from the viewpoint of geomechanics—of national needs for research and development in coastal and offshore engineering (and in related earth sciences). Appendix B, CSIRO: 55 p.

Ginsburg, R. N. (1956). Environmental relationships of grain size and constituent particles in some South Florida carbonate sediments. *Am. Assoc. Pet. Geol. Bull. 40:* 2384–2427.

Ginsburg, R. N. (1960). Ancient analogies of Recent stromatolites. *Internat. Geol. Congr. Pt. XXII, Proc. Internat. Paleontol. Union:* 26–35.

Ginsburg, R. N. (1972). South Florida carbonate sediments. *Sedimenta II, Div. Mar. Geol., Geophys., Rosenstiel School Mar. & Atmos. Sci., Univ. Miami:* 72 p.

Ginsburg, R. N. (Ed.) (1975). "Tidal Deposits: A Casebook of Recent Examples and Fossil Counterparts." Springer–Verlag, New York: 428 p.

Ginsburg, R. N., and James, N. P. (1974). Spectrum of Holocene reef-building communities in the western Atlantic. *In:* Principles of benthic community analysis, Sedimenta IV. *Div. Mar. Geol., Geophys., Rosenstiel School Mar. & Atmos. Sci., Univ. Miami:* 7. 1–7.22.

Ginsburg, R. N., James, N. P., Land, L. S., Moore, C. H., and Neumann, A. C. (1974). Exploration of modern reef and carbonate platform margins by submersibles. *Geol. Soc. Am. Abstr. with Prog., Ann. Mtg. 6(7):* 754–755.

Ginsburg, R. N., Rezak, R., and Wray, J. L. (1972). Geology of the calcareous algae. *Sedimenta I, Div. Mar. Geol., Geophys., Rosenstiel School Mar. & Atmos. Sci., Univ. Miami:* 40 p.

Ginsburg, R. N., and Schroeder, J. H. (1973). Growth and submarine fossilization of algal cup reefs, Bermuda. *Sedimentol. 20:* 575–614.

Gislen, T. (1931). A survey of the marine invertebrates in the Misaki District with notes concerning their environmental conditions. *J. Fac. Sci. Univ. Tokyo, Sec. IV, Zool. 2:* 389–444.

Gjessing, E. T. (1976). Physical and chemical characteristics of aquatic humus. *Ann Arbor Sci. Publ., Ann Arbor, Mich.:* 120 p.

Glaessner, M. F. (1960). The fossil decapod crustacea of New Zealand and the evolution of the Order Decapoda. *Pal. Bull. 31, N. Z. Geol. Survey,* 63 p.

Glaessner, M. F. (1969). Decapoda. *In:* "Treatise on Invertebrate Paleontology, Pt. R, Arthropoda 4(2)." (R. C. Moore, Ed.) Geol. Soc. Am. and Univ. Kansas Press: pp. R399–R533.

Glaessner, M. F., and Daily, B. (1959). The geology and Late Precambrian fauna of the Ediacara Fossil Reserve. *Rec. S. Australian Mus. 13:* 369–401.

Glasby, G. P., and Summerhayes, C. P. (1975). Sequential deposition of authigenic marine minerals around New Zealand; paleoenvironmental significance. *New Zealand J. Geol., Geophys. 18:* 477–490.

Glass, B. P. (1969). Reworking of deep-sea sediments as indicated by vertical dispersion of Australasian and Ivory Coast microtektite horizons. *Earth and Planet. Sci. Lett. 6:* 409–415.

Glenister, B. F., Klapper, G., and Chauff, K. M. (1976). Conodont pearls? *Science 193:* 571–573.

Glude, J. B. (1967). The effect of Scoter Duck predation on a clam population in Dabob Bay, Washington. *Proc. Nat. Shellfisheries Assoc. 55:* 73–86.

Glynn, P. W. (1965). Community composition, structure, and inter-relationships in the marine intertidal *Endocladia muricata-Balanus glandula* association in Monterey Bay, California. *Beaufortia 12:* 1–198.

Glynn, P. W. (1968). Mass mortalities of echinoids and other reef flat organisms coincident with midday, low water exposures in Puerto Rico. *Mar. Biol. 1:* 226–243.

Goethe, F. (1958). Anhäufungen unversehrter Muscheln durch Silbermöwen. *Natur und Volk 88:* 181–187.

Goldring, R. (1964). The trace fossils of the Baggy Beds (Upper Devonian) of North Devon, England. *Paläontol. Zeit. 36:* 232–251.

Goldring, R., and Kazmierczak, J. (1974). Ecological succession in intraformational hardground formation. *Palaeontology 17:* 949–962.

Goldring, R., and Stephenson, D. G. (1972). The depositional environment of three starfish beds. *N. Jb. Geol. Paläontol., Monh.:* 611–624.

Golubic, S. O., Perkins, R. D., and Lucas, K. J. (1975). Boring microorganisms in carbonate substrates. *In:* "The Study of Trace Fossils." (R. W. Frey, Ed.) Springer–Verlag, New York: pp. 229–259.

Goodbody, I. (1961). Mass mortality of a marine fauna following tropical rains. *Ecology 42:* 150–155.

Gordon, D. C., Jr. (1966). The effects of the deposit feeding polychaete *Pectinaria gouldii* on the intertidal sediments of Barnstable Harbor. *Limnol. Oceanogr. 11:* 327–332.

Goreau, T. F. (1964). Mass expulsion of Zooxanthellae from Jamaican reef communities after Hurricane Flora. *Science 145:* 383–386.

Goreau, T. F., and Hartman, W. D. (1963). Boring sponges as controlling factors in the formation and maintenance of coral reefs. *In:* Mechanisms of hard tissue destruction. (R. F. Sognnaes, Ed.) *Am. Assoc. Adv. Sci. Publ. 75:* 25–54.

Gordon, J., and Carriker, M. R. (1978). Growth lines in a bivalve mollusk: Subdaily patterns and dissolution of the shell. *Science 202:* 519–521.

Gould, H. R., and McFarlan, E., Jr. (1959). Geological history of the Chenier Plain, South Western Louisiana. *Trans. Gulf Coast Assoc. Geol. Societies 9:* 261–270.

Gramm, M. N., and Egorov, G. I. (1972). Late Devonian *Cavellina* (Ostracoda) with separate receptacles for eggs. *Nature 238:* 267–268.

Grant, R. E. (1963). Unusual attachment of a Permian linoproductid brachiopod. *J. Paleontol. 37:* 134–140.

Grassle, J. F., and Grassle, J. P. (1974). Opportunistic life histories and genetic systems in marine benthic polychaetes. *J. Mar. Res. 32:* 253–284.

Grassle, J. F., Sanders, H. L., Hessler, R. R., Rowe, G. T., and McLellan, T. (1975). Patterns and zonation: A study of the bathyal megafauna using the research submersible ALVIN. *Deep-Sea Res. 22:* 457–482.

Gray, J. and Boucot, A. J. (1972). Palynological evidence bearing on the Ordovician–Silurian paraconformity in Ohio. *Geol. Soc. Am. Bull. 83:* 1299–1314.

Gray, J., and Boucot, A. J. (1975). Color changes in pollen and spores: A review. *Geol. Soc. Am. Bull. 86:* 1019–1033.

Gray, J. S. (1974). Animal–sediment relationships. *Oceanogr. Mar. Biol. Ann. Rev. 12:* 223–261.

Green, R. H. (1969). Population dynamics and environmental variability. *Am. Zool. 9:* 393–398.

Greenhill, J. F. (1965). New records of marine mollusca from Tasmania. *Papers and Proc. Roy. Soc. Tasmania 99:* 67–69.

Greensmith, J. T., and Tucker, E. V. (1969). The origin of Holocene shell deposits in the Chenier Plain Facies of Essex (Great Britain). *Mar. Geol. 7:* 403–425.

Greensmith, J. T., and Tucker, E. V. (1975). Dynamic structures in the Holocene Chenier Plain setting of Essex, England. *In:* "Nearshore Sediment Dynamics and Sedimentation." (J. Hails and A. Carr, Eds.) Wiley-Interscience, New York:251–272.

Grigg, R. W., and Maragos, J. E. (1974). Recolonization of hermatypic corals on submerged lava flows in Hawaii. *Ecology 55:* 387–395.

Griggs, G. B., Carey, A. G., and Kulm, L. D. (1969). Deep-sea sedimentation and sediment fauna interactions in Cascadia Channel and on Cascadia Abyssal Plain. *Deep-sea Res. 16:* 157–170.

Gripenberg, S. (1934). A study of the sediments of the North Baltic and adjoining seas. *Fennia 20:* 168.

Gripp, K. (1968). Belemniten-Bruch vom Ostseestrand. *Nature und Museum 98:* 274–384.

Groot, S. J. de (1971). On the interrelationships between morphology of the alimentary tract, food and feeding behaviour in flatfishes (Pisces: Pleuronectiformes). *Netherlands J. Sea Res. 5:* 121–196.

Gross, M. G. (1971). Analysis of carbonaceous organic matter in sediments and sedimentary rocks. *In:* "Procedures in Sedimentary Petrology." (R. E. Carver, Ed.) Wiley-Interscience, New York: pp. 573–596.

Gunter, G. (1947). Paleoecological import of certain relationships of marine animals to salinity. *J. Paleontol. 21:* 77–79.

Gunter, G. (1956). Some relationships of faunal distributions to salinity in estuarine waters. *Ecology 37:* 616–619.

Gunter, G. (1957). Temperature. *In:* Treatise on marine ecology and paleoecology. (J. Hedgpeth Ed.) *Geol. Soc. Am. Mem. 67(1) Ecology:* 159–184.

Guinasso, N. L., Jr., and Schink, D. R. (1975). Quantitative estimates of biological mixing rates in abyssal sediments. *J. Geophys. Res. 80:* 3032–3043.

Gundrum, L. E. (1979). Demosponges as substrates: An example from the Pennsylvanian of North America. *Lethaia 12:* 105–119.

Gustavson, T. C. (1976). Paleotemperature analysis of the marine Pleistocene of Long Island, New York, and Nantucket Island, Massachusetts. *Bull. Geol. Soc. Am. 87:* 1–8.

Haas, F. (1940). Ecological observations on the common mollusks of Sanibel Island, Florida. *Am. Midl. Nat. 24:* 369–378.

Hallam, A. (1965). Environmental causes of stunting in living and fossil marine benthonic invertebrates. *Palaeontology 8:* 132–155.

Hallam, A. (1975a). "Jurassic Environments." Cambridge Univ. Press, Cambridge: 258 p.

Hallam, A. (1975b). Evolutionary size increases and longevity in Jurassic bivalves and ammonites. *Nature 258:* 493–496.

Hallam, A. (1975c). Preservation of trace fossils. *In:* "The Study of Trace Fossils." (R. W. Frey, Ed.) Springer-Verlag, New York: pp. 55–63.

Halleck, M. S. (1973). Crinoids, hardgrounds, and community succession: The Silurian Waldron–Laurel contact in southern Indiana. *Lethaia 6:* 239–252.

Hamada, T. (1964). Notes on the drifted *Nautilus* in Thailand. *Contrib. to Geol. and Paleontol. Southeast Asia 21:* 255–278.

Hancock, D. A., and Simpson, A. C. (1962). Parameters of marine invertebrate populations. *In:* "The exploitation of natural animal populations." *Symp. No. 2 Brit. Ecol. Soc.:* 29–50.

Hanley, J. H. (1976). Paleosynecology of nonmarine mollusca from the Green River and Wasatch Formations (Eocene), southwestern Wyoming and northwestern Colorado. *In:* "Structure and Classification of Paleocommunities." (R. W. Scott and R. R. West, Eds.) Dowden, Hutchinson & Ross Publ., Stroudsburg, Pa.: pp. 235–262.

Hansen, T. A. (1978). Larval dispersal and species longevity in Lower Tertiary gastropods. *Science 199:* 885–886.

Hantzschel, W. (1962). Trace fossils and Problematica. *In:* "Treatise on Invertebrate Paleontology." (R. C. Moore, Ed.) Pt. W. Geol. Soc. Am. and Univ. Kansas Press, Lawrence: 259 p.

Hantzschel, W. (1975). Trace fossils and Problematica. *In:* "Treatise on Invertebrate Paleontology." (C. Teichert, Ed.) Pt. W, Suppl. 1. Geol. Soc. Am., and Univ. Kansas Press, Lawrence: 268 p.

Hantzschel, W., El-Baz, F., and Amstutz, G. C. (1968). Coprolites, an annotated bibliography. *Geol. Soc. Am. Mem. 108:* 132 p.

Hardy, A. C. (1956). "The Open Sea, its Natural History: the World of Plankton." Houghton Mifflin, New York: 335 p.

Hardy, J. T., and Hardy, S. A. (1969). Ecology of *Tridacna* in Palau. *Pacific Sci. 23:* 467–472.

Hargens, A. R., and Shabica, S. V. (1973). Protection against lethal freezing temperatures by mucus in an Antarctic limpet. *Cryobiol. 10:* 331–337.

Harger, J. R. E. (1968). The role of behavioural traits in influencing the distribution of two species of sea mussels, *Mytilus edulis* and *Mytilus californianus. The Veliger 12:* 45–49.

Harger, J. R. E. (1969). The effect of wave impact on some aspects of the biology of sea mussels. *The Veliger 12:* 401–404.

Harger, J. R. E. (1970). The effect of species composition on the survival of mixed populations of the sea mussels *Mytilus californianus* and *Mytilus edulis. The Veliger 13:* 147–152.

Harger, J. R. E. (1971). Comparisons among growth characteristics of two species of sea mussel *Mytilus edulis* and *Mytilus californianus. The Veliger 13:* 44–55.

Harger, J. R. E. (1972). Competitive coexistence among intertidal invertebrates. *Am. Sci. 60:* 600–607.

Harper, C. W., Jr. (1975). Standing diversity of fossil groups in successive intervals of geologic time: A new measure. *J. Paleontol. 49:* 752–757.

Harrison, W. B. (1978). The occurrence of larval and young post-larval juvenile mollusca in the Upper Ordovician

Cincinnatian Series. *Geol. Soc. Am. Abstr. with Prog., Ann. Mtg. 10:* 256.

Hart, T. J. (1930). Preliminary notes on the bionomics of the amphipod *Corophium volutator* Pallas. *J. Mar. Biol. Assoc. U. K. 16:* 761–789.

Hartman, O. (1966). Quantitative survey of the benthos of San Pedro Basin, Southern California. *Allan Hancock Pacific Exped. 19:* 144–150.

Hartog, C. den. (1972a). Sea grasses of the world. *Koninkl. Nederl. Akad. Wettensch. 59*(1): 1–38.

Hartog, C. den. (1972b). *In:* "Marine 1(3) environmental factors." (O. Kinne, Ed.) Wiley-Interscience, London: 1283–1285.

Harville, J. P., and Verhoeven, L.A. (1978). Dungeness crab project of the State–Federal Fisheries Management Program. *Pacific Marine Fish. Comm.:* 196 p.

Hartwick, E. B., and Thorarisonsson, G. (1978). Den associates of the giant octopus *Octopus dofleini* (Wulker). *Ophelia 17:* 163–166.

Hauf, B. (1921). Untersuchung der Fossilfundstätten von Holzmaden in Posidonienschiefer des Oberen Lias Württembergs. *Palaeontographica 64:* 1–30.

Haven, D. S., and Morales–Alamo, R. (1966). Use of fluorescent particles to trace oyster biodeposits in marine sediments. *J. Cons. Perm. Int. Explor. Mer. 30:* 267–269.

Hayden, B. P., and Dolan, R. (1976). Coastal marine fauna and marine climates of the Americas. *J. Biogeogr. 3:* 71–81.

Hayward, B. W. (1976). Macropaleontology and paleoecology of the Waitakere Group (Lower Miocene) Waitakere Hills, Auckland. *Tane 22:* 177–206.

Hayward, B. W. (1977). Lower Miocene polychaetes from the Waitakere Ranges, North Auckland, New Zealand. *J. Roy. Soc. N. Z. 7:* 5–16.

Hazel, J. E. (1970). Atlantic continental shelf and slope of the United States—Ostracode zoogeography in the southern Nova Scotian and northern Virginian faunal provinces. *U. S. Geol. Surv. Prof. Paper 529-E:* E1–E21.

Heckel, P. H. (1974). Carbonate buildups in the geologic record: A review. *In:* Reefs in time and space. (L. F. Laporte, Ed.) *Soc. Econ. Paleontol. and Mineral. Spec. Publ. 18:* 90–154.

Hecker, R. F. (1965). "Introduction to Paleoecology." Elsevier, Amsterdam: 166 p.

Hecker, R. Th., Ossipova, A. I., and Belskaya, T. N. (1963). Fergana Gulf of Paleogene Sea of Central Asia, its history, sediments, fauna, and flora, their environment and evolution. *Am. Assoc. Pet. Geol. Bull. 47:* 617–631.

Hedgpeth, J. W. (1953). An introduction to the zoogeography of the Northwestern Gulf of Mexico with reference to the invertebrate fauna. *Publ. Inst. Mar. Sci., Univ. Texas, Port Aransas 3:* 111–224.

Hedgpeth, J. W. (1954). Bottom communities of the Gulf of Mexico. *U. S. Fish & Wildlife Serv., Fish. Bull. 89 55:* 203–214.

Hedgpeth, J. W. (1957a). Classification of marine environments. *In:* Treatise on marine ecology and paleoecology. (J. W. Hedgpeth, Ed.) *Geol. Soc. Am. Mem. 67*(1) *Ecology:* 17–27.

Hedgpeth, J. W. (Ed.) (1957b). Treatise on marine ecology and paleoecology. *Geol. Soc. Am. Mem. 67*(1) *Ecology:* 1296 p.

Hedgpeth, J. W. (1967). Ecological aspects of the Laguna Madre, a hypersaline estuary. *In:* Estuaries. (G. H. Lauff, Ed.) *Am. Assoc. Adv. Sci. Publ. 83:* 408–419.

Hedgpeth, J. W. (1969). Preliminary observations of life between tidemarks at Palmer Station, 64°45'S, 64°05'W. *Antarctic J. U. S. 4:* 106–107.

Hedgpeth, J. W. (1971). Perspectives of benthic ecology in Antarctica. *In:* Research in Antarctica. (L. O. Quam, Ed.) *Am. Assoc. Adv. Sci. Publ. 93:* 93–136.

Heezen, B. C., and Hollister, C. D. (1971). "The Face of the Deep." Oxford Univ. Press, New York and London: 657 p.

Hein, F. J., and Risk, M. J. (1975). Bioerosion of coral heads: Inner patch reefs, Florida Reef Tract. *Bull. Mar. Sci. 25:* 133–138.

Heldt, J. (1925). Sur un cas de trifurcation de l'antenne chez *Palinurus vulgaris* Latr. et sur la persistance de cette malformation apres la mue. *Notes No. 3, Stat. Ocean. de Salammbo:* 3–11.

Henningsmoen, G. (1975). Moulting in trilobites. *In:* "Evolution and Morphology of the Trilobita, Trilobitoidea and Merostomata." (A. Martinsson, Ed.) Fossils and Strata No. 4, Universitetsforlaget, Oslo: pp. 179–200.

Herald, E. S. (1967). "Living Fishes of the World." Doubleday & Co., New York: 303 p.

Herdman, W. A. (1905). The pearl fisheries of Ceylon. *Ann. Rept. Smithsonian Inst. for 1904:* 485–493.

Herdman, W. A. (1906). Report to the government of Ceylon on the pearl oyster fisheries of the Gulf of Manar: Part V. *Roy. Soc. London:* 452 p.

Herr, S. R. (1971). Regeneration and growth abnormalities in *Orthograptus quadrimucronatus* from the Ordovician Maquoketa Formation of Iowa. *J. Paleontol. 45:* 628–632.

Herrnkind, W., and Kanciruk, P. (1978). Mass migration of spiny lobster, *Panulirus argus* (Crustacea: Palinuridae): Synopsis and orientation. *In:* "Animal Migration, Navigation, and Homing." (K. Schmidt-Koenig, and W. T. Keaton, Eds.) Springer–Verlag, New York: 436–439.

Hesse, K. O. (1979). Movement and migration of the queen conch, *Strombus gigas,* in the Turks and Cocos Islands. *Bull. Mar. Sci. 29:* 303–311.

Hiatt, R. W., and Strasburg, D. W. (1960). Ecological relationships of the fish fauna on coral reefs of the Marshall Islands. *Ecol. Monogr. 30:* 65–127.

Hibbert, C. J. (1977). Growth and survivorship in a tidal-flat population of the bivalve *Mercenaria mercenaria* from Southampton Water. *Mar. Biol. 44:* 71–76.

Hickman, C. J. S. (1969). The Oligocene marine molluscan fauna of the Eugene Formation in Oregon. *Mus. Nat. Hist. Univ. Oregon Bull. 16:* 112 p.

Hidu, H. (1969). Gregarious setting in the American oyster *Crassostrea virginica* Gmelin. *Chesapeake Sci. 10:* 85–92.

Hill, B. J. (1976). Natural food, foregut clearance-rate and activity of the crab *Scylla serrata. Mar. Biol. 34:* 109–116.

Hill, D. (1938–1941). Carboniferous rugose corals of Scotland. *Palaeontogr. Soc. Monogr.:* 1–52.

Hill, G. W., and Hunter, R. E. (1976). Interaction of biological and geological processes in the beach and nearshore environments, northern Padre Island, Texas. *In:* Beach and nearshore sedimentation. (R. A. Davis and R. L. Ethington, Eds.) *Soc. Econ. Paleontol. and Mineral. Spec. Publ. 24:* 169–187.

Hoare, R. D. (1978). Annotated bibliography on preservation of color patterns on invertebrate fossils: *The Compass of Sigma Gamma Epsilon 55:* 39–63.

Hoare, R. D., and Steller, D. L. (1967). A Devonian brachiopod with epifauna. *Ohio J. Sci. 67:* 291–297.

Hobson, E. S. (1974). Feeding relationships of teleostean fishes on coral reefs in Kona, Hawaii. *Fish. Bull. 72:* 915–1031.

Hoefs, J. (1973). "Stable Isotope Geochemistry." Springer–Verlag, New York: 140 p.

Hölder, H. (1972). Endo- und epizoen von Belemniten-Rostren (*Megateuthis*) in nordwestdeutschen Bajocium (Mittlerer Jura). *Palaontol. Zeit. 46:* 199–220.

Holland, H. D. (1978). The chemistry of the atmosphere and oceans. John Wiley & Sons, New York: 352 p.

Holland, P. G., and Steyn, D. G. (1975). Vegetational responses to latitudinal variations in slope angle and aspect. *J. Biogeogr. 2:* 179–183.

Holling, C. S. (1966). The functional response of invertebrate predators to prey density. *Mem. Entom. Soc. Canada 48:* 5–86.

Holmes, N. A. (1961). The bottom fauna of the English Channel. *J. Mar. Biol. Assoc. U. K. 41:* 397–461.

Holmes, N. A., and McIntyre, A. D. (Eds.) (1971). Methods for the study of marine benthos. *Internat. Biome Prog. Handbook 16:* 334 p.

Holmes, R. W. (1957). Solar radiation, submarine daylight, and photosynthesis. *In:* Treatise on marine ecology and paleoecology. (J. Hedgpeth, Ed.) *Geol. Soc. Am. Mem. 67(1). Ecology:* 109–128.

Hornell, J. (1922). The Indian pearl fisheries of the Gulf of Manar and Palk Bay. *Madras Fish. Dept. Bull. XVI:* 188 p.

Hough, J. L. (1934). Redeposition of microscopic Devonian plant fossils. *J. Geol. 42:* 646–648.

Howard, J. D. (1975). The sedimentological significance of trace fossils. *In:* "The Study of Trace Fossils." (R. W. Frey, Ed.) Springer–Verlag, New York: pp. 131–146.

Howard, J. D., Reineck, H.-E., and Rietschel, S. (1974). Biogenic sedimentary structures formed by heart urchins. *Senck. Marit. 6:* 185–201.

Howe, H. V., and van den Bold, W. A. (1975). Mudlump ostracoda. *In:* Biology and paleobiology of Ostracoda. (F. M. Swain, Ed.) *Bull. Am. Paleontol. 65:* 303–316.

Howell, B. F. (1957). Vermes. *In:* Treatise on marine ecology and paleoecology. (H. S. Ladd, Ed.) *Geol. Soc. Am. Mem. 67(2), Paleoecology:* 805–816.

Hudson, J. D. (1963). The recognition of salinity-controlled mollusc assemblages in the Great Estuarine Series (Middle Jurassic) of the Inner Hebrides. *Palaeontology 6:* 318–326.

Hunt, A. R. (1885). On the influence of wave-currents on the fauna inhabiting shallow seas. *J. Linn. Soc. 43:* 262–274.

Hupé, P. (1953). Classe des Trilobites. *In:* "Traité de Paleontologie." (J. Piveteau, Ed.) Masson et Cie, Paris: pp. 44–246.

Hutchins, L. W. (1947). The bases for temperature zonation in geographical distribution. *Ecol. Monogr. 17:* 325–335.

Hutchinson, G. E. (1944). Limnological studies in Connecticut. VII. *Ecology. 25:* 3–26.

Hutchinson, G. E. (1950). Survey of contemporary knowledge of biogeochemistry. 3. The biogeochemistry of vertebrate excretion. *Bull. Amer. Mus. Nat. Hist. 96:* 554 pp.

Hylleberg, J. (1970). On the ecology of the sipunculan *Phascolion strombi* (Montagu). *Proc. Internat. Symp. Biol. Sipuncula and Echiura I, Kotor,* June 18–25: 241–250.

Hyman, L. H. (1955). "Echinodermata. the Invertebrates, V. IV." McGraw–Hill, New York: 763 p.

Ingels, J. J. C. (1963). Geometry, paleontology, and petrography of Thornton Reef Complex, Silurian of northeastern Illinois. *Am. Assoc. Pet. Geol. Bull. 47:* 405–440.

Ingle, J. C., Jr. (1972). Stratigraphy and paleoecology of Early Miocene through Early Pleistocene benthonic and planktonic Foraminifera, San Joaquin Hills–Newport Bay, Orange County, California. *Proc. Pacific Coast Miocene Biostrat. Symp. 47th Ann. Pacific Sec., S.E.P.M. Conv., Bakersfield, Calif.:* 255–283.

Isakson, J. S., Simenstad, C. A., and Burgner, R. L. (1971). Fish communities and food chains in the Amchitka area. *Bioscience 21:* 666–670.

Ivanova, E. A. (1949). Usloviya sushestvovaniya, obraz zhisin i istoriya razvitiya nekotorikh brakhiopod srednego i verkhnego Karbona podmoskovnoi kotlovini: *Trudy Paleont. Inst. Akad. Nauk SSSR. T. XXI:* 152 pp.

Jaanusson, V. (1961). Discontinuity surfaces in limestones. *Geol. Inst. Univ. Uppsala Bull. 40:* 221–241.

Jaanusson, V., Laufeld, S., and Skoglund, R. (Eds.) (1979). Lower Wenlock faunal and floral dynamics–Vattenfallet Section, Gotland. *Sveriges Geol. Undersök., Serie C, Nr. 762, Avhandl. Uppsat., Arsbok 73, Nr. 3:* 294 p.

Jackson, J. B. C. (1972). The ecology of molluscs of *Thalassia* communities, Jamaica, West Indies. 2. Molluscan population variability along an environmental stress gradient. *Mar. Biol. 14:* 304–337.

Jackson, J. B. C. (1973). The ecology of molluscs in the *Thalassia*, communities, Jamaica, West Indies. 1. Distribution, environmental physiology, and the ecology of common shallow-water species. *Bull. Mar. Sci. 23:* 310–350.

Jackson, J. B. C. (1974). Biogeographic consequence of eurytopy and stenotopy among marine bivalves and their evolutionary significance. *Am. Nat. 108:* 541–560.

Jackson, J. B. C. (1977a). Habitat area, colonization, and development of epibenthic community structure. *In:* "Biology of Benthic Organisms." (B. F. Keegan, P. O. Ceidigh, and P. J. S. Boadean, Eds.) Pergamon Press, New York: 349–358.

Jackson, J. B. C. (1977b). Competition on marine hard substrata: The adaptive significance of solitary and colonial strategies. *Am. Nat. 111:* 743–767.

Jackson, J. B. C., and Buss, L. (1975). Allelopathy and spatial competition among coral reef invertebrates. *Proc. Nat. Acad. Sci. 72:* 5160–5163.

Jackson, J. B. C., Goreau, T. F., and Hartman, W. D. (1971). Recent brachiopod–coralline sponge communities and their paleoecological significance. *Science 173:* 623–625.

Jacobsen, V. H. (1967). The feeding of the lugworm *Arenicola marina:* Quantitative studies. *Ophelia 4:* 91–109.

James, N. P., Ginsburg, R. N., Marszalek, D. S., and Choquette, P. W. (1976). Facies and fabric specificity of early subsea cements in shallow Belize (British Honduras) reefs. *J. Sed. Pet. 46:* 523–544.

Janssen, A. W. (1967). Beiträge zur Kenntnis des Miocäns von Dingden und seiner Mollusken fauna 1. *Geol. et. Palaontol. 1:* 115–173.

Jarco, S. (Ed.) (1966). "Human Palaeopathology." Yale Univ. Press, New Haven: 182 p.

Jefferies, R. P. S. (1960). Photonegative young in the Triassic lamellibranch *Lima lineata* (Schlotheim). *Palaeontology 3:* 362–369.

Jeppsen, L. (1974). Aspects of Late Silurian conodonts. *Fossils and strata No. 6:* 79 p.

Jeppsen, L. (1976). Autecology of Late Silurian conodonts. *Geol. Assoc. Canada Spec. Paper No. 15:* 105–118.

Jerzmanska, A., and Kotlarczyk, J. (1976). The beginnings of the Sargasso assemblage in the Tethys. *Palaeogeogr., Palaeoclimatol., Palaeoecol. 20:* 297–306.

Jillett, J. B. (1976). Zooplankton associations off Otago Peninsula, southeastern New Zealand, related to different water masses. *N. Z. J. Mar. and Freshwater Res. 10:* 543–557.

Johnson, J. H. (1961). "Limestone-Building Algae and Algal Limestones." Colorado School Mines, Johnson Publ. Co.: pp. 22–38, 251 p.

Johnson, P. T. (1968). "An Annotated Bibliography of Pathology in Invertebrates other than Insects." Burgess Publ. Co.: 322 p.

Johnson, R. G. (1965). Pelecypod death assemblages in Tomales Bay, California. *J. Paleontol. 39:* 80–85.

Johnson, R. G. (1970). Variations in diversity within benthic marine communities. *Am. Nat. 104:* 283–300.

Johnson, R. G. (1974). Particulate matter at the sediment–water interface in coastal environments. *J. Mar. Res. 32:* 313–330.

Johnston, R. F. (1954). The summer food of some intertidal fishes of Monterey County, California. *Calif. Fish. Game 40:* 65–68.

Jones, G. D., and Ross, C. A. (1979). Seasonal distribution of foraminifera in Samish Bay, Washington. *J. Paleontol. 53:* 245–257.

Jones, G. F. (1969). The benthic macrofauna of the mainland shelf of southern California. *Allan Hancock Monogr., Mar. Biol. 4:* 219 p.

Jones, M. L. (1968). On the morphology, feeding and behavior of *Magelona* sp. *Biol. Bull. 134:* 272–297.

Jones, M. L., and Dennison, J. M. (1970). Oriented fossils as paleocurrent indicators in Paleozoic lutites of Southern Appalachians. *J. Sed. Pet. 40:* 642–649.

Joyce, E. A., Jr. (1972). A partial bibliography of oysters, with annotations. *Spec. Sci. Rept. No. 34, Dept. Nat. Res. Florida:* 846 p.

Jumars, P. A. (1978). Spatial autocorrelation with RUM (Remote Underwater Manipulator): Vertical and horizontal structure of a bathyal benthic community. *Deep-Sea Research 25:* 589–604.

Jumars, P. A., and Fauchald, K. (1977). Between-community contrasts in successful polychaete feeding strategies. *In:* "Ecology of Marine Benthos." (B. C. Coull, Ed.) Univ. South Carolina Press, Columbia, S.C.: pp. 1–21.

Kain, J. M. (1962). Aspects of the biology of *Laminaria hyperborea. J. Mar. Biol. Assoc. U. K. 42:* 377–385.

Kaiser, H. E. (1964). Pathological conditions of the soft parts in a Devonian brachiopod species *Stropheodonta. N. Jb. Geol. Paläontol., Monh.:* 196–198.

Kaiser, P., and Voigt, E. (1977). Uber eine als Gastropodenlaich gedeutete Eiablage in einer schale von *Pseudopecten* aus dem Lias von Salzgitter. *Paläontol. Zeit. 51:* 5–11.

Kalyanasundaram, N., Ganti, S. S., and Karande, A. A.

(1974). The habitat and the habitat-selection by *Umbonium vestiarium* L. *INSA Bull. 47, Indian Nat. Sci. Acad. 38:* 273–287.

Kanciruk, P., and Herrnkind, W. (1978). Mass migration of spiny lobster, *Palinurus argus* (Crustacea: Palinuridae): Behavior and environmental correlates. *Bull. Mar. Sci. 28:* 601–623.

Kapp, U. S. (1975). Paleoecology of Middle Ordovician stromatoporoid mounds in Vermont. *Lethaia 8:* 195–207.

Kauffman, E. G. (1967). Coloradoan macroinvertebrate assemblages, Central Western Interior, United States. *In:* Paleoenvironment of the Cretaceous Seaway in the Western Interior: A symposium. (E. G. Kauffman and H. C. Kent, Eds.) *Colorado School Mines Spec. Paper 1:* 67–143.

Kauffman, E. G. (1969). Systematics and evolutionary position of a new Tertiary *Thyasira* from Alaska. *J. Paleontol. 43:* 1099–1110.

Kauffman, E. G. (1973). A brackish water biota from the Upper Cretaceous Harebell Formation of northwestern Wyoming. *J. Paleontol. 47:* 436–446.

Kauffman, E. G. (1975). Dispersal and biostratigraphic potential of Cretaceous benthonic bivalvia in the Western Interior. *Geol. Assoc. Canada Spec. Paper 13:* 163–194.

Kauffman, E. G. (1978). Benthic environments and paleoecology of the Posidonienschiefer (Toarcian). *In:* Paleoecology: Construction, sedimentology, diagenesis and associations of fossils. (A. Seilacher and F. Westphal, Eds.) *N. Jb. Geol. Paläontol., Abh. 157:* 18–36.

Kauffman, E. G., and Kesling, R. V. (1960). An Upper Cretaceous ammonite bitten by a mosasaur. *Contrib. Mus. Paleontol. Univ. Michigan 15:* 193–248.

Kauffman, E. G., and Scott, R. W. (1976). Basic concepts of community ecology and paleoecology. *In:* "Structure and Classification of Paleocommunities." (R. W. Scott, and R. R. West, Eds.) Dowden, Hutchinson & Ross, Stroudsburg, pp. 1–28.

Kauffman, E. G., and Sohl, N. F. (1974). Structure and evolution of Antillean Cretaceous rudist frameworks. *Verhandl. Naturf. Ges. Basel 84:* 399–467.

Keen, M. C. (1977). Ostracod assemblages and the depositional environments of the Headon, Osborne and Bembridge Beds (Upper Eocene) of the Hampshire Basin. *Palaeontology 20:* 405–446.

Keith, M. L., and Parker, R. H. (1965). Isotope contents of mollusk shells. *Mar. Geol. 3:* 115–129.

Keller, T. (1976). Magen- und Darminhalte von Ichthyosauriern des Süddeutschen Posidonienschiefers. *N. Jb. Geol. Paläontol., Monh. 5:* 266–283.

Kennedy, W. J. (1975). Trace fossils in carbonate rocks. *In:* "The Study of Trace Fossils." (R. W. Frey, Ed.) Springer-Verlag, New York: pp. 377–398.

Kennedy, W. J., and Klinger, H. C. (1972). Hiatus concretions and hardground horizons in the Cretaceous of Zululand (South Africa). *Palaeontology 15:* 539–549.

Kern, J. P. (1971). Paleoenvironmental analysis of a Late Pleistocene estuary in Southern California. *J. Paleontol. 45:* 810–823.

Kern, J. P. (1973). Early Pliocene marine climate and environment of the eastern Ventura Basin, southern California. *Univ. Calif. Publ. Geol. Sci. 96:* 117 p.

Kern, J. P. (1978). Paleoenvironment of new trace fossils from the Eocene Mission Valley Formation, California. *J. Paleontol. 52:* 186–194.

Kern, J. P., and Wicander, E. R. (1974). Origin of a bathymetrically displaced marine invertebrate fauna in the upper part of the Capistrano Formation (Late Pliocene) southern California. *J. Paleontol. 48:* 495–505.

Kesling, R. V., and LeVasseur, D. (1971). *Strataster ohioensis,* a new Early Mississippian brittle-star and the paleoecology of its community. *Contrib. Mus. Paleontol. Univ. Michigan 23:* 305–341.

Kidston, R., and Lang, W. H. (1924). On the presence of tetrads of resistant spores in the tissue of *Sporocarpon furcatum* Dawson from the Upper Devonian of America. *Trans. Roy. Soc. Edinburgh 52:* 597–601.

Kier, P. M. (1957). Tertiary echinoidea from British Somaliland. *J. Paleo. 31:* 839–902.

Kier, P. M. (1972). Upper Miocene Echinoids from the Yorktown Formation of Virginia and Their Environmental Significance. *Smithsonian Contribs. Paleobiology,* No. 13: 41 pp.

Kier, P. M. (1977). The poor fossil record of the regular echinoid. *Paleobiology 3:* 168–174.

Kier, P. M., and Grant, R. E. (1965). Echinoid distribution and habits, Key Largo Coral Reef Preserve, Florida. *Smithsonian Misc. Coll. 149:* 68 pp.

Kinne, O. (Ed.) (1970). "Marine Ecology 1(1)." Wiley–Interscience, John Wiley & Sons, Ltd., London: 681 p.

Kinne, O. (1971a). Invertebrates. *In:* "Marine Ecology 1(2)." Wiley–Interscience, John Wiley & Sons, Ltd., London: pp. 821–995.

Kinne, O. (Ed.) (1971b). "Marine Ecology 1(2)." Wiley–Interscience, John Wiley & Sons, Ltd., London: pp. 682–1244.

Kinne, O. (Ed.) (1972). "Marine Ecology 1(3)." Wiley–Interscience, John Wiley & Sons, Ltd., London: pp. 1245–1774.

Kinne, O. (Ed.) (1975). "Marine Ecology 2(1, 2)." Wiley–Interscience, John Wiley & Sons, Ltd., London: pp. 1–449; 451–992.

Kinne, O., Morita, R. Y., Vidaver, W., and Flugel, H. (1971). Pressure. *In:* "Marine Ecology 1(3)." Wiley–Interscience, John Wiley & Sons, Ltd., London: pp. 1323–1450.

Kissling, D. L. (1973). Circumrotary growth form in Recent and Silurian corals. *In:* "Animal Colonies; their Development and Function through Time." (R. S. Boardman, A. H. Cheetham, and W. A. Oliver, Jr., Eds.) Dowden, Hutchinson & Ross, Stroudsburg, Penna. pp. 43–58.

Kitchell, J. A. (1979). Deep-sea foraging pathway: An analysis of randomness and resource exploitation. *Paleobiology 5:* 107–125.

Kitchell, J. A., Kitchell, J. F., Clark, D., and Dangeard, L. (1978a). Deep-sea foraging behavior, its bathymetric potential in the fossil record. *Science 200:* 1289–1291.

Kitchell, J. A., Kitchell, J. F., Johnson, G., and Hunkins, K. L. (1978b). Abyssal traces and megafauna: comparison of productivity, diversity and density in the Arctic and Antarctic. *Paleobiol. 4:* 171–180.

Kitching, J. A. (1937). Studies in sublittoral ecology: II. Recolonization at the upper margin of the sublittoral region with a note on the denudation of *Laminaria* forest by storms. *J. Ecol. 25:* 482–495.

Kitching, J. A. (1972). The effects of pressure on organisms: A summary of progress. *Symp. Soc. Exper. Biol. 26:* 473–482.

Kitching, J. A., and Ebling, F. J. (1967). Ecological studies at Lough Ine. *In:* "Advances in Ecological Research." (J. H. Cragg, Ed.) Academic Press, New York: pp. 197–291.

Kitching, J. A., Macan, T. T., and Gilson, H. C. (1934). Studies in sublittoral ecology. I. A submarine gully in Wembury Bay, South Devon. *J. Mar. Biol. Assoc. U. K. 19:* 677–705.

Klein, G. de V. (1972). Sedimentary models for determining paleotidal range: Reply. *Geol. Soc. Am. Bull. 83:* 539–546.

Klovan, J.E. (1974). Development of western Canadian reefs and comparison with Holocene analogues. *Am. Assoc. Pet. Geol. Bull. 58:* 787–799.

Knight–Jones, E. W., and Quasim, S. Z. (1955). Responses of some marine plankton animals to changes in hydrostatic pressure. *Nature 175:* 941–942.

Knox, G. A. (1970). Antarctic marine ecosystems. *In:* "Antarctic Biology," V. 1. (M. W. Holdgate, Ed.) Academic Press, New York: pp. 69–96.

Kobluk, D. R., and James, N. P. (1979). Cavity-dwelling organisms in Lower Cambrian patch reefs from southern Labrador. *Lethaia 12:* 193–218.

Kobluk, D. R., James, N. P., and Pemberton, S. G. (1978). Initial diversification of macroboring ichnofossils and exploitation of the macroboring niche in the lower Paleozoic. *Paleobiol. 4:* 163–170.

Koch, D. L., and Strimple, H. L. (1968). A new Upper Devonian cystoid attached to a discontinuity surface. *Iowa Geol. Surv. Rept. Invest. 5:* 1–49.

Koehler, R. (1922). Anomalies et irregularites du test des Echinides. *Bull. Inst. Oceanogr., Monaco, No. 419:* 1–158.

Kohn, A. J. (1959). The ecology of *Conus* in Hawaii. *Ecol. Monogr. 29:* 47–90.

Kohn, A. J. (1971). Diversity, utilization of resources, and adaptive radiation in shallow-water marine invertebrates of tropical oceanic islands. *Limnol. and Oceanogr. 16:* 332–348.

Kohn, A. J., and Nybakken, J. M. (1975). Ecology of *Conus* on Eastern Indian Ocean fringing reefs; Diversity of species and resource utilization. *Mar. Biol. 29:* 211–234.

Kornas, J. (1972). Corresponding taxa and their ecological background in the forests of temperate Eurasia and North America. *In:* "Taxonomy, Phytogeography, and Evolution." (D. H. Valentine, Ed.) Academic Press, New York: pp. 37–59.

Kornicker, L. S., Wise, C. D., and Wise, J. M. (1963). Factors affecting the distribution of opposing mollusk valves. *J. Sed. Pet. 33:* 703–712.

Korringa, P. (1951). The shell of *Ostrea edulis* as a habitat. *Neerland. Zool. 10:* 32–52.

Kozlowski, R. (1974). Decouverte des oeufs des Polychaetes dans l'Ordovicien. *Acta Pal. Polonica 19:* 437–442.

Kranz, P. M. (1974a). Computer simulation of fossil assemblage formation under conditions of anastrophic burial. *J. Paleontol. 48:* 800–808.

Kranz, P. M. (1974b). The anastrophic burial of bivalves and its paleoecological significance. *J. Geol. 82:* 237–265.

Kristensen, I. (1957). Differences in density and growth in a cockle population in the Dutch Wadden Sea. *Arch. Neerland. Zool. 12:* 351–453.

Kriz, J., (1979). Silurian Cardiolidae (Bivalvia). *Sbornik geol. ved, Paleontologie 22:* 157 pp.

Krone, R. B. (1976). Engineering interest in the benthic boundary layer. *In:* "The Benthic Boundary Layer." (I. N. McCave, Ed.) Plenum Press, New York: pp. 143–156.

Kruger, F. (1959). Zur Ernährungsphysiologie von *Arenicola marina*. *Zool. Anz. Suppl. 22:* 115–120.

Krumbein, W. C. (1936). Application of logarithmic moments to size–frequency distribution of sediments. *J. Sed. Pet. 6:* 35–47.

Kummel, B., and Raup, D. (Eds.) (1965). "Handbook of Paleontological Techniques." W. H. Freeman & Co., San Francisco: 852 p.

Kuris, A. M. (1974). Trophic interactions: Similarity of parasitic castrators to parasitoids. *Quart. Rev. Biol. 49:* 129–148.

LaBarbera, M. (1974). Larval and post-larval development of five species of Miocene bivalves (Mollusca). *J. Paleontol. 48:* 256–277.

Ladd, H. S. (Ed.) (1957). Treatise on marine ecology and paleoecology, 2 Paleoecology. *Geol. Soc. Am. Mem. 67:* 1077 p.

Ladd, H. S. (1966). Chitons and gastropods (Haliotidae through Adeorbidae) from the Western Pacific Islands. *U. S. Geol. Surv. Prof. Paper 531:* 98 p.

Ladd, H. S. (1970). Existing reefs—geological aspects. *In: Proc. N. Am. Paleontol. Conv. 2* (E. Yochelson, Ed.): 1273–1300.

Lagaaij, R. (1973). Shallow-water bryozoa from deep-sea sands of the Principe Channel, Gulf of Guinea. *In:* Living and fossil bryozoa. (G. P. Larwood, Ed.) Academic Press, New York: pp. 139–150.

Lamb, I. M., and Zimmermann, M. H. (1977). Benthic marine algae of the Antarctic Peninsula, a preliminary guide to the common benthic marine algae of the Antarctic Peninsula and adjacent islands. *In:* Biology of the Antarctic Seas, V, Paper 4 (D. L. Pawson, Ed.) *Antarc. Res. Ser. 23,* no. 4: 130–229.

Landenberger, D. E. (1968). Studies on selective feeding in the Pacific starfish *Pisaster* in Southern California. *Ecology 49:* 1062–1075.

Lander, W. S. (1954). Notes on the predation of the hard clam, *Venus mercenaria,* by the mud crab, *Neopanope texana. Ecology 35:* 422.

Lander, W. S. (1967). Infestation of the hard clam, *Mercenaria mercenaria,* by the boring polychaete worm, *Polydora ciliata. Proc. Nat. Shellfisheries Assoc. 57:* 63–66.

Lane, N. G. (1971). Crinoids and reefs. *Proc. North Am. Paleontol. Conv. Pt. J:* 1430–1443.

Lane, N. G. (1973). Paleontology and paleoecology of the Crawfordsville fossil site (Upper Osagian: Indiana). *Univ. California Publ. Geol. Sci. 99:* 1–141.

Lang, J. C. (1974). Biological zonation at the base of a reef. *Am. Sci. 62:* 272–281.

Larsson, K. (1979). Silurian tentaculitids from Gotland and Scania. *Fossils and Strata 11:* 180 pp.

Lawrence, D. L. (1968). Taphonomy and information losses in fossil communities. *Geol. Soc. Am. Bull. 79:* 1315–1350.

Leckwijck, W. P. van, and Chesaux, C. H. (1962). Vertical and lateral variation in the lithology and the fauna of the Petit Buisson Band in the Norimage Coalfield, Southern Belgium. *Paläontol. Zeit., H.* Schmidt–Festband: 140–153.

Lecompte, M. (1970). Die Riffe im Devon der Ardennen und ihre Bildungsbedingungen. *Geol. Palaeontol. 4:* 25–71.

Leighton, D. L. (1960). An abalone lacking respiratory apertures. *The Veliger 3:* 48.

Leppakoski, E. (1969). Transitory return of the benthic fauna of the Bornholm Basin after extermination by oxygen insufficiency. *Cah. Biol. Mar. 10:* 163–172.

Lessertisseur, J. (1955). Traces fossiles d'activité animale et leur signification paleobiologique. *Géol. Soc. France, Mém. 74:* 150 p.

Lever, J. (1958). Quantitative beach research. I. The left-right phenomenon: Sorting of lamellibranch valves on sandy beaches. *Basteria 22:* 21–51.

Lever, J. (1961). Quantitative beach research. II. The "hole effect". *Netherlands J. Sea Res. 1:* 339–358.

Levinton, J. S. (1970). The paleoecological significance of opportunistic species. *Lethaia 3:* 69–78.

Levinton, J. S., and Bambach, R. K. (1975). A comparative study of Silurian and Recent deposit-feeding bivalve communities. *Paleobiol. 1:* 97–124.

Levinton, J. S., and Lopez, C. R. (1977). A model of renewable resources and limitation of deposit-feeding benthic populations. *Oceologia (Berl.) 31:* 177–190.

Lewis, D. B. (1968). Feeding and tube-building in the Fabriciinae (Annelida, Polychaeta). *Proc. Linn. Soc. London 179:* 37–49.

Lewis, J. B. (1965). A preliminary description of some marine benthic communties from Barbados, West Indies. *Can. J. Zool. 43:* 1049–1074.

Lewis, J. R. (1964). "The Ecology of Rocky Shores." English Univ. Press, London: 323 p.

Liddell, W. D. (1975). Recent crinoid biostratinomy. *Geol. Soc. Am. Abstr. with Prog., Ann. Mtg.:* 1169.

Lie, U., and Evans, R. A. (1973). Long-term variability in the structure of subtidal benthic communities in Puget Sound, Washington, U. S. A. *Mar. Biol. 21:* 122–126.

Lim, H.-K., and Heyneman, D. (1972). Intramolluscan intertrematode antagonism: A review of factors influencing the host-parasite system and its possible role in biological control. *Adv. Parasitol. 10:* 192–268.

Lindström, M. (1963). Sedimentary folds and the development of limestone in an Early Ordovician sea. *Sedimentology 2:* 243–292.

Lindström, M. (1979). Diagenesis of Lower Ordovician hardgrounds in Sweden. *Geol. et Paleontol. 13:* 9–30.

Livingstone, D. A. (1963). Data of geochemistry, sixth edition, Ch. G, Chemical composition of rivers and lakes. *U. S. Geol. Surv. Prof. Paper 440–G:* 64 p.

Lloyd, R. M. (1972). Isotopic record of circulation gradient. *In:* South Florida carbonate sediments. Sedimenta II (R. N. Ginsburg, Ed.) *Div. Mar. Geol., Geophys., Rosenstiel School Mar. & Atmos. Sci., Univ. Miami:* 15–19.

Lochman, C. (1941). A pathologic pygidium from the Upper Cambrian of Missouri. *J. Paleontol. 15:* 324–325.

Lockwood, S. (1884). An oyster on a crab. *Am. Nat. 18:* 200.

Loesch, H. C. (1953). Studies of the ecology of two species of *Donax* on Mustang Island, Texas. *Publ. Inst. Mar. Sci. Univ. Texas, Port Aransas 3:* 201–227.

Logan, B. W., Hoffman, P., and Gebelstein, C. D. (1974). Algal mats, cryptalgal fabrics and structures, Hamelin Pool, Western Australia. *In:* Evolution and diagenesis of Quaternary carbonate sequences, Shark Bay, Australia. *Am. Assoc. Pet. Geol. Mem. 22:* 140–195.

Lohman, K. E. (1957). Diatoms. *In:* Treatise on marine ecology and paleoecology. (H. S. Ladd, Ed.) *Geol. Soc. Am. Mem.* 67(2), *Paleoecology:* 731–736.

Longbottom, M. R. (1970). The distribution of *Arenicola marina* (L.) with particular reference to the effects of particle size and organic matter of the sediments. *J. Exper. Mar. Biol. Ecol. 5:* 138–157.

Longhurst, A. R. (1958). An ecological survey of the West African marine benthos. *Colonial Off., Fish. Publ. 11:* 102 p.

Longhurst, A. R. (1964). A review of the present situation in benthic synecology. *Bull. Inst. Oceanogr. Monaco 63:* 1–54.

Longhurst, A. R. (1969). Species assemblages in tropical demersal fisheries. *In:* "Proc. Symp. on the Oceanogr. Fish. Res. Tropical Atlantic: Review Papers and Contributions, UNESCO," Paris: pp. 147–168.

Loosanoff, V. L. (1956). Two obscure oyster enemies in New England waters. *Science 123:* 1119–1120.

Loosanoff, V. L. (1962). Effects of turbidity on some larval and adult bivalves. *Proc. Gulf Carib. Fish. Inst. 14:* 80–95.

Loosanoff, V. L., and Davis, H. C. (1963). Rearing of bivalve mollusks. *Adv. Mar. Biol. 1:* 1–135.

Loosanoff, V. L., and Engle, J. (1940). Spawning and setting of oysters in Long Island Sound in 1937, and discussion of the method for predicting intensity and time of oyster setting. *U. S. Bur. Fish. Bull. 33, 49:* 218–255.

Lowe (McConnell), R. H. (1962). The fishes of the British Guiana continental shelf, Atlantic coast of South America, with notes on their natural history. *J. Linn. Soc. London 44:* 669–700.

Lowenstam, H. A. (1973). Biogeochemistry of hard tissues, their depth and possible pressure relationships. *In:* "Barobiology and the Experimental Biology of the Deep Sea." (R. W. Brauer, Ed.) Univ. North Carolina, Chapel Hill: pp. 19–32.

Loya, Y. (1972). Community structure and species diversity of hermatypic corals at Eilat, Red Sea. *Mar. Biol. 13:* 100–123.

Lubchenco, J., and Menge, B. A. (1978). Community development and persistence in a rocky intertidal zone. *Ecol. Mono. 59:* 67–94.

Luders, K., and Trusheim, F. (1929). Beitrage zur Ablagerung mariner Mollusken in der Flachsee: Entstehung und Aufbau von Grossrucken mit Schillbedeckung in Flutbezv. Ebbetricktern der Aussenjade. *Senck. 11:* 123–141.

Ludvigsen, R. (1977). Rapid repair of traumatic joint injury by an Ordovician trilobite. *Lethaia 10:* 179–266.

Ludvigsen, R. (1979a). Fossils of Ontario: Part I: The trilobites. *Life Sci. Misc. Publ. Roy. Ontario Mus.:* 96 p.

Ludvigsen, R. (1979b). A trilobite zonation of Middle Ordovician rocks, southwestern District of Mackenzie: *Geol. Surv. Canada Bull. 312:* 99 pp.

Lund, E. J. (1957). Self-silting, survival of the oyster as a closed system, and reducing tendencies of the environment. *Publ. Inst. Mar. Sci.* (Texas) *4:* 313–319.

Lunz, G. R., Jr. (1947). *Callinectes* vs. *Ostrea. J. Elisha Mitchell Sci. Soc. 63:* 81.

Lutz, R. A., and D. Jablonski. (1978a). Cretaceous bivalve larvae. *Science 199:* 439–440.

Lutz, R. A., and Jablonski, D. (1978b). Larval bivalve shell morphometry: A new paleoclimatic tool? *Science 202:* 51–53.

Lutz, R. A., and Rhoads, D. (1977). Anaerobiosis and a theory of growth line formation. *Science 198:* 1222–1227.

Lützen, J. (1972). Studies on parasitic gastropods from echinoderms II. On *Stilifer* Broderip, with special reference to the structure of the sexual apparatus and reproduction. Kong. *Danske Videnskab. Selsk. Biol. Skr. 19:* (6): 18 pp.

Lützen, J., and Nielsen, K. (1975). Contributions to the anatomy and biology of *Echineulima* n. g. (Prosobranchia: Eulimidae), parasitic on sea urchins: Vidensk. Meddr. dansk naturh. Foren. *38:* 171–199.

MacArthur, R. H. (1957). On relative abundance of bird species. *Proc. Nat. Acad. Sci. U. S. A. 43:* 293–295.

MacArthur, R. H. (1960). On relative abundance of species. *Am. Nat. 94:* 25–36.

MacArthur, R. H. (1971). Patterns of terrestrial bird communities. *In:* "Avian Biology." (D. S. Farner, J. R. King, and K. C. Parks, Eds.) Academic Press, New York: pp. 189–221.

MacDonald, K. B. (1976). Paleocommunities: Toward some confidence limits. *In:* "Structure and Classification of Paleocommunities." (R. W. Scott and R. R. West, Eds.) Dowden, Hutchinson & Ross, Stroudsburg, Penna. pp. 87–106.

MacGeachy, J. K., and Stearn, C. (1976). Boring by macroorganisms in the coral *Monastrea annularis* on Barbados Reef. *Int. Revue ges. Hydrobiol. 61:* 715–745.

MacGillavry, H. J. (1978). Foraminifera and parallel evolution—how or why? *Geol. Mijn. 57:* 385–394.

MacGinitie, G. E. (1930). Natural history of the mud shrimp *Upogebia pugettensis* (Dana). *Ann. Mag. Nat. Hist. 6(31):* 36–41.

MacGinitie, G. E. (1934). The natural history of *Callianassa californiensis* (Dana). *Am. Midl. Nat. 15:* 166–177.

MacGinitie, G. E., and MacGinitie, N. (1949). "Natural History of Marine Animals." McGraw–Hill, New York: 473 p.

MacKay, T. F. C., and Doyle, R. W., (1978). An ecological genetic analysis of the settling behaviour of a marine polychaete. *Heredity 40:* 1–12.

MacNae, W. (1968). A general account of the fauna and flora of mangrove swamps and forests in the Indo–West-Pacific region. *Adv. Mar. Biol. 6:* 73–270.

Macurda, D. B., Jr., and Meyer, D. L. (1975). The behavior and ecology of the crinoids of the East Indies. *Geol. Soc. Am. Abstr. with Prog., Ann. Mtg.:* 1184.

Malkowski, K. (1975). Attachment scars of the brachiopod *Coenothyris vulgaris* (Schlotheim, 1820) from the Muschelkalk of Upper Silesia: *Acta Geol. Polonica 25:* 275–283.

Malkowski, K. (1976). Regeneration of some brachiopod shells. *Acta Geol. Polonica 26:* 439–442.

Manger, G. E. (1963). Porosity and bulk density of sedimentary rocks. *U. S. Geol. Surv. Bull. 1144-E:* E1–E55.

Mani, M. S. (1968). "Ecology and Biogeography of High Altitude Insects." Junk, The Hague: 527 p.

Manning, R. B. (1962). A striking abnormality in *Squilla bigelowi* Schmitt (Stomatopoda). *Crustaceana 4:* 243–244.

Manton, S. M., and Stephenson, T. A. (1931). Ecological surveys of coral reefs. *British Mus.* (Nat. Hist.) *Great Barrier Reef Exped.* 1928–29. *Sci. Rept. 3(1):* 274–312.

Mariscal, R. N., Conklin, E., and Bigger, C. (1977). The

ptychocyst, a major new category of cnida used in tube construction by a cerianthid anemone. *Biol. Bull. 152:* 392–405.

Marliave, J. B. (1977). Substratum preferences of settling larvae of marine fishes reared in the laboratory: *J. Exper. Mar. Biol. Ecol. 27:* 47–60.

Marshall, S. M., and Orr, A. P. (1960). Feeding and nutrition. *In:* "The Physiology of Crustacea. Vol. I. Metabolism and Growth." (T. H. Waterman, Ed.) Academic Press, New York: pp. 227–258.

Martin–Kaye, P. (1951). Sorting of lamellibranch valves on beaches in Trinidad. *Geol. Mag. 88:* 432–434.

Martinsson, A. (1956). Ontogeny and development of dimorphism in some Silurian ostracodes. *Bull. Geol. Inst. Univ. Uppsala 37:* 1–42.

Martinsson, A. (1962). Ostracodes of the family Beyrichiidae from the Silurian of Gotland. *Bull. Geol. Inst. Univ. Uppsala 41:* 369 pp.

Martinsson, A. (1965a). Aspects of a Middle Cambrian thanatotope on Öland. *Geol. Foren. I. Stockholm Förhand. 87:* 171–230.

Martinsson, A. (1965b). Phosphatic linings in bryozoan zooecia. *Geol. Fören. Stockholm Förh. 86:* 404–408.

Martinsson, A. (1970). Toponomy of trace fossils. *In:* Trace fossils. (T. P. Crimes and J. C. Harper, Eds.) *Geol. J. Spec. Issue 3:* 323–330.

Marwick, J. (1971). An ovoviviparous gastropod (Turritellidae, *Zeocolpus*) from the upper Miocene of New Zealand. *N. Z. J. Sci. Tech. 14:* 66–70.

Matthews, J. V., Jr. (1970a). Quaternary environmental history of interior Alaska: Pollen samples from organic colluvium and peats. *Arctic and Alpine Res. 2:* 241–251.

Matthews, J. V., Jr. (1970b). Two new species of *Micropeplus* from the Pliocene of western Alaska with remarks on the evolution of Micropeplinae (Coleoptera: Staphylinidae). *Can. J. Zool. 48:* 779–788.

Matthews, J. V., Jr. (1974a). A preliminary list of insect fossils from the Beaufort Formation, Meighen Island, District of Franklin. *Geol. Surv. Canada Paper* 74–1(A): 203–206.

Matthews, J. V., Jr. (1974b). Quaternary environments at Cape Deceit (Seward Peninsula, Alaska): Evolution of a tundra ecosystem. *Geol. Soc. Am. Bull. 85:* 1353–1384.

Matthiessen, G. C. (1960). Intertidal zonation in populations of *Mya arenaria*. *Limnol. and Oceanogr. 5:* 381–388.

Mattison, J. E., Trent, J. D., Shanks, A. L., Akin, T. R. and Pearse, J. S. (1977). Movement and feeding activity of red sea urchins (*Strongylocentrotus franciscanus*) adjacent to a kelp forest. *Mar. Biol. 39:* 25.

Mattox, N. T. (1955). "Observations on the Brachiopod Communities Near Santa Catalina Island. Essays in Natural Sciences in Honor of Capt. A. Hancock." U. S. C. Press, Los Angeles: pp. 73–86.

May, R. M. (1973). "Stability and Complexity in Model Ecosystems." Princeton Univ. Press, Princeton: 235 p.

May, R. M. (1975). Patterns of species abundance and diversity. *In:* "Ecology and Evolution of Communities." (M. L. Cody and J. M. Diamond, Eds.) Belknap Press of Harvard Univ. Press: Cambridge, Mass. pp. 81–120.

Mayr, E. (1974). Behavior programs and evolutionary strategies. *Am. Sci. 62:* 650–659.

McAlester, A. L., and Rhoads, D. C. (1967). Bivalves as bathymetric indicators. *Mar. Geol. 5:* 383–388.

McAlester, A. L., Speden, I. G., and Buzas, M. A. (1964).

Ecology of Pleistocene molluscs from Martha's Vineyard—a reconsideration. *J. Paleontol. 38:* 985–990.

McCave, I. N. (1976). "The Benthic Boundary Layer." Plenum Press, New York: 323 p.

McDermott, J. J. (1960). The predation of oysters and barnacles by crabs of the family Xanthidae. *Proc. Pennsylvania Acad. Sci. 34:* 199–211.

McGowan, J. A. (1971). Oceanic biogeography of the Pacific. *In:* "The Micropaleontology of Oceans." (B. M. Funnell and W. R. Riedel), Cambridge Univ. Press, London and New York pp. 3–74.

McIntosh, G. C. (1978). Pseudoplanktonic crinoid colonies attached to Upper Devonian (Frasnian) logs. *Geol. Soc. Am. Abstr. with Prog. Ann. Mtg. 10*(7): 451.

McIntosh, R. P. (1963). Ecosystems, evolution and relational patterns of living organisms. *Am. Sci. (Mar.)* 246–267.

McIntyre, A. D. (1961). Quantitative differences in the fauna of boreal mud associations. *J. Mar. Biol. Assoc. U. K. 41:* 599–616.

McIntyre, A. D. (1964). Meiobenthos of sublittoral muds. *J. Mar. Biol. Assoc. U. K. 44:* 665–674.

McIntyre, A. D., Be, A. W., and Preikstas, R. (1967). Coccoliths and the Pliocene–Pleistocene boundary. *In: Progress in oceanography* (M. Sears, Ed.) 4: 3–26.

McIntyre, A. D., and Eleftheriou, A. (1968). The bottom fauna of a flatfish nursery ground. *J. Mar. Biol. Assoc. U. K. 48:* 113–142.

McKee, E. D., Chronic, J., and Leopold, E. B. (1959). Sedimentation belts in lagoon of Kapinganmarangi Atoll. *Am. Assoc. Pet. Geol. Bull. 43:* 501–562.

McKnight, D. G. (1968). Recent and relict molluscan faunas from off Cape Farewell. *N. Z. J. Mar. Freshwater Res. 2:* 708–711.

McKnight, D. G. (1969a). An outline distribution of the New Zealand shelf fauna. *N. Z. Dept. Sci. Industr. Res. Bull. 195:* 86 p.

McKnight, D. G. (1969b). A Recent, possibly catastrophic burial in a marine molluscan community. *N. Z. J. Mar. Freshwater Res. 3:* 177–179.

McLusky, D. S., Nair, S. A., Stirling, A., and Bhargava, R. (1975). The ecology of a Central West Indian beach, with particular reference to *Donax incarnatus*. *Mar. Biol. 30:* 267–276.

McNulty, J. K., Work, R. C., and Moore, H. B. (1962). Level sea bottom communities in Biscayne Bay and neighboring areas. *Bull. Mar. Sci. Gulf and Caribbean 12:* 204–233.

Meadows, P. S. (1964a). Substrate selection by *Corophium* species: The particle size of substrates. *J. Animal Ecol. 33:* 387–394.

Meadows, P. S. (1964b). Experiments on substrate selection by *Corophium volutator* (Pallas): Depth selection and population density. *J. Exper. Biol. 41:* 677–687.

Meadows, P. S. (1964c). Experiments on substrate selection by *Corophium* species: Films and bacteria on sand particles. *J. Exper. Biol. 41:* 499–511.

Meadows, P. S., and Anderson, J. G. (1968). Microorganisms attached to marine sand grains. *J. Mar. Biol. Assoc. U. K. 48:* 161–175.

Meadows, P. S., and Campbell, J. I. (1972a). Habitat selection and animal distribution in the sea: The evolution of a concept. *Proc. Roy. Soc. Edinburgh*(B) 73: 145–157.

Meadows, P. S., and Campbell, J. I. (1972b). Habitat selection by aquatic invertebrates. *Adv. Mar. Biol.* 10: 271–382.

Meadows, P. S., and Ried, A. (1966). The behavior of *Corophium volutator* (Crustacea: Amphipoda) *J. Zool. Soc. London* 150: 387–399.

Medcof, J. C. (1950). Burrowing habits and movements of soft-shelled clams. *Fish. Res. Bd. Canada Progress Rept. of the Atlantic Coast Sta. No. 50:* 17–21.

Medcof, J. C. (1952). The winter flounder—a clam enemy. *Fish. Res. Bd. Canada, Progress Rept. of the Atlantic Coast Sta. No. 52:* 3–8.

Meijer, J. J., De, (1969). Fossil non-calcareous algae from insoluble residues of algal limestones. *Leidse Geol. Meded.* 44: 235–263.

Mellett, J. S. (1974). Scatological origin of microvertebrate fossil accumulations. *Science 185:* 349–350.

Menard, H. W., and Boucot, A. J. (1951). Experiments on the movement of shells by water. *Am. J. Sci. 249:* 131–151.

Menge, B. A. (1972). Competition for food between two intertidal starfish species and its effect on body size and feeding. *Ecology 53:* 635–644.

Menzel, R. W., and Nichy, F. E. (1958). Studies of the distribution and feeding habits of some oyster predators in Alligator Harbor, Florida. *Bull. Mar. Sci. Gulf Carib. 8:* 125–145.

Menzies, R. J., George, R. Y., and Rowe, G. T. (1973). "Abyssal Environment and Ecology of the World Oceans." Wiley–Interscience, New York: 488 p.

Merriam, C. W. (1931). Notes on a brittle-star limestone from the Miocene of California. *Am. J. Sci. 221:* 304–310.

Meyerhoff, A. A. (1970). Continental drift, II: High-latitude evaporite deposits and geologic history of Arctic and North Atlantic oceans. *J. Geol. 78:* 406–444.

Michael, R. (1894). Ammoniten-Brut mit Aptychen in der Wohnkammer von *Oppelia steraspis* Oppel sp. *Zeit. Deutsch. Geol. Gesell. 46:* 697–702.

Middlemiss, F. A. (1962). Brachiopod ecology and Lower Greensand palaeogeography. *Palaeontology 5:* 253–267.

Mileikovsky, S. A. (1971). Types of larval development in marine bottom invertebrates, their distribution and ecological significance: A re-evaluation. *Mar. Biol. 10:* 193–213.

Miles, R. S. (1969). VI-Features of placoderm diversification and the evolution of the arthrodire feeding mechanism. *Trans. Roy. Soc. Edinburgh 68(6):* 123–170.

Miller, B. A., and Croker, R. A. (1972). Distribution and abundance of an isolated population of *Terebra gouldi* (Gastropoda: Terebridae) on an Hawaiian subtidal sand flat. *Ecology 53:* 1120–1126.

Miller, B. S. (1967). Stomach contents of adult starry flounder and sand sole in East Sound, Orcas Island, Washington. *J. Fish. Res. Bd. Canada 24:* 2515–2526.

Miller, B. S. (1970). Food of flathead sole *Hippoglossoides elassodon* in East Sound, Orcas Island, Washington. *J. Fish. Res. Bd. Canada 27:* 1661–1665.

Miller, C. D. (1961). The feeding mechanism of fiddler crabs with ecological considerations of feeding adaptations. *Zoologica 46:* 89–100.

Miller, W., III, and Brown, N. A. (1979). The attachment scars of fossil balanids. *J. Paleontol. 53:* 208–210.

Milliman, J. D. (1977). Role of calcareous algae in Atlantic continental margin sedimentation. *In:* "Fossil Algae."

(E. Flugel, Ed.) Springer–Verlag, New York: pp. 232–247.

Mills, E. L. (1969). The community concept in marine zoology, with comments on continua and instability in some marine communities: A review. *J. Fish. Res. Bd. Canada* 26: 1415–1428.

Mitchell, C. E., and Bergström, S. M. (1977). Three-dimensionally preserved Richmondian graptolites from southwestern Ohio and the graptolite correlation of the North American Upper Ordovician Standard. *Boll. Soc. Paleont. Italiana 16:* 257–270.

Mitchell, D. F. (1953). An analysis of stomach contents of California tide pool fishes. *Am. Midl. Nat. 49:* 862–871.

Mitchell, S. W. (1975). Variation in the ontogenetic development of radial ornamentation in pelecypods and brachiopods. *Biol. Bull. 149:* 437.

Miyadi, D. (1941a). Ecological survey of the benthos of the Ago-wan. *Annot. Zool. Japan. 20:* 169–180.

Miyadi, D. (1941b). Indentation individuality in the Tanabe-wan. *Mem. Imp. Mar. Observ. Kobe, Japan 7:* 471–482.

Mobius, K. (1877, translation 1883). The oyster and oyster-culture. *Misc. Doc. No. 29, U. S. Senate, 46th Congr., 3rd Session, U. S. Comm. of Fish and Fisheries, Rept. of Commissioner for 1880:* 683–751.

Molander, A. R. (1928). Investigations into the vertical distribution of the fauna of the bottom deposits in Gullmar fjord. *Svenska Hydrog.-Biol. Komm. Skrifter, Ser. Hydrografi 6:* 1–5.

Moodie, R. L. (1923). "Paleopathology: An Introduction to the Study of Ancient Evidences of Disease." Univ. Chicago Press, Chicago: 567 p.

Moore, H. B. (1931). The muds of the Clyde Sea area. III. Chemical and physical conditions: rate and nature of sedimentation; and fauna. *J. Mar. Biol. Assoc. U. K. 17:* 325–358.

Moore, H. B. (1936). The biology of *Purpura lapillus*. I. Shell variation in relation to environment. *J. Mar. Biol. Assoc. U. K. 21:* 61–89.

Moore, H. B. (1972). An estimate of carbonate production by macrobenthos in some tropical soft-bottom communities. *Mar. Biol. 17:* 145–148.

Moore, H. B., Davies, L. T., Fraser, T. H., Gore, R. H., and Lopez, N. R. (1968). Some biomass figures from a tidal flat in Biscayne Bay, Florida. *Bull. Mar. Sci. 18:* 261–279.

Moore, H. B., Jutare, R., Bauer, J. C., and Jones, J. A. (1961). The biology of *Lytechinus variegatus*. *Bull. Mar. Sci. Gulf Carib. 13:* 23–53.

Moore, H. F. (1913). Enemy of the oyster, the drill, investigated by U. S. experts. *New Jersey Bd. of Shellfish. Rept. for 1912:* 71–75.

Moore, H. F., and Pope, T. E. B. (1910). Oyster culture experiments and investigations in Louisiana. *Bur. Fish. Doc. No. 731, Dept. Commerce & Labor:* 1–52.

Moore, J. A. (1940). Stenothermy and eurythermy of animals in relation to habitat. *Am. Nat. 74:* 188–192.

Moore, P. G. (1973). Bryozoa as a community component on the northeast coast of Britain. *In:* "Living and Fossil Bryozoa." (G. Larwood, Ed.) Academic Press, New York: pp. 21–36.

Moore, T. C., and Heath, G. R. (1966). Manganese nodules, topography, and thickness of Quaternary sediments in the central Pacific. *Nature 212:* 983–985.

Moore, T. C., and Scrutton, P. C. (1957). Minor internal

structures of some Recent unconsolidated sediments. *Am. Assoc. Pet. Geol. Bull. 41:* 983–985.

Moorman, F. R. (1963). Acid sulfate soils (cat-clays) of the tropics. *Soil Sci. 95:* 271–275.

Morgan, E. (1972). The pressure sensitivity of marine invertebrates—a resume after 25 years. *Proc. Roy. Soc. Edinburgh*(B) *73:* 287–299.

Mori, S. (1938). Characteristic tidal rhythmic migration of a mussel *Donax semignosus* Dunker, and the experimental analysis of its behavior at flood tide. *Zool. Mag.* (Japan) *50:* 1–12.

Morris, S. C. (1977). Fossil priapulid worms. *Spec. Papers in Palaeontol. No. 20, Palaeontol. Assoc.:* 95 p.

Morse, E. S. (1902). Observations on living Brachiopoda. *Mem. Boston Soc. Nat. Hist. 5:* 313–386.

Morton, J. E. (1954). The crevice fauna of the upper intertidal zone at Wembury. *J. Mar. Biol. Assoc. U. K. 33:* 187–224.

Morton, J. E., and Challis, D. A. (1969). The biomorphology of Solomon Island shores with a discussion of zoning patterns and ecological terminology. *Phil. Trans. Roy. Soc.* (B) *255:* 459–516.

Morton, J. E., and Miller, M. (1968). "The New Zealand Sea Shore." Collins, London: 638 p.

Mosher, C. (1980). Distribution of *Holothuria arenicola* Semper in the Bahamas with observations on habitat, behavior and feeding activity. *Bull. Marine Science 30:* 1–12.

Müller, A. H. (1957). "Lehrbuch der Palaozoologie. Band I, Allegemeine Grundlagen." Veb. Gustav Fischer Verlag, Jena: 322 p.

Müller, A. H., (1970). Uber Ichnia vom Typ *Ophiomorpha* und *Thalassinoides* (*Vestigia invertebratorum*, Crustacea). *Deutsch. Akad. Wiss. Berlin, Monatsber. 12:* 775–787.

Müller, A. H. (1979). Fossilization (Taphonomy). *In:* "Treatise on Invertebrate Paleontology." (R. A. Robison and C. Teichert, Eds.) Pt. A, Introduction. Geol. Soc. Am. and Univ. Kansas Press, Lawrence: A2–A78.

Müller, C. D. (1966). Seltene Bryozoen-Kugelform in einem Spülsaum. *Natur und Museum 96:* 176–179.

Muller, J. (1959). Palynology of Recent Orinoco delta and shelf sediments. Micropaleontology 5: 1–32.

Muller, J. (1966). Montane pollen from the Tertiary of N. W. Borneo. *Blumea 14:* 231–235.

Müller, K. J. (1969). Börstenbildung bei Conodonten. *Paläontol. Zeit. 43:* 64–71.

Muller, P. J., and Seuss, E. (1977). Interaction of organic compounds with calcium carbonate. III. Amino acid composition of sorbed layers. *Geochim. et Cosmochim. Acta 41:* 941–949.

Multer, H. G., and Milliman, J. D. (1967). Geologic aspects of sabellarian reefs, southeastern Florida. *Bull. Mar. Sci. 17:* 257–267.

Murray, J. W. (1973). "Distribution and Ecology of Living Benthic Foraminiferids." Crane, Russak & Co., Inc., New York: 274 p.

Murray, J. W. (1976). Comparative studies of living and dead benthic foraminiferal distributions. *In:* "Foraminifera." (R. H. Hedley and C. G. Adams, Eds.) V. 2, Academic Press, New York: pp. 45–109.

Murray, J. W., and Wright, C. A. (1974). Palaeogene foraminifera and palaeoecology, Hampshire and Paris Basins and the English Channel. *Spec. Papers in Palaeontol. 14, Palaeontol. Assoc.:* 129 p.

Muus, K. (1967). The fauna of Danish estuaries and lagoons. *Meddr. Danm. Fisk.-og. Havunders 5:* 1–316.

Muus, K. (1973). Settling, growth and mortality of young bivalves in the Oresund. *Ophelia 12:* 79–116.

Myers, A. C. (1972). Tube-worm–sediment relationships of *Diopatra cuprea* (Polychaetes: Onuphidae). *Mar. Biol. 17:* 350–356.

Myers, A. C. (1973). Sediment reworking, tube building, and burrowing in a shallow subtidal marine bottom community: Rates and effects. Ph.D. dissert., Univ. Rhode Island, Kingston: 117 p.

Myers, A. C. (1977a). Sediment processing in a marine subtidal sandy bottom community: I. Physical aspects. *J. Mar. Res. 35:* 609–632.

Myers, A. C. (1977b). Sediment processing in a marine subtidal sandy bottom community: II. Biological consequences. *J. Mar. Res. 35:* 633–647.

Nagle, J. S. (1967). Wave and current orientation of shells. *J. Sed. Pet. 37:* 1124–1138.

Natland, M. L. (1963). Presidential address: Paleoecology and turbidites. *J. Paleontol. 37:* 946–951.

Naylor, E., and Atkinson, R. J. A. (1972). Pressure and the rhythmic behavior of inshore marine animals. *In:* "The Effects of Pressure on Organisms." Symp. Soc. Exper. Biol. 26. Academic Press, New York: pp. 395–415.

Neall, V. E. (1970). Notes on the ecology and paleoecology of *Neothyris*, an endemic New Zealand brachiopod. *N. Z. J. Mar. Freshwater Res. 4:* 117–125.

Needler, A. W. H. (1933). Mortality of scallops in the southern Gulf of St. Lawrence. *Canada Fish. Res. Bd. Ann. Rept.* 30–33.

Nelson, J. R. (1931). Trapping the oyster drill. *Ann. Rept. Dept. of Sewage Disposal, Bull. 521, New Jersey Agricul. Exper. Sta., New Brunswick, N. J.:* 2–12.

Neudecker, S. (1979). Effects of grazing and browsing fishes on the zonation of corals in Guam. *Ecol. 60:* 666–672.

Neuffer, O. F. (1971). Nachweis von Färbungsmustern an tertiären Bivalven unter UV-Licht. *Abh. Hess. Landesamt. Bodenforsch. 60:* 121–130.

Neumann, A. C., Kofoed, J. W., and Keller, G. H. (1977) Lithoherms in the Straits of Florida. *Geology 5:* 4–10.

Neumann, A. C., and Land, L. S. (1975). Lime mud deposition and calcareous algae in the Bight of Abaco, Bahamas: A budget. *J. Sed. Pet. 45:* 763–786.

Neushul, M. (1967). Studies of subtidal marine vegetation in Western Washington. *Ecology 48:* 83–94.

Newcombe, C. L. (1935). A study of the community relationships of the sea mussel *Mytilus edulis* L. *Ecology 16:* 234–243.

Newell, N. D. (1971). An outline history of tropical coral reefs. *Am. Mus. Nat. Hist. Novitates 2465:* 37 p.

Newell, N. D., Imbrie, J., Purdy, E. G., and Thurber, D. L. (1959). Organism communities and bottom facies, Great Bahamas Bank. *Bull. Am. Mus. Nat. Hist. 117*(4): 181–228.

Newell, R. C. (1965). The role of detritus in the nutrition of two marine deposit feeders, the prosobranch *Hydrobia ulvae* and the bivalve *Macoma baltica*. *Proc. Zool. Soc. London 144:* 25–45.

Newell, R. C. (1970). "Biology of Intertidal Animals." Elsevier, Amsterdam: 555 p.

Newman, G. G., and Pollock, D. E. (1974). A mass stranding

Trueman, E. R. (1968). Burrowing habit and the early evolution of body cavities. *Nature 218:* 96–98.

Trueman, E. R., and Ansell, A. (1969). The mechanics of burrowing into soft substrate by marine animals. *In: Oceanogr. Mar. Biol. Ann. Rev.* (H. Barnes, Ed.) *7:* 315–366.

Turekian, K. K., Cochran, J. K., and DeMaster, D. J. (1978). Bioturbation in deep-sea deposits: rates and consequences. *Oceanus 21*(1): 34–41.

Turner, H. J., Jr. (1950). Third report on investigations of methods of improving the shellfish resources of Massachusetts. *Contrib. No. 564 WHOI:* 31 p.

Turner, H. J., Jr. (1958). The effect of nutrition on the color of the callus of *Polinices duplicatus. The Nautilus 72:* 1–3.

Turner, H. J., Jr., Ayers, J. C., and Wheeler, C. L. (1948a). Appendix II, The horseshoe crab and boring snail as factors limiting the abundance of the soft-shell clam. *WHOI Contrib. No. 462, Report on Investigation of the propagation of the soft-shell clam,* Mya arenaria: 43–45.

Turner, H. J., Jr., Ayers, J. C., and Wheeler, C. L. (1948b). Appendix III, Further observations on predators of the soft-shell clam. *WHOI Contrib. 462, Report on Investigation of the propagation of the soft-shell clam,* Mya arenaria: 47–49.

Turner, R. D. (1973). Wood-boring bivalves, opportunistic species in the deep sea. *Science 180:* 1377–1379.

Turney, W. J., and Perkins, B. F. (1972). Molluscan distribution in Florida Bay, Sedimenta III, Div. Mar. Geol., Geophys., *Rosenstiel School Mar. & Atmos. Sci., Univ. Miami:* 37 p.

Tyler, A. V. (1971). Periodic and resident components in communities of Atlantic fishes. *J. Fish. Res. Bd. Canada 28:* 935–946.

Tyler, A. V. (1972). Food resource division among northern, marine, demersal fishes. *J. Fish. Res. Bd. Canada 29:* 997–1003.

Urey, H. C., Lowenstam, H. A., Epstein, S., and McKinney, C. R. (1951). Measurement of paleotemperatures and temperatures of the Upper Cretaceous of England, Denmark, and the southeastern United States. *Geol. Soc. Am. Bull. 62:* 399–416.

Ursin, E. (1952). Change in the composition of the bottom fauna in the Dogger Bank area. *Nature 170:* 324.

Valencia, M. J. (1977). Pleistocene stratigraphy of the western equatorial Pacific. *Bull. Geol. Soc. Am. 88:* 143–150.

Valentine, P. C. (1976). Zoogeography of Holocene Ostracoda off western North America and paleoclimatic implications. *U. S. Geol. Surv. Prof. Paper 916:* 47 p.

van Leckwijck, W. P., and Chesaux, C. H. (1962). Vertical and lateral variation in the lithology and fauna of the Petit Buisson Band in the Norimage Coalfield, Southern Belgium. *In Pal. Zeit.* (H. Schmidt-Festband, Ed.): pp. 140–153.

van Steenis, C. G. G. J. (1960). *In:* M. S. van Meeuwen, H. P. Nooteboom, and C. G. G. J. van Steenis, Preliminary revisions of some genera of Malaysian Papilionaceae. *Reinwardtia 5:* 419–456.

van Straaten, L. M. J. U. (1950). Environment of formation and facies of the Wadden Sea sediments. *Tijdschr. Hiet Koninkl. Nederl. Aarddij. Genootsch. 67:* 94–108.

van Straaten, L. M. J. U. (1952a, b). Biogene textures and the formation of shell beds in the Dutch Wadden Sea. *Proc. Koninkl. Nederl. Acad. Wettensch. Amsterdam: Proc. Ser. B:* 500–516.

van Straaten, L. M. J. U. (1956). Composition of shell beds formed in tidal flat environment in the Netherlands and in the Bay of Arcachon (France). *Geol. en. Mijnbouw 18* (N. S.) (7): 209–226.

van Straaten, L. M. J. U. (1960). Marine mollusc shell assemblages of the Rhone Delta. *Geol. en. Mijnbouw 39:* 105–129.

Vance, R. R. (1979). Effects of grazing by the sea urchin, *Centrostephanus coronatus,* on prey community composition. *Ecol. 60:* 537–546.

Vella, P. (1964). Foraminifera and other fossils from late Tertiary deep-water coral thickets, Wairarapa, New Zealand. *J. Paleontol. 38:* 916–929.

Veevers, J. J. (1959). Size and shape variations in the brachiopod *Schizophoria* from the Devonian of Western Australia. *J. Paleontol. 33:* 888–901.

Venema, S. C., and Creutzberg, F. (1973). Seasonal migration of the swimming crab *Macropipus holsatus* in an estuarine area controlled by tidal streams. *Netherl. J. Sea Res. 7:* 94–102.

Vermeij, G. J. (1975). Marine faunal dominance and molluscan shell form. *Evolution 28:* 656–664.

Vermeij, G. J. (1977a). Patterns in crab claw size: The geography of crushing. *Syst. Zool. 26:* 138–151.

Vermeij, G. J. (1977b). The Mesozoic marine revolution: Evidence from snails, predators and grazers. *Paleobiology 3:* 245–258.

Vermeij, G. J. (1978). "Biogeography and Adaptation." Harvard Univ. Press, Cambridge, Mass. 332 p.

Veron, J. E., and Done, T. J. (1979). Corals and coral communities of Lord Howe Island. *Austral. J. Mar. Freshwater Res. 30:* 203–236.

Villadolid, D. V., and Villaluz, D. K. (1938). Animals destructive to oysters in Bacoor Bay, Luzon. *Philippine J. Sci. 67:* 393–399.

Virnstein, R. W. (1977). The importance of predation by crabs and fishes on benthic infauna in Chesapeake Bay. *Ecology 58:* 1199–1217.

Voigt, E. (1956). Der Nachweis des Phytals durch Epizoen als Kriterium der Tiefe Vorzeitlicher Meere. *Geol. Rundschau 45:* 97–119.

Voigt, E. (1973). Environmental conditions of bryozoan ecology of the hardground biotope of the Maastrichtian Tuff-Chalk, near Maastricht (Netherlands). *In:* "Living and Fossil Bryzoa." (G. Larwood, Ed.) Academic Press, New York: pp. 185–197.

Voigt, E. (1977). On grazing traces produced by the radula of fossil and Recent gastropods and chitons. *In:* Trace fossils 2. (T. P. Crimes and J. C. Harper, Eds.) *Geol. J. Spec. Issue No. 9,* Seel House Press, Liverpool: 335–346.

Voigt, E., and Soule, J. D. (1973). Cretaceous burrowing bryozoans. *J. Paleontol. 47:* 21–23.

Vokes, H. E., and Vokes, E. H. (1968). Variation in the genus *Orthaulax* (Mollusca: Gastropoda). *Tulane studies in Geology 6:* 71–79.

von Gaertner, H. R. (1958). Vorkommen von Serpelriffen nordlich des Polarkreises an der Norwegischen Kuste. *Geol. Rundschau 47:* 72–73.

Wagner, H. (1936). Die Wandermuschel (*Dreissenia*) erobert den Platten-See. *Natur u. Volk 66:* 37–41.

Wahlman, G. P. (1974). Stratigraphy, structure, paleontology, and paleoecology of the Silurian reef at Montpelier, Indiana. Indiana Univ. MS thesis.

Waksman, S. A., and Hotchkiss, M. (1937). On the oxidation of organic matter in marine sediments by bacteria. *J. Mar. Res. 1:* 101–118.

Walcott, C. D. (1883). Injury sustained by the eye of a trilobite at the time of moulting of the shell. *Am. J. Sci. 26:* 302.

Walker, K. R., and Alberstadt, L. P. (1975). Ecological succession as an aspect of structure in fossil communities. *Paleobiology 1:* 238–257.

Walker, K. R., and Bambach, R. K. (1974). Feeding by benthic invertebrates: Classification and terminology for paleoecological analysis. *Lethaia 7:* 67–78.

Walker, K. R., and Parker, W. C. (1976). Population structure of a pioneer and a later stage species in an Ordovician succession. *Paleobiology 2:* 191–201.

Wallace, R. J., and Frost, S. H. (1976). A comparison of an Oligocene–Miocene reef tract in Jamaica with modern Caribbean and Pacific reefs. *Geol. Soc. Am. Abstr. with Prog., Ann. Mtg:* 1159.

Waller, T. R. (1969). The evolution of the *Argopecten gibbus* stock (Mollusca: Bivalvia), with emphasis on the Tertiary and Quaternary species of eastern North America. *Paleontol. Soc. Mem. 3:* 125 p.

Walne, P. R., and Dean, G. J. (1972). Experiments on predation by the shore crab *Carcinus maenas* L., on *Mytilus* and *Mercenaria*. *J. Cons. int. Explor. Mer 34:* 190–199.

Warburton, F. E. (1958). The effects of boring sponges on oysters. *Fish. Res. Bd. Canada, Prog. Rept. Atlantic Coast Sta. No. 68:* 3–8.

Warme, J. E. (1969). Live and dead mollusks in a coastal lagoon. *J. Paleontol. 43:* 141–150.

Warme, J. E. (1971a). Paleoecological aspects of a modern coastal lagoon. *Univ. California Publ. Geol. Sci. 87:* 110 p.

Warme, J. E. (1971b). Recent advances in paleoecology and ichnology. Am. Geol. Inst. Short Course Lecture Notes, 30–31 Oct., 1971. Washington, D. C. Biological energy in erosion and sedimentation, and animal–sediment interrelationships: pp. 55–80.

Warme, J. E. (1975). Borings as trace fossils, and the processes of marine bioerosion. *In:* "The Study of Trace Fossils." (R. W. Frey, Ed.) Springer–Verlag, New York: pp. 181–227.

Warme, J. E., Ekdale, A. A., Ekdale, S. F., and Peterson, C. H. (1976). Raw material of the fossil record. *In:* "Structure and Classification of Paleocommunities." (R. W. Scott and R. R. West, Eds.) Dowden, Hutchinson & Ross, Stroudsburg, Pa: pp. 143–169.

Warn, J. M. (1974). Presumed myzostomid infestation of an Ordovician crinoid. *J. Paleontol. 48:* 506–513.

Warner, W. W. (1976). Winter 'drudging' lifts crabs from Chesapeake mud. *Smithsonian 6*(11): 82–89.

Warren, A. E. (1936). An ecological study of the sea mussel (*Mytilus edulis* Linn.). *J. Biol. Bd. Canada 2:* 89–94.

Warren, P. S. (1948). Chimaeroid fossil egg capsules from Alberta. *J. Paleontol. 22:* 630–631.

Waterhouse, J. B. (1973). Communal hierarchy and significance of environmental parameters for brachiopods: The New Zealand Permian model. *Roy. Ontario Mus. Life Sci. Contrib. 92:* 49 p.

Watkins, R. (1974a). Molluscan paleobiology of the Miocene Wimer Formation, Del Norte County, California. *J. Paleontol. 48:* 1264–1282.

Watkins, R. (1974b). Carboniferous brachiopods from northern California. *J. Paleontol. 48:* 304–325.

Watkins, R. (1975). British Ludlow paleoecology and its bearing on the Silurian marine ecosystem. Univ. Oxford, Wolfson College and Dept. of Geol. & Mineral., Oxford, England, Ph.D. thesis.

Watkins, R., and Boucot, A. J. (1975). Evolution of Silurian brachiopod communities along the southeastern coast of Acadia. *Geol. Soc. Am. Bull. 86:* 243–254.

Watkins, R., and Hurst, J. (1977). Community relations of Silurian crinoids at Dudley, England. *Paleobiology 3:* 207–217.

Watt, K. (1973). "Principles of Environmental Science." McGraw–Hill, New York.

Webb, G. R., Logan, A., and Noble, J. P. A. (1976). Occurrence and significance of brooded larvae in a Recent brachiopod, Bay of Fundy, Canada. *J. Paleontol. 50:* 869–871.

Weihe, S. C., and Gray, I. E. (1968). Observations on the Biology of the Sand Dollar *Mellita quinquiesperforata* (Leske). *J. Elisha Mitchell Scientific Soc. 84:* 315–321.

Weiler, W. (1929). Über das Vorkommen isolierter Köpfe bei fossilen Clupeiden. *Senckenberg. 11:* 40–47.

Weimer, R. J., and Hoyt, J. H. (1964). Burrows of *Callianassa major* Say, geologic indicators of littoral and shallow neritic environments. *J. Paleontol. 38:* 761–767.

Weis, J. S. (1968). Fauna associated with pelagic *Sargassum* in the Gulf Stream. *Am. Midl. Nat. 80:* 554–558.

Weiss, C. M. (1948). The seasonal occurrences of sedentary marine organisms in Biscayne Bay, Florida. *Ecology 29:* 153–172.

Welch, J. R. (1976). *Phosphannulus* on Paleozoic crinoid stems. *J. Paleontol. 50:* 218–225.

Weller, S. (1899). Kinderhook fauna studies. I. The fauna of the vermicular sandstone at Northview, Webster County, Missouri. *Trans. Acad. Sci. St. Louis 9:* 9–51.

Wells, C. (1964). "Bones, Bodies, and Disease." Praeger, New York: 288 p.

Wells, G. P. (1944). Mechanism of burrowing in *Arenicola marina*. *Nature 154:* 396.

Wells, H. W. (1961). The fauna of oyster beds, with special reference to the salinity factor. *Ecol. Monogr 31:* 239–266.

Wells, H. W., and Wells, M. J. (1962). The polychaete *Ceratonereis tridentata* as a pest of the scallop *Aequipecten gibbus*. *Biol. Bull. 122:* 149–159.

Wells, H. W., Wells, M. J., and Gray, I. E. (1964). The calico scallop community in North Carolina. *Bull. Mar. Sci. Gulf and Carib. 14:* 561–593.

Wells, J. W. (1941). Crinoids and Callixylon. *Am. J. Sci. 239:* 454–456.

Wells, J. W. (1947). Provisional paleoecological analysis of the Devonian rocks of the Columbus region. *Ohio J. Sci. 47:* 119–126.

Wells, J. W. (1957). Coral reefs. *In:* Treatise on marine ecology and paleoecology, V. 1 Ecology. (J. Hedgpeth, Ed.) *Geol. Soc. Am. Mem. 67:* 609–631.

Wells, J. W. (1967a). Corals as bathometers. *Mar. Geol. 5:* 349–365.

Wells, J. W. (1967b). The Devonian coral *Pachyphyllum*

vagabundum, a necroplotic *P. woodmani? J. Paleontol.* *41:* 1280.

Wells, J. W. (1976). Eocene corals from Eua, Tonga. *U. S. Geol. Surv. Prof. Paper 640-G:* G1–G13.

Wells, M. J., and Wells, J. (1970). Observations on the feeding, growth rate and habits of newly settled *Octopus cyanea. J. Zool. 161:* 65–74.

Wendt, J. (1970). Stratigraphische Kondensation in triadischen und jurassischen Cephalopodenkalken der Tethys. *N. Jb. Geol. Paläontol., Monh.:* 433–448.

Weymouth, F. W., McMillin, H. C., and Rich, W. H. (1931). Latitude and relative growth in razor clam, *Siliqua patula. J. Exper. Biol. 8:* 228–249.

Whatley, R. C., and Wall, D. R. (1975). The relationship between ostracoda and algae in littoral and sublittoral marine environments. *In:* Biology and paleobiology of Ostracoda. (F. M. Swain, Ed.) *Bull. Am. Paleontol. 65:* 173–203.

Wherry, E. T. (1916). Glauberite crystal–cavities in the Triassic rocks of eastern Pennsylvania. *Am. Mineral. 1*(3): 37–43.

Whitehouse, F. W. (1973). Coral reefs of the New Guinea region. *In:* "Biology and Geology of Coral Reefs." V. 1 Geology 1. (O. A. Jones and R. Endean, Eds.) Academic Press, New York: pp. 169–186.

Whitlach, R. B. (In preparation). Patterns of resource utilization and coexistence in intertidal deposit feeding communities.

Whitlach, R. B., and Johnson, R. G. (1974). Methods of staining organic matter in marine sediments. *J. Sed. Pet. 44:* 1310–1312.

Wiedmann, H. Y. (1972). Shell deposits and shell preservation in Quaternary and Tertiary estuarine sediments in Georgia, U. S. A. *Sed. Geol. 7:* 103–125.

Wieser, W. (1960). Benthic studies in Buzzards Bay. II. The Meiofauna. *Limnol. and Oceanogr. 5:* 121–137.

Wigley, R. L., and Stinton, F. C. (1973). Distribution of macroscopic remains of Recent animals from marine sediments off Massachusetts. *U. S. Fish & Wildlife Serv. Fish. Bull. 71:* 1–40.

Wilde, G. L. (1965). Abnormal growth conditions in fusulinids. Contrib. Cushman Foundation. *Foraminiferal Res. 16:* 121–124.

Willard, B. (1926). A six-rayed *Devonaster eucharis* (Hall). *J. Geol. 34:* 85–87.

Williams, A., and Rowell, A. J. (1965). Morphology. *In:* "Treatise on Invertebrate Paleontology, Part H, Brachiopoda." (R. C. Moore, Ed.) Geol. Soc. America & Univ. Kansas Press, Lawrence: pp. H57–H155.

Williams, M. E. (1972). The origin of "spiral coprolites". *Univ. Kansas Paleontol. Contrib. Paper 59:* 19 p.

Williams, M. J. (1978). Opening of bivalve shells by the mud crab *Scylla serrata* Forskål. *Aust. J. Mar. Freshwater Res. 29:* 699–702.

Williams, T. C. (1976). Paleosalinities within a Pliocene bay, Kettleman Hills, California: A study of the resolving power of isotopic and faunal techniques: Discussion and reply. *Geol. Soc. Am. Bull. 87:* 158–160.

Wills, L. J. (1963). *Cyprilepas holmi* Wills 1962, a pedunculate cirripede from the Upper Silurian of Oesel, Esthonia. *Palaeontology 6:* 161–165.

Wilson, D. P. (1937). The influence of the substratum on the metamorphosis of *Notomastus* larvae. *J. Mar. Biol. Assoc. U. K. 22:* 227–243.

Wilson, D. P. (1948). The relation of the substratum to the metamorphosis of *Ophelia* larvae. *J. Mar. Biol. Assoc. U. K. 27:* 723–759.

Wilson, D. P. (1952). The influence of the nature of the substratum on the metamorphosis of the larvae of marine animals, especially the larvae of *Ophelia bicornis* Savigny. *Ann. L'Inst. Oceanogr. 27:* 49–156.

Wilson, D. P. (1953). The settlement of *Ophelia bicornis* Savigny larvae. The 1951 experiments. *J. Mar. Biol. Assoc. U. K. 31:* 413–438.

Wilson, E. C. (1975). Light show from beyond the grave. *Terra 13:* 10–13.

Wilson, J. B. (1967). Palaeoecological studies on shell-beds and associated sediments in the Solway Firth. *Scottish J. Geol. 3:* 329–371.

Wilson, J. L. (1975). "Carbonate Facies in Geologic History." Springer–Verlag, New York: 471 p.

Windle, T. M. F. (1979). Reworked Carboniferous spores: An example from the Lower Jurassic of northeast Scotland. *Rev. Palaeobot. Palynol. 27:* 173–184.

Wodinsky, J. (1969). Penetration of the shell and feeding on gastropods by *Octopus. Am. Zool. 9:* 997–1010.

Womersley, H. B. S., and Edmonds, S. J. (1958). A general account of the intertidal ecology of South Australian coasts. *Australian J. Mar. Freshwater Res. 9:* 217–260.

Woodin, S. A. (1977). Algal "gardening" behavior by nereid polychaetes: Effects on soft-bottom community structure. *Mar. Biol. 44:* 39–42.

Woodring, W. P. (1928). Miocene mollusks from Bowden, Jamaica: Part II. Gastropods and discussion of results. *Carnegie Inst. Washington Publ. No. 385:* 564 p.

Woodring, W. P. (1931). A Miocene *Haliotis* from southern California. *J. Paleontol. 5:* 34–39.

Woodring, W. P. (1965). Endemism in Middle Miocene Caribbean molluscan fauna. *Science 148:* 961–963.

Woods Hole Oceanographic Institution. (1952). The fouling community. *In:* "Marine Fouling and its Prevention." U. S. Naval Inst., Annapolis: pp. 37–76; 91–101.

Worsley, D. (1971). Faunal anticipation in the Lower Llandovery of the Oslo region, Norway. *Norsk Geol. Tidsskr. 51:* 161–167.

Wray, J. (1971). Algae in reefs through time. *Proc. North. Am. Paleontol. Conv. 1969 2:* 1358–1373.

Yamaguchi, M. (1975). Sea level fluctuations and mass mortalities of reef animals in Guam, Mariana Islands. *Micronesica 11:* 227–243.

Yamanouchi, T. (1927). Some preliminary notes on the behavior of the holothurian, *Caudina chilensis.* (J. Muller, Ed.). *Sci. Rept. Tohoku Univ., Ser. 4, 2:* 85–91.

Yamanouchi, T. (1941). Study of the useful holothurians from Palau. *Kagaku Nanyo* (Science of the South Seas) *4:* 36–52. In Japanese. (Translation prepared by Military Geology Branch of U. S. Geol. Surv. for Intelligence Div. Office of the Engineer Headquarters Army Forces Far East, Tokyo, 1953).

Yingst, J. (1974). The utilization of organic detritus and associated micro-organisms by *Parastichopus parvimensis,* a benthic deposit feeding holothurian. Ph.D. thesis, Univ. Southern California, Los Angeles: 154 p.

Yonge, C. M. (1928). Feeding mechanisms in the invertebrates. *Biol. Rev. 3:* 19–76.

Yonge, C. M. (1937). Evolution and adaptation in the digestive system of the metazoa. *Biol. Rev. 12:* 87–115.

Yonge, C. M. (1949). On the structure and adaptations of the

Tellinacea, deposit feeding Eulamellibranchia. *Phil. Trans. Roy. Soc., Ser. B 234:* 29–76.

Yonge, C. M. (1955). Adaptation to rock boring in *Botula* and *Lithophaga* (Lamellibranchia, Mytilidae) with a discussion on the evolution of this habit. *Quart. J. Microscop. Sci.* 96: 383–410.

Yonge, C. M. (1960). "Oysters." The New Naturalist Series; Collins, London: 209 p.

Yonge, C. M. (1963a). The biology of coral reefs. *In:* "Advances in Marine Biology 1." (F. Russell, Ed.) Academic Press, London: pp. 209–260.

Yonge, C. M. (1963b). Rock-boring organisms. *In:* Mechanisms of hard tissue destruction. (R. F. Sognnaes, Ed.) *Am. Assoc. Adv. Sci. Publ.* 75: 1–24.

Young, D. K. (1971). Effects of infauna on the sediment and seston of a subtidal environment. *Vie et Milieu. Suppl.* 22: 557–571.

Young, D. K., Buzas, M. A., and Young, M. W. (1976). Species densities of macrobethos associated with seagrass: A field experimental study of predation. *J. Mar. Res. 34:* 577–592.

Young, D. K., and Rhoads, D. C. (1971). Animal–sediment relations in Cape Cod Bay, Massachusetts. I. A transect study. *Mar. Biol. 11:* 242–254.

Zach, R. (1978). Selection and dropping of whelks by northwestern crows: *Behaviour 67:* 134–148.

Zakharov, V. A. (1966a). Bespozvonochnie, prizhiznenno zakhoronennie v valanzhinskikh peskakh khatanskoi vpadini. *In:* Organism i Sreda v Geologicheskom proshlom. (R. F. Gekker, Ed.) *Akad. Nauk. Otdel Biologii:* 31–54.

Zakharov, V. A. (1966). "Pozdneoorskie i Rannemelovoe Dvustvorchatie mollooski severa Sibiri i Ustoviya ikh sushestvovaniya (Otryad Anisomyaria)." Izdatel. "Nauka": 189 p.

Zangerl, R., and Richardson, E. S., Jr. (1963). The paleo-ecological history of two Pennsylvanian black shales. *Fieldiana, Geology Mem. 4:* 352 p.

Zenkevitch, L. (1963). "Biology of the Seas of the USSR." George Allen and Unwin Publ., London: 955 p.

Zidek, J. (1976). A new shark egg capsule from the Pennsylvanian of Oklahoma and remarks on the chondrichthyan egg capsule in general. *J. Paleontol. 50:* 907–915.

Ziegelmeier, E. (1963). Das Makrobenthos in Ostteil der Deutschen Bucht nach qualitativen und quantitativen Bodengreiferuntersuchungen in der Zeit von 1949–1960. Veroffentlichungen Inst. Meerforsch., Bremerhaven. *Meeresbiol. Symp. 23:* 101–114.

Ziegler, A. M. (1965). Silurian marine communities and their environmental significance. *Nature 207:* 270–272.

Ziegler, A. M. (1966). Unusual stricklandiid brachiopods from the Upper Llandovery beds near Presteigne, Radnorshire. *Palaeontology 9:* 346–350.

Ziegler, A. M., Boucot, A. J., and Sheldon, R. P. (1966). Silurian pentamerid brachiopods preserved in position of growth. *J. Paleontol. 40:* 1032–1036.

Index

447

LOURDES LIBRARY
GWYNEDD-MERCY COLLEGE
GWYNEDD VALLEY, PA. 19437